CONE PENETRATION TESTING OF SOILS IN GEOTECHNICS

Igor B. Ryzhkov
Oleg N. Isaev

ASV Construction
Stockholm, Sweden
2016

Translation from English:

Sheshukova S.V.

Reviewers:

Doctor of Technical Sciences, Professor Amaryan L.S.
(PNIIIS, Moscow, Russia)
Doctor of Technical Sciences, Professor Bakholdin B.V.
(NIIOSP, Moscow, Russia)

Editor:

Dr. Nikolay Volkov, P.Eng., (Fugro Consultants Inc., Houston, USA)

Ryzhkov I.B., Isaev O.N.

Cone penetration testing of soils in geotechnics. Stockholm, Sweden: ASV Construction, 2016. – 408 p.

ISBN 978-91-982223-4-0

The book reviews research and analysis of cone penetration testing of soils practical applications both in Russia and worldwide during the last five decades. The book describes penetrometer designs and CPT rigs as well as CPT techniques, data obtained interpretation and penetration data application techniques when solving various practical problems.

The book is addressed to engineers, technicians and scientists involved in research, design, construction and science; it is also a helpful source for students of civil engineering and geological specialties.

ISBN 978-91-982223-4-0

Signed for printing 12.01.2016. Format 60×90 1/16.
Offset paper. Times type. Offset printing.
Conventional 24 printed sheets.

ASV Constraction, Sweden,
Mårdvägen 16 131 50 Saltsjö-Duvnäs

LIST OF CONTENTS

INTRODUCTION

Nowadays cone penetration testing (CPT) is one of the principal methods of soil conditions investigation. During past 20th century it had greatly evolved from the simplest sounding rods penetrated by hand to the powerful mobile units with highly mechanized control and automatic measurements. Applying CPT we can estimate soil in-situ with maximum efficiency. The principal merits of CPT are the quickness and simplicity, i.e. the possibility to do lots of measurements in the shortest time possible. The necessity of doing it depends upon two factors.

Firstly, the soil under study is always heterogeneous; its properties are different in every location of the examined site. Few field and laboratory tests of soil do not guarantee necessary completeness of the obtained information no matter how accurate they are. There are always some concerns about the unnoticed "weak" or "strong" lentils of the soil between boreholes or locations of field tests, the factual bounds between strata (soil layers) differ from the accepted ones, i.e. indicated on lithological section, etc. Such holes often cause failures and even collapse of the structures. Thus, to prevent these concerns it is necessary to estimate soil conditions of the site on the possible number of locations; this requires quick and cheap methods to be applied such as cone penetration testing.

Secondly, it is rational application of quick and cheap methods of soil examination ("express-methods") that reduce the duration of investigation, cone penetration testing is relevant. Partial replacement of boreholes and costly soil testing by CPT allow reducing terms of investigation even if the number of CPT locations significantly increases.

Great attention to the rates of investigation is associated with its economic role in construction investment cycle in whole (i.e. in the cycle "investigation – designing – building and assembly – balancing and commissioning"). Investigation and designing are stages that demand significant time costs but not material costs. For this reason, duration of investigation and designing may have greater economic value than their cost. For an investor increase in the time of investment cycle only one month equals rise in the cost of construction about 1...2 % (at the cost of additional payments on the loan and reduction of the turnover of investment rates). On the contrary, reduction in the time of the cycle equals reduction in price of construction. With regard to the cost of engineering-geological surveys of 0.5...1 % from estimated total cost of construction it is obvious that rates of investigation are of primary importance.

In practice when applying highly productive sounding units it is possible to do a great number of measurements on the investigated site but it is practically impossible with traditional costly methods of soil estimation. "Thickness" of penetration locations location on the site can easily be brought to the level when there is no the slightest fear of weak soils lentils missing or any significant deformations of bounds between different strata. Besides, CPT allows estimating piles' bearing capacity in every distinctive place of the site, in every depth which is of great interest for a design engineer yielding to static pile load testing in accuracy of estimation.

Vast research on CPT has been done in the developed countries during last decades. A great number of publications have been devoted to this method, international symposiums on its application are being held. National and international standards regulating CPT application have been developed. A number of firms specialize on manufacturing, selling and using of CPT equipment. The professional engineers of the firms have been working in many countries and last time including Russia.

In the former USSR great attention was paid to CPT. In 1960s…1980s great interest to CPT was considerably connected with pile foundations application, the designing of the foundations appeared to be the most efficient when applying CPT. In the country's regions where this method was applied extensively and professionally, higher quality of pile foundations construction was observed, i.e. their cost efficiency. Errors in choosing the piles' length were infrequent, volume of cutting and incomplete driving decreased.

In the USSR special state commission on CPT coordinated research and practical application of this method in the scientific, designing, research and industrial institutions. Industry manufactured small numbers of different CPT rigs which were successfully used by designing and research bureaus and in some regions – by construction institutions as well. Up to 1990s development of CPT in the USSR was intensive. In 1990s its growth recession was due to the total decrease in scientific- technical work in Russia, but not to changed attitude to the method. Nowadays, the situation is nearly normalized.

In the second half of the 20th century the most outstanding non-Russian works generalizing CPT were as follows: Sanglerat (1971); Meigh A.C. (1987); Lunne, Robertson & Powell (1997, republished in 2001, 2001, 2004); Burns & Mayne (1998). Unfortunately, the last two have not been translated into Russian. In Russia the most outstanding works are the following: Bondarick (1964), Ferronsky (1969), Trofimenkov & Vorobkov (1974, republished in 1981), Razorenov (1980) et al. In 1990s detailed generalization of non-Russian experience was done by Trofimenkov (1995), in the first half of the 21st century – by M.S. Zaharov (2007).

Nevertheless, the experience, theoretical knowledge and practical methods of CPT applications have not been formed into an integral system and they need to be additionally systemized and generalized. The data published are often incomplete and sometimes contradictory. There are only a few works on analysis of CPT global application, systematization and generalization of the data obtained. The majority of publications, especially the non-Russian ones show illustrations of different contradictory, empirical dependencies without any attempts to explain their physics. Nowadays it is difficult to find a sphere in soil mechanics where empiricism would play the principal role as in CPT application. Herewith, achievements of Russian (Soviet) professional engineers in CPT are practically ignored in most outstanding non-Russian books (Sanglerat 1971, Lunne et al. 2004, Burns & Mayne 1998).

Up to present time possibilities of CPT have not been used in practice in full scale. Despite a lot of publications and normative documents instructions on CPT applications, professional engineers of many organizations underestimate this method. It is a common practice when CPT is not used even if it is of primary importance or it is used inefficiently to check the box. Here it duplicates "standard" tests. To a great extent, it is due to data obtained interpretation difficulty, specialists' lack of knowledge on the issue and sometimes lack of necessary equipment. CPT application efficiency is defined by experience and theoretical knowledge of a specialist since a lot of geotechnical tasks solutions cannot be schematized as prepared instructions and recommendations.

The book gives results of research and CPT practical application generalization. The book, as we hope, is to show that CPT is not equal to empirical dependencies application, reliability of which cannot always be forecast and the results are usually corrected by senior engineers highly experienced in CPT. The greatest attention is paid to both highly productive penetrometer units with friction sleeve (type II of standard penetrometers) applications, and research on variety and contradiction of empirical dependencies linking CPT results with standard soil properties or pile resistance in soils of different kinds and origins. The book presents issues on CPT application in specific soils (permafrost, collapsible, rubble soils, etc.) as well as principles of joint application of CPT together with other methods of soil investigation. The book gives details of research on CPT application in space objects surface exploration (Lunar soil serves as an example).

The book does not show CPT applications in structural design under specific conditions: on the shelf, deep under the ground, etc., since the issues in Russia are not so urgent and infringe on specialists' interests. In a few words the book tells us about equipping the penetrometers with additional sensors and other equipment to test the following: blade shear, pressiometry, ecological measurements (e.g. occurrence of carbohydrates in the soil), etc.

We thank scientific workers of BashNIIstroy (in particular Candidate of Technical Sciences Kolesnick G.S.), NIIOSP and other institutions for providing us with necessary materials.

We appreciate Doctor of Technical Sciences, Professor Amaryan L.S., Doctor of Technical Sciences, Professor Bakholdin B.V., Doctor of Technical Sciences, Professor Petrukhin V.P., Doctor of Technical Sciences, Professor Shvarev V.V., Doctor of Geological-Engineering Sciences Minkin M.A. and Candidate of Technical Sciences Kolybin I.V. their valuable criticisms and recommendations which have helped improve this book.

The authors are grateful to hear the any critical notes, comments and recommendations. Please send them on one of the addresses: Dr. Ryzhkov I.B., BGAU, 34, 50 let Oktyabrya str., Ufa, 450001; E-mail: ig-ryzhk@yandex.ru; Dr. Isaev O.N., Gersevanov N.M. NIIOSP, 2nd Institutskaya str. 6, Moscow, 109428; E-mail: geotechnika2008@gmail.com.

LIST OF MAIN SYMBOLS

a	=	coefficient of compressibility
a	=	area ratio of the cone $= A_n/A_c$
a_{max}	=	maximum horizontal acceleration of ground surface, induced by earthquake
a_0	=	coefficient of relative compressibility $= a/(1+e_0)$
A	=	cross-sectional area of a pile (boulder)
A_b	=	cross-sectional area of the leader hole
A_c	=	area of the cone base
A_f	=	side surface area of a pile
$A_{f,i}$	=	area of i-sector of pile side surface
A_i	=	side surface area of the penetrometer fitting i-engineering-geological element
A_n	=	cross-sectional area of load cell ore shaft
A_p	=	cross-sectional area of the pile end
$A_{pf,i}$	=	freezing surface area of frozen soil i-layer with pile side surface
A_s	=	area of friction sleeve (type II penetrometer)
b, B	=	width of foundation bottom
B	=	"boulder density" = ratio of total volume of boulders to total volume of ground massif
B_q	=	pore pressure parameter $= (u_2 - u_0)/(q_c - \sigma_{vo})$
c	=	cohesion
c_{eq}	=	long-term value of equivalent cohesion in frozen soil (ball plate test)
c_h	=	horizontal coefficient of consolidation
c_v	=	vertical coefficient of consolidation
C	=	coefficient of soil subgrate reaction
d	=	depth of foundation
d_b	=	diameter of the ball plate
d_c	=	diameter of the cone basis
d_{eq}	=	"equivalent" pile diameter
d_p	=	diameter of the pile
d_s	=	diameter of the friction sleeve
D	=	boulder size
D, D_{eq}	=	diameter (side) of the pile cross-section
D_k	=	"impassible" boulder size

D_{50}	=	size such that 50 % (by weight) of the sample consists of particles having a smaller nominal diameter (average diameter of particles)
e	=	void ratio
e_L	=	void ratio corresponding to water content close to fluidity
e_{max}	=	maximum void ratio
e_{min}	=	minimum void ratio
e_o	=	initial void ratio
E, E_{st}	=	modulus of deformation, drained (stamp)
E_f	=	compressive modulus of deformation of frozen soil
E_b	=	modulus of elasticity of pile concrete
E_s	=	modulus of elasticity of pile helmet layer
E_u	=	undrained modulus of deformation
ER_r	=	rod energy ratio in standard penetration test (SPT)
f, f_p	=	pile unit averaged side friction
f_i	=	unit pile side bearing capacity (in i-layer)
f_{pf}	=	unit long-term resistance of frozen soil to shear along frozen surface of the pile
$f^{(0)}{}_{pf}$	=	unit long-term resistance of frozen soil to shear along frozen surface of the driven pile embedded without the leader hole, $A_b = 0$
$f^{(1)}{}_{pf}$	=	unit ultimate resistance of frozen soil to shear along frozen surface of the boring pile embedded in the leader hole in $A_b \cong A_p$
f_{pi}	=	unit averaged resistance of soil in i-sector of pile side surface
f_s	=	unit sleeve friction resistance
f_{ss}, f_{cm}	=	unit sleeve friction resistance at relaxation-creep CPT
f_{sv}, f_v	=	unit sleeve friction resistance at pushing the penetrometer with constant velocity (standard CPT)
f_{si}	=	unit sleeve friction resistance in i-layer of soil
f_{si}	=	unit sleeve friction resistance at the beginning of pushing after relaxation-creep CPT
$f_{s,wet}$	=	unit sleeve friction resistance after wetting of soil
f_t	=	unit sleeve friction resistance corrected for pore pressure effects
F_d	=	pile bearing capacity
F_{du}	=	pile bearing capacity in CPT location
F_t	=	normalized friction ratio $= f_s/(q_t - \sigma_{vo})$
F_u	=	particular value of ultimate pile resistance to vertical load
F_{uf}	=	particular value of long-term pile resistance in frozen soils to vertical load

$F_{u,i}$	=	particular value of ultimate pile resistance in i-location of the site
$F_{u,n}$	=	normative value of ultimate pile resistance
$F_{u,wet.}$	=	ultimate pile resistance obtained by static tests after wetting of soil
$F_{u,pile}$	=	ultimate pile resistance obtained by static tests
$F_{u,CPT}$	=	ultimate pile resistance obtained by CPT
g	=	gravity acceleration
G	=	weight of striking part of hammer
G	=	shear modulus
G_0	=	shear modulus in small deformations (initial shear modulus)
h	=	depth from the ground surface
h_i	=	thickness of i-layer of soil
H	=	head of striking part of hammer
H_0, H_1	=	entropy before and after the test correspondingly
I	=	information quantity $= H_0 - H_1$
I_D	=	density index of sands $= (e_{max} - e)/(e_{max} - e_{min})$
I_f	=	friction index $= (q_c / f_s) \times 100\,\%$
I_L	=	liquidity index
I_p	=	plasticity index
I_r	=	rigidity index $= G/s_u$
I_z	=	"settlement influence factor"
k, k_{per}	=	coefficient of permeability
k_α	=	angular coefficient of calibration curve
k_c	=	pile end bearing capacity factor
k_h	=	coefficient of permeability in horizontal direction
k_v	=	coefficient of permeability in vertical direction
$k_{\theta F}, k_{\theta R}$	=	coefficient of difference in permafrost soils of pile basement and isolated foundation during CPT procedure and life cycle of the designed structure
K_o	=	coefficient of side earth pressure at rest $= \dfrac{\sigma'_{ho}}{\sigma'_{vo}}$
K_d	=	parameter of durability showing resistivity of pile material to repeated dynamic loads
K_{ds}	=	parameter of dynamic strengthening $= R_d/R_{st}$
K_w	=	coefficient of cone resistance decrease in wetting $= q_c/q_{c,wet}$
$K_{w,f}$	=	coefficient of sleeve friction resistance decrease in wetting $= f_s/f_{s,wet}$
$K_{pile-penetr}$	=	parameter showing the ratio of boulder volume "impassible" for piles to the analogous volume for the penetrometer
l, L	=	depth of pile end location (pile length);

xi

m	=	gradient of curve initial sector "$\Delta u_t / \Delta u_i \sim \sqrt{t}$"
$m`$	=	coefficient of relative compressibility $= m_0/(1+e_o)$
m_f	=	coefficient of frozen soil compressibility
m_0	=	coefficient of compressibility
m_θ	=	coefficient of penetrometer temperature stabilization $= \Delta\theta_{cs}/\Delta t_s$
M, E_c	=	odometric modulus of deformation of soil loading (compressive)
M	=	gradient of pore pressure corresponding to theoretical curve of pore pressure dissipation with regard to piezocone of specified geometry
M	=	earthquake magnitude
n	=	creep index
n_i	=	anticipated number of hammer blows in one meter immersion of the pile
N	=	total number of blows necessary to drive in the pile
N	=	number of blows necessary to immerse the sampler in 30 cm
N	=	cone factor
N'	=	angle of resistance $= \text{arctg}\,(\tau/\sigma)$
N_b	=	load on the ball plate
N_c	=	theoretical cone factor
N_k	=	empirical cone factor $= \dfrac{q_c - \sigma_{vo}}{s_u}$
N_{ke}	=	empirical cone factor $= \dfrac{q_t - u_2}{s_u}$
N_{kt}	=	empirical cone factor $= \dfrac{q_t - \sigma_{vo}}{s_u}$
N_t	=	allowable number of blows by condition of the required productivity of pile-driving works
$N_{strength}$	=	allowable number of blows by condition of prevention of the pile from destruction
OCR	=	overconsolidation ratio $= \dfrac{\sigma'_p}{\sigma_{zg}}$
p	=	probability
p_a	=	atmospheric pressure
P_{ic}	=	initial critical pressure under standard stamp
$p_{max,base}$	=	ultimate (maximum) pile end bearing capacity
$p_{max,shaft}$	=	ultimate (maximum) pile side bearing capacity
P_n	=	negative friction
p_r	=	radial pressure on walls of expanded cavity

LIST OF MAIN SYMBOLS

q	=	weight of striking part of hammer
q_c	=	unit cone resistance $= Q_c/A_c$
$\overline{q}_c$	=	unit average cone resistance
$(q_c)_{cr}$	=	unit critical cone resistance corresponding to liquefied sands
q_{ca}	=	unit average (corrected by definite rules) cone resistance in pile end
q_{ci}	=	unit cone resistance at the beginning of pushing after relaxation-creep CPT
q_{cs}, q_{ce}	=	unit cone resistance at relaxation-creep CPT
$q_{cs;5}$	=	unit cone resistance at relaxation-creep CPT in "stabilization" of the penetrometer $t_s = 5$ min
q_{cv}, q_v	=	unit cone resistance at pushing the penetrometer with constant velocity (standard CPT)
$q_{cv;0,5}$	=	unit cone resistance at pushing the penetrometer with constant velocity $v = 0{,}5$ m/min
$q_{c;I;mean}$, $q_{c;II;mean}$, $q_{c;III;mean}$	=	unit mean cone resistances in the adjacent zones of pile end
q_{cl}	=	unit normalized cone resistance
$(q_{cl})_{cr}$	=	unit critical normalized cone resistance corresponding to liquefied sands
$q_{c,wet}$	=	unit cone resistance after wetting of soil
q_p	=	unit pile end resistance
q_t	=	unit corrected cone resistance $= q_c + (1 - a)\,u_2$
q_{tl}	=	unit normalized cone resistance
q_{ult}	=	tip bearing capacity of piles
Q	=	pile weight
Q_c	=	resistance of soil to cone pushing
Q_f	=	resistance of soil on pile side surface
Q_i	=	total side friction resistance on i-engineering-geological element
Q_p	=	pile end bearing capacity
Q_s	=	total side friction resistance (for type II cone penetrometer)
Q_t	=	normalized cone resistance $= \dfrac{q_t - \sigma_{vo}}{\sigma'_{vo}}$
Q_{ult}	=	ultimate pile bearing capacity
r	=	roughness of sleeve friction surface
r	=	coefficient of correlation
r_o	=	radius of the penetrometer
r_o	=	radius of expanded spherical cavity or cylindrical one

R	=	design pile end bearing capacity
R, R_o	=	design foundation soil resistance of shallow foundations
R_c	=	long-term strength of frozen soil in uniaxial compression
R^c_{vs}	=	stabilization factor of cone resistance = q_{cv}/q_{cs}
R_f	=	friction ratio = (f_s / q_c)x100 %
R_{pf}	=	unit long-term resistance of frozen soil under pile end
R_s	=	unit pile end bearing capacity
R_{sf}	=	unit long-term resistance of frozen soil under isolated foundation bottom
R_d	=	dynamic strength of pile material
R_{st}	=	static strength of pile material
s	=	settlement of foundation
$s_{a,i}$	=	anticipated settlement per blow of the pile immersed in i-depth
s_b	=	depth of ball plate penetration in soil at the end of the test
s_u	=	allowable mean settlement for the designed structure;
s_u	=	undrained shear strength
s_u'	=	remolded undrained shear strength
S_t	=	sensitivity = s_u/s_u'
t	=	time
t_s	=	time after relaxation-creep CPT beginning
t_α	=	student coefficient
t_{50}	=	time for 50 % dissipation of excess pore water pressure ($\Delta u_t/\Delta u_i = 0.5$)
t^*	=	unconditioned small value of time
T^*	=	"modified time factor"
u	=	pore water pressure
u	=	perimeter of pile cross-section
u_i	=	initial pore pressure at time $t = 0$
u_t	=	pore pressure at time = t
u_0	=	in-situ pore pressure (equilibrium pore pressure before penetration);
u_1	=	pore pressure measured on the cone
u_2	=	pore pressure measured behind the cone (between the cone and friction sleeve)
u_3	=	pore pressure measured behind the friction sleeve
U	=	normalized excess of pore pressure $=\dfrac{u_t-u_o}{u_i-u_o}\times100\%$
υ	=	coefficient of variation
v	=	velocity of penetration

v_s	=	velocity of seismic shear wave
V	=	volume
w_l	=	water content at liquidity limit
w_p	=	water content at plastic limit
w_{tot}	=	total water content in frozen soil
Π	=	collapsibility criterion
$\Im_p$	=	calculated hammer capacity

Greek

α	=	transition coefficient from s_u to f_p
α	=	transition coefficient from $\overline{q}_c$ to q_p
α	=	confidence probability
α	=	cone tip angle
α'	=	half apex angle of the cone
α'	=	ratio of reinforced pile side bearing capacity to unit sleeve friction resistance
α_i	=	transition coefficient from $(q_t - \sigma_{vo})$ to M
α_{LCPC}	=	friction coefficient dependent on the types of piles and soils
α_m	=	transition coefficient from q_c to M
β	=	coefficient considering the absence of shear expansion of soil in the compressive device
β	=	transition coefficient from p_r to q_c
β	=	transition coefficient from f_s to f_p
β_i	=	transition coefficient from f_{si} to f for type II cone penetrometer
β_1	=	transition coefficient from q_c to R_s
β_2	=	transition coefficient from f_s to f_i for type II cone penetrometers, from f_s to f - for type I ones
γ	=	unit weight of soil
γ_c	=	application coefficient
γ_{cf}	=	application coefficient of soil behavior on pile side surface
γ_{cR}	=	application coefficient of soil behavior under pile end
γ_g	=	reliability coefficient of soil
γ_k	=	reliability coefficient
γ_s	=	unit weight of solid particles
γ_w	=	unit weight of water
γ_I	=	design value of unit weight of soil
δ	=	angle of soil friction upon the cone

LIST OF MAIN SYMBOLS

φ	=	angle of internal friction of soil
Δ	=	change (correction)
Δp	=	additional pressure on foundation = total pressure under foundation footing minus overburden pressure at the depth of footing location
Δu_i	=	excess pore pressure = $u_i - u_0$
Δu_t	=	excess pore pressure = $u_t - u_0$
Δz	=	thickness of foundation layer
$\Delta \theta_c$	=	alteration of penetrometer temperature = $(\theta_{cv} - \theta_{cs})$
ε	=	relative linear deformation
ε_{sl}	=	relative soil collapsibility
λ	=	transition coefficient from q_c to $\sigma_p{}'$
ν	=	coefficient of lateral deformation (Poisson ratio)
π	=	pi
θ_c	=	penetrometer temperature
θ_{cv}	=	penetrometer temperature in immersion
θ_{cs}	=	temperature of immovable penetrometer
θ_s	=	natural temperature of soil
θ_{sp}	=	prognostic temperature of soil corresponding to life cycle of the designed structure
ρ	=	density of soil
ρ	=	radius of limit equilibrium zone
σ	=	standard deviation
σ	=	normal total stress
σ'	=	normal effective stress
σ'_a	=	effective stress equal to atmospheric one
σ_c	=	soil consolidation stress = $c/tg\varphi$
σ_{hg}, σ_{ho}	=	natural overburden total horizontal stress in soil
$\sigma_{hg}{}'$, $\sigma_{ho}{}'$	=	natural overburden effective horizontal stress in soil;
$\sigma_{mean}{}'$	=	natural overburden effective octahedral stress in soil
$\sigma_p{}'$	=	maximum overburden effective vertical stress during formation of soil (overconsolidation stress)
σ_r	=	radial stress of soil on pile side surface
σ_{zg}, σ_{vo}, σ_g	=	natural overburden total vertical stress in soil

σ_{zg}', σ_{vo}'	=	natural overburden effective vertical stress in soil
σ_θ	=	tangential normal stress
τ	=	tangential stress
τ	=	shear stress
τ_d	=	cyclic shear stress amplitude
τ_n	=	coefficient dependent on magnitude
τ_g	=	shear stress of soil in overburden stress
ξ	=	coefficient of lateral stress of soil
ψ	=	coefficient in the equation of elastic-plastic model of expanding opening in soil

Footnote – When writing the book, the authors tried to leave the symbols as they are given in publications presented. Therefore, the symbols often mean several terms and vice versa – a number of terms have several alphabetic symbols.

GLOSSARY ON CPT TERMINOLOGY

CPT – Cone Penetration Test (static sounding)

The pushing of a cone penetrometer at the end of a series of cylindrical push rods into the ground at a constant rate of penetration.

CPT continuous

The penetration is regarded as continuous even if the penetration is stopped regularly for a new stroke or mounting of a new push rod.

CPT discontinuous

The penetration is regarded as discontinuous if larger stops are introduced, such as dissipation, relaxation-creep, quasi-static tests or due to unforeseen malfunctions of the equipment.

CPT rig (equipment or system)

The system including a penetrometer, pushing and extraction equipment, support-anchor equipment, a measuring system. It is mounted in a movable or floating vessel or it can be used independently.

Cone

The part of the penetrometer tip on which the end bearing is developed. The cone has an apex angle of $60°$ and forms the bottom part of the cone penetrometer.

Cone penetrometer (penetrometer tip)

The lower part of the penetrometer.

Cone resistance, q_c

Measured cone resistance, q_c, is found by dividing the measured force on the cone, Q_c, by the cross-sectional area, A_c ($q_c = Q_c/A_c$)

Data channel

The device to signal from a penetrometer to recording apparatuses.

Dissipation Test (DT)

A test when the decay of the pore water pressure is monitored during a pause in penetration.

Friction ratio, R_f

The ratio, expressed as a percentage, of the sleeve friction, f_s, to the cone resistance, q_c, both measured at the same depth [$R_f = (f_s/q_c \cdot 100)$].

Friction reducer

A local enlargement on the push-rod surface, placed at a distance above the cone penetrometer, and provided to reduce the friction on the push rods.

Friction sleeve

The section of the cone penetrometer upon which the sleeve friction is measured.

Incremental loading cone penetration test mode

After stopping the penetrometer is tested by stepwise - increasing static load, the load being constant on each step.

Measuring system

The system includes the measuring devices themselves and the means of transmitting information from the penetrometer tip to where it can be seen or recorded. It consists of sensors indicating soil resistivity to pushing (for special penetrometers and other sensors), data channel and recording apparatuses.

Penetration depth

Depth of the base of the cone, relative to a fixed horizontal plane.

Penetration length

Sum of the length of the push rods and the cone penetrometer, reduced by the height of the conical part, relative to a fixed horizontal plane.

Penetrometer (probe)

The device immersed in soil during CPT. It consists of a push rod (rods) and a cone penetrometer (penetrometer tip).

Penetrometer electric

The penetrometer with a penetrometer tip consisting of the cone, friction sleeve, any other sensors and measuring systems as well as the connection to the push rods.

Penetrometer hydraulic or pneumatic

The penetrometer in which uses hydraulic or pneumatic devices built into the penetrometer tip.

Penetrometer mechanical

The penetrometer in which uses a set of inner rods to operate the penetrometer tip.

Penetrometer special

The term "special penetrometer" embraces the group of penetrometers capable to measure, besides soil resistance to penetration, some additional characteristics of soil and/or control CPT procedure. Pore water pressure, temperature, radiation, electrical resistivity, seismic, inclinometer and other sensors may be used as additional measuring devices.

Pore pressure, u

The pore pressure, u, is the fluid pressure measured during penetration and dissipation or relaxation-creep penetration testing.

Push rod

The push rods are a string of rods for transfer of compressive and tensile forces to the cone penetrometer.

Pushing and extraction equipment

The load-bearing mechanical, hydraulic or pneumatic device to push or extract a penetrometer.

Quasi-Static Test (QST)

Penetrometer testing is done in constant, slow (below the standard one), controlled speed; as a rule, a series of tests with stepwise increasing speeds is applied at a depth.

Recording apparatus

The device used for recording the data obtained by a penetrometer (soil resistivity, etc.).

Relaxation-Creep Test (RCT)

In relaxation-creep test when the penetrometer stops at the given depth, the load acting on it and the speed of its immersion gradually decrease with reduced intensity due to neighboring soil relaxation and creep. The test is carried out by oil supply shutdown to penetrometer pushing hydro-jacks. During the test one can measure penetrometer setting, temperature, pore pressure, etc. As a rule, the length of the test is about 5-10 min. It is determined by the specified conditional stabilization criterion for one of the measured parameters or specified stabilization time.

Sleeve friction (friction resistance), f_s

Measured sleeve friction, f_s, is found by dividing the measured force acting on the friction sleeve, F_s, by the area of the sleeve, A_s ($f_s = F_s/A_s$)

Support-anchor equipment

The equipment incorporating pushing and extraction equipment.

ABBREVIATIONS IN PUBLICATIONS ON CPT

ASCE	=	American Society of Civil Engineers
ASTM	=	American Society for Testing and Materials
CCE	=	Cylindrical cavity expansion
CCPT	=	Conductivity CPT
CCPTU	=	Conductivity CPTU
CEN	=	Comité Européen de Normalisation
CHT	=	Crosshole test
CID	=	Consolidated Isotropic Drained
CIU	=	Consolidated Isotropic Undrained
CP	=	Cone Pressuremeter
CPT	=	Cone Penetration Test
CPTT	=	Cone Penetration Test with Temperature Measurement
CPTU	=	Cone Penetration Test with Pore Pressure Measurement (Piezocone Test)
DMT	=	Flat Dilatometer Test
DP	=	Dynamic Probing Test
DPT	=	Direct Push Technologies
DSS	=	Direct Simple Shear
EN	=	European Standard
FC	=	Fines Content
FFD	=	Fuel Fluorescence Detector
EN	=	Euronorm (European standard)
ENV	=	European pre-standart
ERT	=	Electrical Resistivity Test
ESOPT	=	European Symposium on Penetration Testing
FDP	=	Full Displacement Pressuremeter
FDT	=	Flexible Dilatometer Test
FVT	=	Field Vane Test
HIM-probe	=	High-frequency Impedance Measuring probe
ICSMFE	=	International Conference of Soil Mechanics and Foundation Engineering
IRTP	=	International Reference Test Procedure
ISOPT	=	International Symposium on Penetration Testing
ISSMFE	=	International Society of Soil Mechanics and Foundation

		Engineering
ISSMGE	=	International Society for Soil Mechanics and Geotechnical Engineering
LCPC	=	Laboratoire Central des Ponts et Chaussées
LIF	=	Laser Induced Fluorescence
MPM	=	Ménard Pressuremeter
NAPL	=	Non Aqueous Phase Liquid
NC	=	Normally Consolidated
ND	=	Nuclear Density (probe)
NDT	=	Nuclear Density Test
NM	=	Neutron Moisture (probe)
OC	=	Overconsolidated
OCR	=	Overconsolidation Ratio
OED	=	Odometer Test
PBP	=	Pre-Bored Pressuremeter
PLT	=	Plate Loading Test
PMT	=	Pressuremeter Test
RCPT	=	Resistivity CPT
RCPTU	=	Piezocone with Resistivity Module
SBP	=	Self Boring Pressuremeter
SCAPS	=	Site Characterization and Analysis Penetrometer System
SCE	=	Spherical cavity expansion
SCPT	=	Seismic CPT
SCPTU	=	Seismic CPTU
SPT	=	Standart Penetration Test
TC	=	Triaxial Compression
UCT	=	Unconfined Compression Test
UU	=	Unconsolidated Undrained
UV	=	Ultra Violet
Video-CPT	=	Cone Penetration Test with miniature video camera
WST	=	Weight Sounding Test

1 BASIC CONCEPTS, APPLIED EQUIPMENT, SHORT RETROSPECT OF CPT

1.1 BASIC CONCEPTS

Under **penetration** we understand the process of special equipment immersion into the soil – that is the penetrometer – and cone resistance indexes measurement in relation to that immersion. This technique allows quick investigation of in-situ soil usually 10…20 m in depth. In accordance with the CIS International standards – the GOST 19912-2012 – penetration is called **static** (Cone Penetration Test – CPT) if it occurs under the static pushing load and it is called **dynamic** (dynamic probing – DP) in case the load is impact or impact-vibrational. **Penetrometer** is a device made up of a "push rod" (metal rod) and a penetrometer tip ("cone") which is fixed on the rod's end. Pushing load (static or dynamic) is transformed to the cone through the push rod which is composed of several sections built-up during penetration, or of only one section. Figure 1.1 shows dimensions and forms of the cones.

The book carefully describes just only static CPT, as to dynamic probing test the books (Trofimenkov & Vorobkov 1981, Rubinshtein & Kulachkin 1984) could be recommended. Depending on the cone penetrometer design CPT penetrometers fall into two types in accordance with the CIS Standard GOST 19912-2012 (fig. 1.1a):

Type I (mechanical CPT) – penetrometers with a penetrometer tip consisting of a cone and a mantle,

Type II (electrical CPT) – penetrometers with a penetrometer tip consisting of a cone and a friction sleeve.

In type I penetrometers part of the tip located above the cone is called **the mantle**, in type II – **the friction sleeve**. The mantle is firmly connected to the cone, the friction sleeve is not.

In CPT the following values are usually evaluated:

- **Measured cone resistance** q_c (MPa or kPa) – cone resistance strength to cone penetration related to cone's area of base (the term "frontal penetrometer resistance" is also used.

- **Unit sleeve friction resistance** f_s (MPa or kPa) is evaluated only in type II penetrometers; it is the soil side friction resistance on a short sector to the friction sleeve, related to the friction sleeve side area.

- **Total side friction resistance** Q_s (kN) is evaluated only in type I penetrometers; it is the soil side friction resistance to the embedded part of the push rod.

In practice the values f_s & Q_s are sometimes referred to one and the same term – "side cone resistance" or "side friction resistance".

It is important to underline that cone resistance to the mantle of type I penetrometer is not evaluated as such but is considered to be a part of the value q_c, i.e. cone resistance. Thus, the indexes being measured with types I & II penetrometers differ not only in specifying side friction resistance but that of cone

1

resistance which in type I penetrometers appear to be higher than that of type II (see section 1.2 for details).

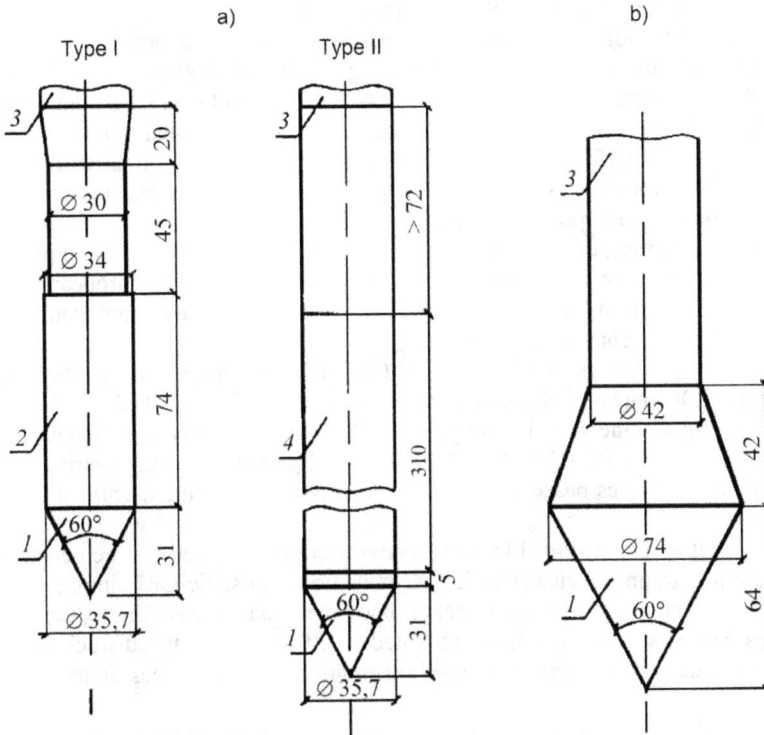

Fig. 1.1 Schematic design of cone penetrometer in accordance with the CIS Standard GOST 19912–2012:

a – for CPT; b – for dynamic probing test (impact); 1 – cone, 2 – mantle, 3 – push rod, 4 – friction sleeve

In dynamic probing test **"parameter"** n – that is a number of hammer impacts after which the penetration depth is measured – proves to be the measured index. **"Dynamic cone resistance"** p_d, is also used in a number of cases; it represents specific cone penetration strength related to penetration depth unit (m).

For years works on standardization of equipment and CPT techniques have been done worldwide. Non-Russian standards do not greatly differ from those of CIS mentioned earlier as the cone dimensions as well as the techniques of penetrating are the same in most countries. The executive committee of International Society of Soil Mechanics and Geotechnical Engineering (ISSMGE) ratified the first International Standard for CPT in 1977. International reference test procedure for CPT (IRTP) was adopted in 1988. The content aims to complete and specify the International Standard (Proposed European Standard 1977) mentioned above without changing it significantly. Improved IRTP for the CPT and the CPTU (Cone Penetration Test with pore pressure) was adopted in 1999 and revised 2001.

In 1997 Comite Europeen de Normalisation (CEN) adopted the temporary Normative Document ("pre-standard") on geotechnical designing – ENV-1997-3 being the part of the system Eurocode 7. ENV-1997-3 included CPT equipment and technology requirements almost the same as that mentioned above (Proposed European Standard 1977, IRTP 1989). As compared to these Normative Documents ENV-1997-3 presented issues on CPT applications completely and in detail. Issues on data obtained interpretation, tables and formulae for soil testing, pile bearing capacity calculations, etc. were shown in EVN-1997-3 equally to CPT equipment and technology requirements, but these were not shown in (Proposed European Standard 1977, IRTP 1989). In 2006 CEN "converted" ENV-1997-3 into European Standard EN 1997-2 having changed it slightly.

Great attention is being paid to CPT data interpretation in the Russian Normative Documents. Necessary tables are given in the Federal Normative Document on Engineering Investigations SP 11-105-97 (1998); National Building Code of Russia (SNiP 2.02.03-85* & SP 50-102-2003) describe pile bearing capacity calculation techniques based on CPT data. The issues are being discussed further in detail.

Penetration is performed by **CPT systems** comprising besides the penetrometer a special penetrating device (pushing & extraction – in static CPT, impact one – in dynamic probing test), support-anchor equipment and measuring systems. CPT systems can be self-propelled, i.e. mounted in a truck or a tracked truck or mobile ones, i.e. mounted in a trailer, portable demountable or as pieces attached to drill units.

Depending on thrusts necessary to push down the penetrometer as well as the evaluated cone resistance indexes range, CPT systems are divided into light, medium and heavy ones in accordance with the CIS Standard GOST 19912–2012.

Table 1.1 CPT systems classification by Standard GOST 19912–2012

Type of equipment	Penetrometer pushing & extraction thrust capacity	Cone resistance indexes range		
		q_c (MPa)	f_s (kPa)	Q_s (kN)
Light	up to 50 kN inclusively	0.5…10	2…100	0.5…10
Medium	over 50 up to 100 kN inclusively	1…30	5…200	1…30
Heavy	over 100 kN	1…50	10…500	2…60

In practice, largely worldwide, besides mechanized CPT systems there exist the simplest portable units aiming at soil investigation by manual pushing (driving) the special rod with a tip on it. The units are generally called "manual penetrometers", "manual probes", "probes", etc. They are generally used in technological control of earth bulk density, soil investigations in pits when expertizing structures or in engineering survey of weak soils. Here the depth of penetration is not very high. It can be 0.3...0.7 m (according to thickness of filled layer) when controlling the bulk density, and it can achieve 7...10 m in weak soils, e.g. silt or sapropel.

The penetrometer parameters (tips' forms & diameters, penetration technique, etc.) do not agree with the Standards (GOST 19912-2012, Proposed European Standard 1977, IRTP 1999) that is why some professional engineers consider probe application as an independent method of testing, apart from CPT. As a rule, the empirical dependencies for soil properties or pile bearing capacity testing by CPT given in the Codes are not relevant to the manual probes. Decoding of such kind of simplified sounding (i.e. pushing down of probes with cone resistance evaluation) requires the following to be taken into account: application of own dependencies showing the specific features of such tests, specific features of the probes applied, specific features of locality. Hence, quantitative estimation of soil conditions based on pushing down the probes appears to be quite efficient if it is followed by calibration of the data obtained related to the soil under study. Such a gauging presupposes local empirical dependencies determination between standard properties of investigated soil and the indexes of its resistance to the probe penetration.

Due to its simplicity and cheapness, application of probes is believed to be a useful addition to standard tests; however the range of their application is very small.

1.2 PRINCIPAL PROBLEMS RESOLVED WITH CPT

Standard GOST 19912–2012 reveals the following problems being resolved with CPT (jointly with other kinds of engineering-geological works or separately):

- Identification of engineering-geological elements (thickness of layers and lentils, bounds of different soil types distribution).
- Dimensional changeability estimation of soil structure and properties.
- Rocky soils and large fragmental rocks occurrence depth determination.
- Quantitative estimation of physical-mechanical properties of soils (density, modulus of deformation, angle of internal friction, adherence of soils, etc.).
- Determination of both coextensive compactness degree and strengthening of soils.
- Possibility of pile driving estimation and their penetration depth determination.
- Data collection for pile foundations calculation.
- Identification of experimental sites location and depth of excavation during field tests as well as identification of soil samplers location for laboratory testing.
- Quality control of geotechnical works.

In the International Standards (Proposed European Standard 1977, IRTP 1999) CPT is generally understood in the same way.

In fact, both in Russia and worldwide the range of problems being resolved with CPT is wider than that given in the Standards. For example, in Russian publications there are books on CPT application for collapsible soils estimation, soil conditions testing (frozen or thawed), a share of rubbles in moraine deposits identification, evaluation of coefficient of soil reaction when applying the model "slab (or beam) on cushion course", selection of experimental sites with" the "strongest" or "the weakest" soils, etc.

In their technical report at the symposium CPT-95 in Linkoping (1995) Lunne & Keaveny (1995) analyzed subjects of reports of 32 member countries and defined 8 trends in CPT applications:

- CPT data general interpretation, soil stratification investigation, soil testing – in the reports of almost all 32 member countries of CPT-95.
- Excavation works, determination of underground working, bulking – in the reports of 53 % of member countries.
- Natural bed foundations calculations, their bearing capacity and settings – 75 %.
- Pile bearing capacity – 75 %.
- Estimation of pile behavior under horizontal load, settings of pile foundations – 12 %.
- Foundation soil reinforcement control – 3.4 %.
- Soil liquefaction estimation – 2.8 %.
- Ecological issues (estimation of environmental contamination) – 25 %.

Thus, the problems resolved with CPT have extended and probably will be extending in future. Nevertheless, such problems as soil stratification investigation, soil testing, and pile bearing capacity testing will probably be of primary importance.

The problems resolved with CPT may be divided into two groups:

Engineering-geological problems connected with experimental site (section) lithological structure testing, i.e. different strata distribution bounds testing, their appearance, condition, dimensional variability estimation, etc.

Constructional problems, connected with quantitative data obtaining to calculate concrete foundations (especially the pile ones), quality control of bulking, stability of slopes, etc.

It is obvious that the given division is quite conditional as engineering-geological and constructional problems are closely interrelated and cannot be separated.

When considering constructional problems, i.e. CPT data application in designing it is necessary to define two principally different approaches. M. Jamiolkowski (one of the Presidents of ISSMGE) called them "direct" and "indirect" (Jamiolkowski 1995).

In *direct approach* correlation of CPT data is used with foundation system behavior without standard soil testing (angle of internal friction φ, specific adherence c, modulus of deformation E, etc.). Applying this method it is possible to test pile bearing capacity, pressure under shallow foundation corresponding to bearing capacity depletion or other condition, etc. As a rule, the formulae used here do not consider non-standard properties of soil (φ, c, E, v), but more concrete indexes of cone resistance, e.g. cone resistance under pile foot or on its side face zones which are specified with CPT data (q_c, f_s, Q_s).

Indirect approach means preliminary soil testing (φ, c, E), formula and model evaluation of soils possessing the following properties is carried out later. The more complicated calculations connected with modeling of stress-strain condition of the ground massif are being done by means of this method.

Direct approach, empirical as it is, may result in more accurate data in the limited series of conditions. However, it demands precautions to be taken beyond the mentioned series of conditions. Additional adjustment is usually necessary here.

Indirect approach considers physical essence of the processes and it is better adapted to different mathematical models application (design models), i.e. the range of regularities obtained application is wider than that in direct approach, but its accuracy is usually lower.

As it was mentioned above, in accordance with the Standard GOST 19912-2012 CPT can be applied jointly with other kinds of engineering-geological works or separately. The greatest effect is achieved when using CPT together with other methods of soil investigation. To a considerable extent it is the method of *relative* estimation; it is desirable to "anchor" the results to a reliable "reference", that is the most reliable traditional methods of soil investigation. When estimating lithological structure of the site, boreholes may serve as the "reference", when estimating pile bearing capacity – static tests of piles, when estimating properties of soil – standard field and laboratory tests results, etc. It is natural that this CPT feature should not be understood as a primitive requirement to accompany CPT with parallel works on drilling, static testing of piles, etc. Calculations on CPT data may be "anchored" to the results of investigations carried out earlier on adjacent sites in case they are analogous to the conditions under study. If there is a lack of any possibility to anchor, it is necessary to introduce relevant reliability coefficients in the calculations which compensate anticipated inaccuracy. Here, reliability coefficients must be applied in the least favorable situations of the designed structure. Principles of adjustment are given in section 6.2 in detail.

There are many possibilities in CPT application but in a number of cases it faces the limitations. The static penetrometer may fail "to pick" dense sands, gravels, pebbles or even layers or lentils of such deposits. Here, data on drilling or dynamic probing test must be taken into account.

1.3 APPLIED EQUIPMENT

At present several dozens of penetrometer types are used worldwide. Considering the measurement techniques they fall into three groups:
- Mechanical cone penetrometers in which the system of inner push rods to transfer thrust onto the cone tip is used.
- Hydraulic (pneumatic) cone penetrometers in which hydraulic (pneumatic) devices are mounted in the tip.
- Electric (tensometric, etc.) cone penetrometers in which measurements are done with electric (tensometric, etc.) load cells.

In accordance with the International Standards (Proposed European Standard 1977, IRTP 1989) it is important to show on the graphs whether the penetrometer and

CPT technique are standard and the type of the applied penetrometer as well. For this, special indexing is defined. In the column "type of penetration" letter "S" is written if penetration technique agrees with the Standard (Proposed European Standard 1977), then the type of penetrometer is given: mechanical cone penetrometer – "M", hydraulic cone penetrometer – "H", electric one – "E". In some well-known penetrometer models the number is written after the indexes "M", "H" or "E". It shows specific design of the penetrometer (in accordance with the Standard specified numeration) , e.g. Danish mechanical cone penetrometer without a friction sleeve is marked as M1, the same one with a friction sleeve – M2, tensometric cone penetrometer Degebo – E2, etc.

1.3.1 Mechanical cone penetrometers

Until recently mechanical cone penetrometers were the most popular ones but during last 20...30 years they have been displacing by tensometric cone penetrometers. Geometries of these cone penetrometers are presented in a number of Russian and non-Russian books (Bondarick 1964, Sanglerat 1971, Trofimenkov & Vorobkov 1981, Cosstay & Sanglerat 1981, Trofimenkov 1995). Figure 1.2 shows the most typical mechanical cone penetrometers.

Mechanical cone penetrometers are easy to manufacture, mount, repair and they are quite cheap. However, the problems being resolved with these penetrometers appear to be quite limited. They include pile bearing capacity testing, soil testing, lithological origin of soils (clays, sands, etc.) must be known beforehand (according to the results of drilling). It is more difficult to supply these cone penetrometers with additional adjustments in order to control verticality of penetrometer position, pore pressure measurement or geophysical procedures.

Application of friction sleeve in these penetrometers is also more difficult than that in the tensometric ones despite the fact that friction sleeve was firstly suggested by Begemann (Netherlands) in 1953 just for mechanical penetrometers (Begemann 1953). Taking all these additional adjustments into account, mechanical cone penetrometers lack their principal advantage – simplicity.

It is necessary to point out that it is more difficult to provide high productivity of CPT applying mechanical cone penetrometers since the measurement procedure by means of manometers and dynamometers is always considered to be less suitable and more prolonged than in tensometric penetrometers with automatic reporting of results.

High probability of significant inaccuracies is considered to be one of the disadvantages of mechanical cone penetrometers. It is due to friction initiation between the inner rod (mark 2, fig. 1.2a or mark 3, fig. 1.2c) and inner surface of the tube that is the outer rod where the inner rod is placed (mark 3, fig. 1.2a or mark 4, fig. 1.2c). The additional friction may significantly tamper data on cone resistance q_c.

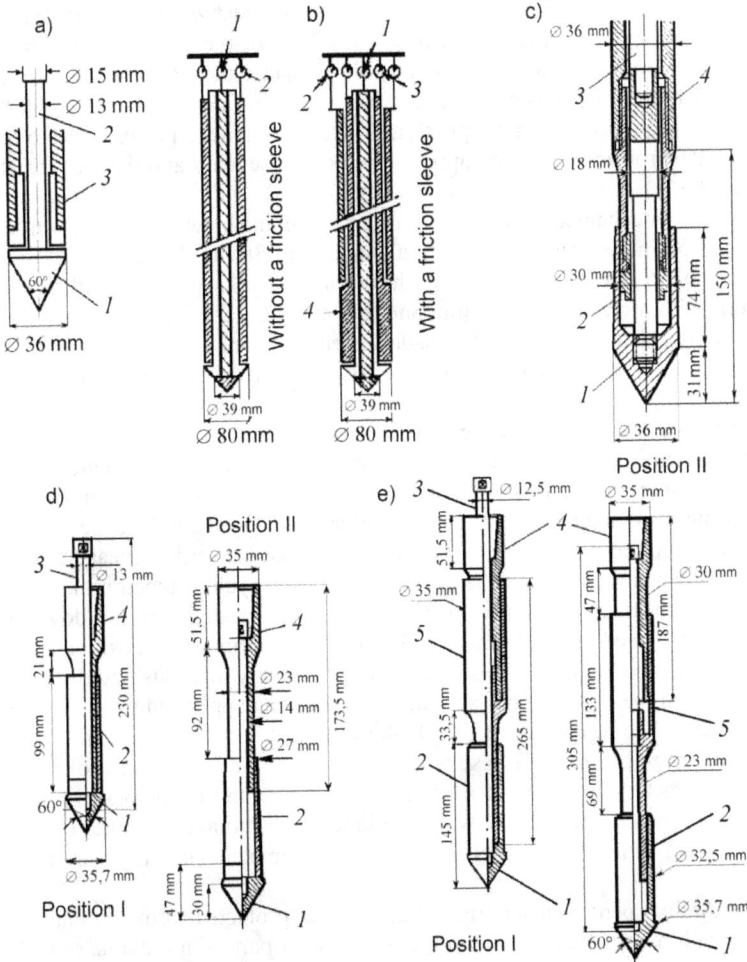

Fig.1.2 Schematics of mechanical cone penetrometers:

a – ordinary penetrometer without a mantle and friction sleeve: 1 – cone, 2 – inner push rod, 3 – outer push rod; *b* – Andina penetrometer (two variants): 1 – cone resistance measurement, 2 – the same one up to the total length of push rod; 3 – the same one up to the friction sleeve, 4 – friction sleeve, *c* – Fundamentproject penetrometer in Soviet (Russian) systems S-979, SP-59: 1 – cone, 2 – mantle, 3 – rod (lower end of the inner push rod), 4 – lower end of the outer push rod; *d, e* – Danish penetrometer without a friction sleeve and with it correspondingly: 1 - cone, 2 – mantle, 3 – rod, 4 – lower end of the outer push rod, 5 – friction sleeve

8

Besides, it is possible of the soil to penetrate contact zone of mantle 2 and outer rod 4, compressive deformations of push rods, bent of the outer push rod tampering total side friction resistance Q_s may have an impact either. Chapter 2 describes the issues in detail. If the expected inaccuracies are significant (e.g. 200...300 %) and they appear unevenly, it is not a problem for an experienced professional to discover them in CPT procedure. If they appear more or less evenly and the indexes are high (e.g. 20...60 %), it is not easy to discover them.

The first mechanical cone penetrometers were analogues to that shown in figure 1.2a where ordinary metal cone 1 with a diameter of 36 mm fixed on push rod 2 with a diameter of 15 mm was used. The push rod (inner one) is placed into the tube (outer push rod) with a diameter of 36 mm and this excludes friction between soil and the inner push rod. Both push rods are made of sections 1 m in length. Such penetration procedure may be carried out in two ways: step-by-step (discontinuous) and continuous.

Under discontinuous CPT the cone penetrometer is firstly penetrated 20 cm deep into the soil; pushing thrust capacity is measured, i.e. total cone resistance to pushing. Then the cone is pushed down 6...8 cm deep (in fixed outer push rod) by means of the inner push rod and cone resistance is evaluated. Then again the cone penetrometer is penetrated 20 cm deep into the soil, pushing thrust capacity is measured; the cone is pushed down, cone resistance is evaluated, etc. (the cycle is repetitive).

Under continuous CPT the cone with both push rods is pushed down into the soil with even speed of 1...2 cm/sec. The cone is in a projected position (in relation to the outer push rod); this allows measuring of not only total pushing thrust capacity but its share on the cone. To measure the share one can apply additional device, e.g. a dynamometer. The results are recorded after every 20 cm of penetration.

Here, reliability of CPT results is not high. The disadvantages of mechanical cone penetrometers are fully manifested. In the lower end of the penetrometer (near the cone) soil enters the zone between the inner push rod and the outer one, the rods bend, etc. Due to these, at present ordinary penetrometers (fig. 1.2a) are not widely used.

The Andina penetrometer (fig. 1.2b) is better protected from soil entering; in it the cone with a diameter of 80 mm has a projected part with a diameter of 39 mm. The penetrometer may have a more complicated geometry – with a friction sleeve. In the former USSR and in Russia the Andina penetrometers were not applied, but the analysis of their geometries allows assuming that most of the disadvantages of mechanical cone penetrometers are typical for the Andina penetrometers. Here, the technique of cone resistance evaluation with the projected cone causes additional problems in pile bearing capacity calculations. The soil will penetrate in the cavity being formed over the projected part of the cone one way or another. This will decrease its resistance compared to resistance under the pile foot. In any case special technique is necessary in these pile bearing capacity calculations. It is important to underline that dimensions of the Andina penetrometer differ from those given in the International Standard.

Mantle cone penetrometers are considered to be more upgraded (fig. 1.2c, d and e); hazards of soil penetration into the outer push rod are reduced to minimum for them. Despite their name (Danish) these penetrometers have been widely applied not in Denmark but in Netherlands (Holland) where they have been used since 1947.

Russian mantle cone penetrometer (fig. 1.2c) is intended to be used in continuous CPT which is considered in the Standard GOST 19912-2012 as compulsory. Penetration shall be carried out with the speed of 1.2±0.3 m/min; intervals in penetration are permitted if only to lengthen push rods.

Tests with Danish cone penetrometer are usually carried out step-by-step, although it can be applied in continuous CPT. Step-by-step CPT is carried out in the same way as with the ordinary cone penetrometer (fig. 1.2a). The cone penetrometer in position I (fig. 1.2d) is penetrated 20 cm deep into the soil, pushing thrust capacity is measured, i.e. total cone resistance to pushing. After that the mantle cone is pushed down 6...7 cm deep (transition to position II, fig. 1.2d) and cone resistance is evaluated. Then the outer push rod is penetrated gripping the mantle and the penetrometer as well, again the whole system is pushed down 20 cm deep and pushing thrust capacity is measured. Side friction resistance is evaluated as the difference between two measurements. Then the cone is pushed down, cone resistance is evaluated, etc. (the cycle is repetitive).

The Dutch specialist Begemann (1953) suggested placing an additional device behind the mantle – "friction sleeve" in the form of a cylindrical zone where "soil side friction" is evaluated simultaneously (fig. 1.2e).

Begemann cone penetrometer tests (fig. 1.2e) are carried out step-by-step in three stages (phases). Position I is a starting location for the penetrometer (fig. 1.2e). At first, mantle cone is penetrated 4 cm deep and cone resistance is evaluated. Then the cone and friction sleeve are penetrated and total cone resistance is evaluated (transition to position II, fig. 1.2e). Unit sleeve friction resistance is evaluated as the difference between these two measurements. After that the outer push rod is projected gripping the friction sleeve (the penetrometer is again in position I), the system is penetrated 20 cm deep. At the last stage only the outer push rod is penetrated 20 cm deep; the friction sleeve passes the distance of 16 cm, the mantle cone – 12 cm. The cycle is repetitive. Begemann cone penetrometer tests provide with the additional information that is used in soil type estimation (sand, clay). It is also advantageous in side face pile resistance calculations.

Rather complicated measurement procedure (three stages), unequal depth of the cone and friction sleeve placement (the cone is placed 25 cm lower than the friction sleeve) are considered to be the disadvantages of Begemann cone penetrometers (besides the disadvantages of mechanical cone penetrometers mentioned above). Nevertheless, the idea of friction sleeve application appeared to have been quite promising and later it was developed in many countries, the USSR among them. It is necessary to point out that merits of the device application refer to tensometric cone penetrometers but not to mechanical ones. In most penetrometers the friction sleeve was adjusted to the cone, i.e. the mantle and the interstitial zone were excluded. The issue is discussed in detail in the section on tensometric penetrometers.

Manual probes are a version of mechanical cone penetrometers (Terzaghi & Peck 1958, Bondarick 1964, Sanglerat 1971, Cosstay & Sanglerat 1981, Amaryan 1990). Their cones' designs prove to be simple without any built-in adjustments (fig. 1.3). Designs and dimensions of the cones can be non-standard. Most of the probes are provided with removable cone tips of different diameters. To minimize the pushing thrust which shall be done by hand by one or two workers, the diameter of the push rod is usually made 1.5…3 times smaller than that of the cone.

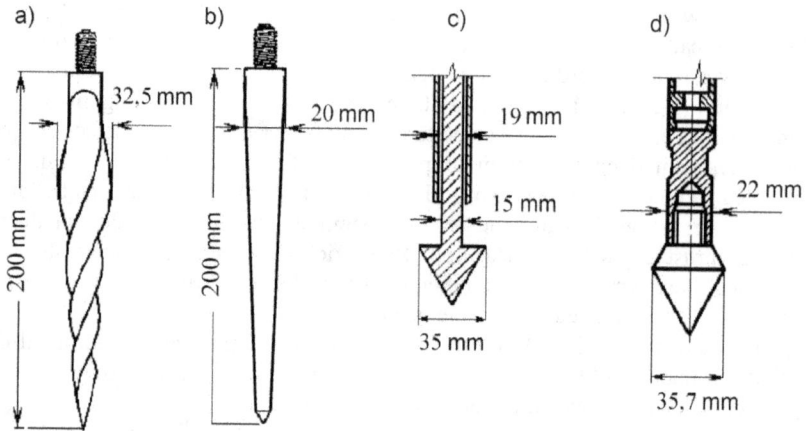

Fig.1.3 Schematics of probe cone tips:

a – screw-shaped (Swedish), *b* – oblong cone (Danish), *c* – cone with a protective tube (Dutch), *d* – Russian cone probe P-5

Diameters of most cones vary between 15…36 mm but they can be larger either, e.g., Soviet (Russian) probe "Penetrometer P-5" intended to investigate soft soils has removable cones with diameters of 35.6 mm, 50.8 mm and 71.4 mm (Amaryan 1990).

1.3.2 Hydraulic (pneumatic) cone penetrometers

Hydraulic penetrometers as well as pneumatic ones are not widely used in practice and in Russia they have hardly been ever applied. As it was said earlier, special hydraulic (pneumatic) devices are mounted in these penetrometers. Hydraulic penetrometers do not have central (inner) push rod. Geometries of the penetrometers are shown in figure 1.4 (Sanglerat 1971, Trofimenkov 1995).

There is a hydro cylinder with a piston in the hydraulic penetrometer cone tip. During CPT soil pressure on the cone is transferred through piston 4 to oil 5 in the cylinder. Pressure measurement is done with the manometer being mounted in the ground surface (in the aerial part of CPT system). Instead of oil there is air in the pneumatic cone penetrometer cylinder; the air is compressed under soil pressure on the cone. Thus, cone resistance is evaluated.

Figure 1.4 shows hydraulic cone penetrometer Sol-Esais that with regard to the Standards (Proposed European Standard 1977, IRTP 1989) has the enlarged diameter of 45 mm. It is due to technological difficulties in hydraulic devices of small sizes manufacture and maintenance.

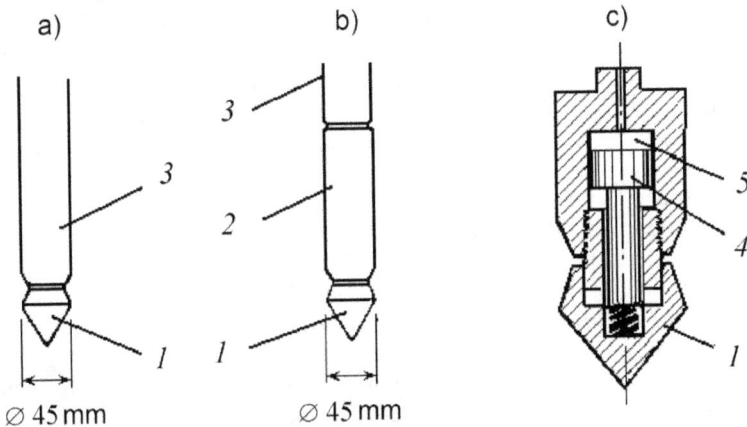

Fig.1.4 Schematics of hydraulic cone penetrometers:

a, b – designs of Parez cone penetrometers without a friction sleeve and with it correspondingly, *c* – design of a hydraulic cone penetrometer tip (Sol-Esais – France): 1 – cone, 2 – friction sleeve, 3 – push rod, 4 – piston, 5 – liquid to estimate soil resistance by its pressure

The authors failed to find any books on generalization of hydraulic cone penetrometers applications to consider their factual reliability, convenience in maintenance, etc. Nevertheless, the hydraulic cone penetrometers are expected to match the mechanical cone penetrometers in maintenance convenience; their manufacture is more complicated but they shall excel the mechanical cone penetrometers in reliability of results. It is obvious that these will depend upon the geometries of the penetrometers as well as the quality of their manufacture.

1.3.3. Tensometric cone penetrometers

At present tensometric (load cell) cone penetrometers are considered by specialists as the most promising. Lunne T., Robertson P.K. and Powell J.J.M. (2004) put forward the following characteristics of tensometric cone penetrometers:

- The elimination of possible erroneous interpretation of test results caused by friction between inner rods and the outer tubes.
- A continuous testing with a continuous rate of penetration without the need for alternative movements of different parts of the penetrometer tip and no possibility for undesirable soil movements.
- The simpler and more reliable electrical measurement of the cone resistance with the possibility for continuous readings and easy recording of the results.
- High sensitivity allowing obtaining accurate readings in very soft soils.

Non-Russian specialists underline the simplicity in additional adjustments to be mounted in tensometric cone penetrometers, firstly – friction sleeves and then a lot of other adjustments as well in order to control verticality of penetrometer position, pore pressure measurement, radiation temperature measurements, etc. Tensometric cone penetrometers are good in CPT of water reservoirs' bottoms or offshore. When using other cone penetrometers the given problems appear hard to be resolved.

The principal disadvantages of tensometric cone penetrometers are as follows: high cost (their cost is 10…20 times higher than that of mechanical cone penetrometers), availability of highly skilled personnel being able to maintenance the device.

Lunne et al. (2004) describe three main design types of tensometric cone penetrometers (fig. 1.5):

- Cone resistance (q_cA_c) and sleeve friction (f_sA_s) are evaluated by two independent load cells both in compression (q_c, f_s – cone resistance and unit sleeve friction resistance, A_c, A_s – areas of cone base & friction sleeve side).
- The sleeve friction compressive load cell of the previous one is replaced by one in tension.
- The sleeve friction load cell in compression (q_cA_s) records the total load from both the cone resistance and sleeve friction ($q_cA_c + f_sA_s$).

In the third type sleeve friction is obtained from the difference in thrust between friction and cone resistance load cells; this cone is often referred to as the *subtraction cone penetrometers.*

When applying tensometric cone penetrometers the results obtained are recorded automatically as "transformed" specific values q_c, f_s (kPa, MPa).

Lunne et al. (2004) classification does not embrace all types of tensometric cone penetrometers applied, some of them are well-known and widely used. E.g. tensometric cone penetrometer S-832 has been in use in the former USSR (now in Russia) since 1963 but it does not agree with the given classification as both of its load cells are in tension (cone and friction sleeve).

Lunne et al. (2004) present penetrometers with friction sleeve fixed to the cone. In some countries "the old" penetrometers with friction sleeve fixed 20…25 cm from the cone have still been applied together with those ones.

Figure 1.6 shows some typical designs of tensometric cone penetrometers being in use (Proposed European Standard 1977, Trofimenkov & Vorobkov 1981, IRTP 1989, Lunne et al. 2004, Ryzhkov & Goncharov 2006).

13

In the USSR tensometric cone penetrometers of CPT equipment PIKA-5 developed in NIIOSP were applied together with BashNIIstroy cone penetrometer (fig. 1.6d). The first PIKA-5 was manufactured in 1970s.

In accordance with International Standard IRTP 1999 the cone shall have a nominal apex angle of $60°$, the cross-sectional area of the cone shall nominally be 10 cm² (Ø 35.7 mm). Cones with a diameter between 25 mm ($A_c = 5$ cm²) and 50 mm ($A_c = 20$ cm²) are permitted for special purposes. The friction sleeve shall have a nominal surface area shall be 150 cm² (length 133.8 mm). In the CIS Standard GOST 19912-2012 the cone also shall have a nominal apex angle of $60°$, the cross-sectional area of the cone shall nominally be 10 cm² (Ø 35.7 mm). The friction sleeve shall have a nominal surface area shall be 350 cm² (length 310.0 mm). Thus, requirements of both Standards for the cone are almost the same, but for friction sleeve are different.

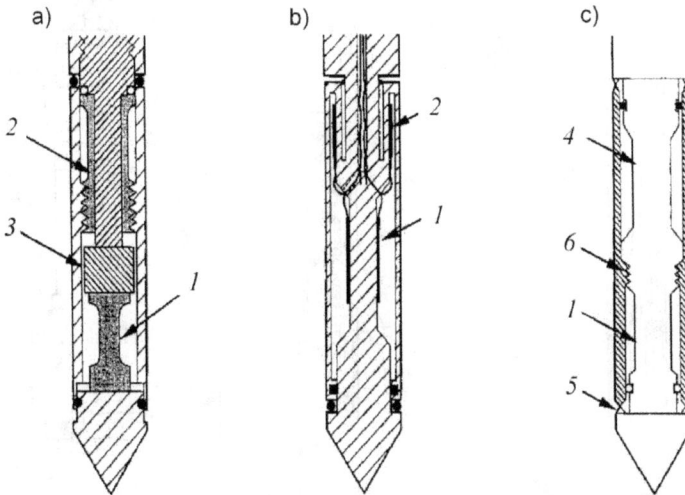

Fig. 1.5 Lunne et al. (2004) schematics of tensometric cone penetrometers: a – cone resistance and sleeve friction are evaluated by two independent load cells both in compression, b – the sleeve friction compressive load cell of (a) is replaced by one in tension, c – the sleeve friction load cell, in compression, records the summation of the loads from both the cone resistance and sleeve friction: 1 – cone load cell, 2 – sleeve load cell, 3 – point load cell overload protection device, 4 – cone+sleeve load cell, 5 – soil seal, 6 – thread

Designs of tensometric cone penetrometers are often more complex than those of mechanical ones. Although principles of tensometric measurements may be realized in different ways, complex electronic devices (bridge circuits, resistance strain gages, etc.) shall be applied, penetrometer waterproofness, its workability in a wide range of temperatures and its durability to shots occurred in soil shall be provided. Tensometric cone penetrometer assembly always requires high accuracy and strict technological instructions to be followed.

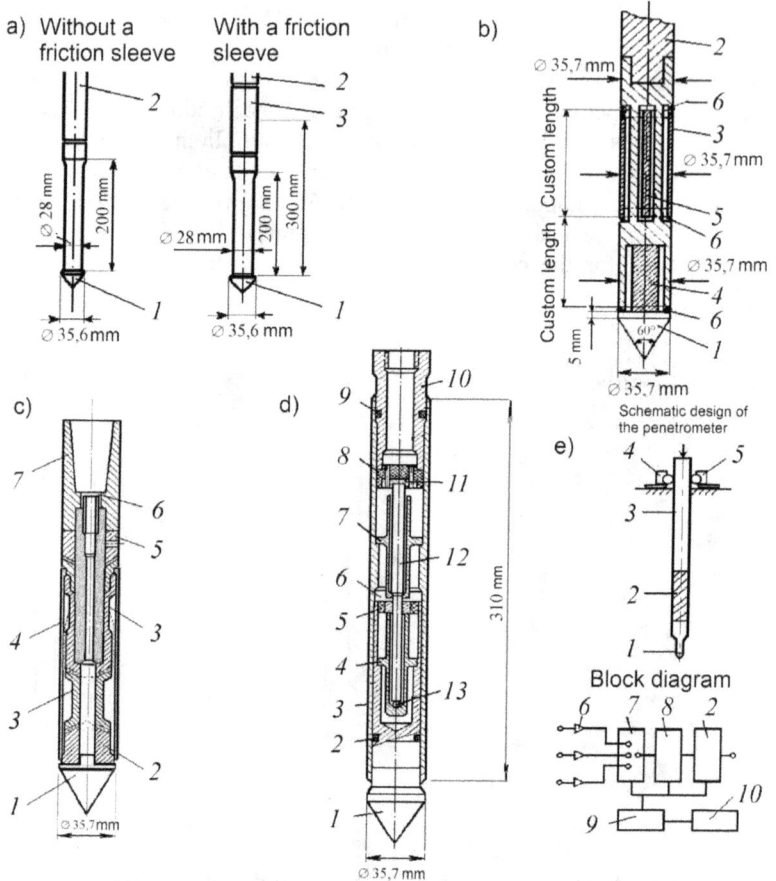

Fig.1.6 Schematics of tensometric cone penetrometers:

a – Delft electric penetrometer: 1 – cone, 2 – push rod, 3 – friction sleeve; *b* – Degebo electric penetrometer: 1 – cone, 2 – push rod, 3 – friction sleeve, 4 – cone load cell, 5 – sleeve load cell, 6 – seal; *c* – Fugro electric penetrometer: 1 – cone, 2 – seal, 3 – load cells (cone & friction sleeve), 4 – friction sleeve, 5 – adjusting ring, 6 – damp-proof member, 7 – push rod connection lock; *d* – BashNIIstroy S-832 penetrometer (USSR-Russia): 1 – cone, 2,9 – seal, 3 – friction sleeve, 4 – cone load cell, 5 – guiding disc, 6 – restricting ring, 7 – sleeve load cell, 8 – abut, 10 – push rod connection lock, 11 – thrust, 12 – cone and friction sleeve pushing load transmission rod, 13 – supporting ball; *e* – Japanese penetrometer with cable-free transmission of signals: 1 – cone, 2 – memory capsule, 3 – push rod, 4 – depth tester, 5 – microcomputer with a clock, 6 – pickup amplifiers, 7 – distributor, 8 – transformer, 9 – device operation automatic controller, 10 – batteries

15

Nevertheless, the manufacturing complexity is abundantly compensated by tensometric cone penetrometers ease of operation. Measurement and recording procedures may be done automatically; reliability of the obtained information is higher than that in other penetrometer types. The workers operate in comfortable environment. It is important to point out that the conditions described must agree with quality control since it is neither abolished nor ignored. When applying mechanical cone penetrometers manometer calibration procedure is to be done not less than once in 6 months in accordance with the International Standard (Proposed European Standard 1977), but when applying tensometric cone penetrometers dynamometer calibration procedure is to be done more often - not less than once in 3 months. Besides, CPT equipment requires regular in-situ control. According to the Standard GOST 19912-2012 gauging of measuring apparatuses is to be done not less than once in 3 months or 100 CPT location.

Tensometric cone penetrometers were proved to be quite efficient after many years of application. In particular, forty years of S-832 tensometric CPT rig operation (100...120 rigs in different regions of the former USSR and contemporary Russia) revealed that principal causes of the equipment failures are mechanical breaking of the thrust machine and its repairing but as to penetrometer failure occurrence it accounted for less than 10 % of the whole failures.

Occurrence of a cable is a disadvantage of tensometric (electric) cone penetrometers; its placement has always been a problem when designing the penetrometers; violation of its safety causes problems in maintenance and operation of penetrometers. Wireless penetrometers possessing built-in electronic memory are believed to be the most promising. Figure 1.6e gives example of one of them. These penetrometers are not widely used due to their high cost but they may be applied in future.

Cone resistance q_c in sands may be significantly higher than that in clayey soils. Nevertheless, any penetrometer needs to be suited to operate in both sandy and clayey soils. This causes definite problems as high measurement accuracy as well as wide range of loads seems to be hardly ever combined. Therefore, the constructors have to decrease the penetrometer measurement system accuracy providing its validity to operate in much denser soils. There are various approaches of the problem to be resolved, e.g. two-stage penetrometer was designed. Here, both small and larger cone resistance recordings are realized with different load cells (Trofimenkov 1995). In advanced cone resistance small resistance load cells become disconnected but those of larger resistance become connected. The simpler solution of the problem is realized in Russian CPT rig – S-832. The same load cells are used in the whole range of resistance, the penetrometer signals "are roughened" due to additional electric resistance operation when cone resistance is advanced. The system allows high accuracy of measurements in small cone resistance and lower accuracy (but acceptable) in larger cone resistance. The operator chooses necessary cone resistance measurement scale q_c and f_s depending on their values (the equipment allows to use three scales).

16

1.3.4 Special cone penetrometers with additional capabilities

Besides cone resistance and side friction resistance evaluation (or sleeve friction resistance), i.e. q_c, Q_s or f_s, some additional measurements are sometimes carried out in CPT. The latter may not refer to CPT (i.e. be independent tests), or they may be means of CPT methodical control.

Both in Russia and worldwide attempts to equip penetrometers with additional measurement devices have been undertaken. To a large extent these refer to tensometric cone penetrometers improvement. "Complicated" (special) penetrometers are usually applied in particular soils studies (silt, collapsible, permafrost, etc.). In non-Russian practice Cone Penetration Test with Pore Pressure Measurement (Piezocone Test) – CPTU as well as Seismic CPTU – SCPTU have been widely applied. Great interest depends on pore pressure theory popularity in most non-Russian countries.

Non-Russian experience in cone penetrometers with additional capabilities application is presented in Burns & Mayne (1998) in detail.

There is no united solution to the principal problem whether it is reasonable to combine devices for numerous tests conducting in one unit. In accordance with theory of reliability any complication of equipment, in particular addition of some parts has a negative effect on equipment reliability as a system. This makes special measures be taken in order to provide reliability of equipment (reservation, protection from malfunctioning, special operating duty application, etc.), but manufacture, maintenance and operation of the equipment increase in cost. Every time the problem comes to a reasonable limit of such complications to be economically and technologically justified. It is a common practice that in penetrometers only a few adjustments are required in a limited range of conditions. It is realized when applying special removable penetrometers with additional removable adjustments being chosen with regard to concrete conditions.

The additional removable adjustments and, hence, measurement procedures are as follows:

- Cone penetrometer verticality measurement.
- Pore pressure measurement.
- Additional mechanical tests (rotary shear, pressiometry, etc.).
- Geophysical measurements including seismic exploration, gamma radiation, neutron radiation, etc.
- Ecological measurements connected with exposition and estimation of contaminants' concentration in the soil (oil products, oils, acids, etc.).
- Measurement of soils' temperature.

There exist a lot of other adjustments but they are hardly ever applied and are not discussed in the book.

Here we consider just only equipment for additional measurements conducting. Expedience of cone penetrometers equipping with these adjustments as well as the technique of "complicated" cone penetrometers application are discussed in the following chapters.

Cone penetrometer verticality control is carried out by means of special miniature devices – *inclinometers* which are placed in the tensometric cone tip. T. Lunne et al. (2004) point out that placement and application of inclinometers are the simplest technical problem among numerous variants of penetrometers' equipping with additional measurement devices. Electromagnetic measurement technique is usually realized in inclinometers. Inclinations of one of the elements functioning as a plumb maintaining verticality of the penetrometer in any position are recorded. The inclinometer is usually placed in the center of the cone tip, sometimes behind the tip (behind the friction sleeve). In Russian CPT rig PIKA-17 verticality control sensor (inclinometer) is placed in special friction reducer located between the tensometric cone tip and the push rod.

In designs of cone penetrometers with inclinometers the automatic switch usually connects the inclinometer with recording device at every interception of the push rod (usually in 1 m). The device cancels penetrometer verticality inclination graph depending on the penetration depth. Figure 1.7 shows inclinometer location in one of the Furgo's (Trofimenkov & Vorobkov 1981) penetrometers:

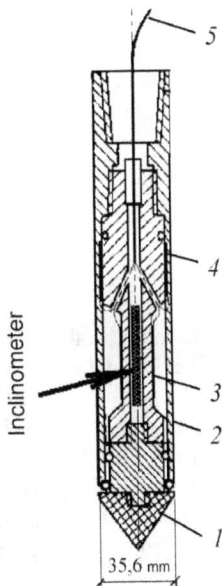

Fig. 1.7 Inclinometer location in cone penetrometer of Fugro:

1 – cone, 2 – friction sleeve, 3 – cone load cell, 4 – sleeve load cell, 5 – cable

Pore pressure measurement is done with special sensors located in the penetrometer. Lunne et al. (2004) consider pore pressure measurement device placement in the penetrometers to be a complex technical problem as the devices should meet a number of requirements being to a certain extent contradictory.

Thus, to ensure quickness of equipment reaction to pore pressure changes it is important that the filters applied would be of as small sizes as possible, pores should be totally filled with liquid possessing minimal viscosity. It is necessary that the filters would possess high permeability, their thickness to area ratio should be as small as possible, and flexibility of power transformers should also be minimal. At the same time, in order to ensure pores to be filled with liquid larger resistivity of the filter to air entrainment is necessary, this requires reduced permeability of filters and increased viscosity of liquid as well. A compromise should be found to combine the requirements and guarantee acceptable optimization.

Pore pressure measurement filter can be located on the cone penetrometer in one, two or three locations shown in fig. 1.8 (Lunne et al. 2004, Burns & Mayne 1998): on the cone, behind the cone and behind the friction sleeve. Measured values of pore pressure are defined as u_1, u_2 & u_3.

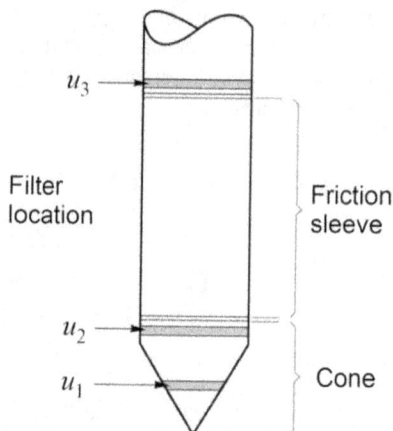

Filter location

Friction sleeve

Cone

Fig.1.8 Pore pressure measurements filter locations:

u_1 – pore pressure is measured on the cone, u_2 – pore pressure is measured behind the cone, u_3 – pore pressure is measured behind the friction sleeve

The Standard GOST 19912 permits pore pressure sensors placement in the cone penetrometer on special assignment, here pore pressure sensors filters may be placed just on the cone or just behind it.

In worldwide practice pore pressure measurement units have been applied, finely-dispersed filters made up of ceramics and/or porous bronze are used in their bodies. In fact, to measure pore pressure in soft water-saturated soils (silts, peats) sensors (filters) made up of ceramic compound are used. The diameters of such filters' pores are less than 0.001 mm; this prevents clayey particles from penetration into the filters.

Figure 1.9 illustrates the piezocone used in Sweden (the design of this piezocone is shown in Swedish Standard for cone testing) (Lunne et al. 2004).

In Russia piezocones are not widely used. In the former USSR and in contemporary Russia application of this equipment at mechanized units has been restricted with separate experimental Russian patterns during last 30...40 years; non-

Russian piezocones have been applied during last 10 years. Piezometer probes have been widely used is weak water-saturated soils studies (silts, sapropels, etc.). Pore piezometer probe PP-2 designed by Amaryan (1990) can serve as an example of such equipment (fig. 1.10).

PP-2 consists of an operational cone 1 with its body 2, connective push rods 4 and a pressure-and-vacuum gage 6 with a division value of 1...2 kPa. The pressure-and-vacuum gage (or simply manometer or vacuum gage) 6 is joined to a ceramic sensor (filter) by means of a polyvinyl chloride (P.V.C.) tube 3. It is mounted in the push rod 4 with a nut 5. The push rods of PP-2 are unified with those of other devices including the manual probe P-5 mentioned earlier (see fig. 1.3d). This allows to reduce the quantity and, hence, weight of transported parts; it is of principal importance when investigating difficult swampy terrain.

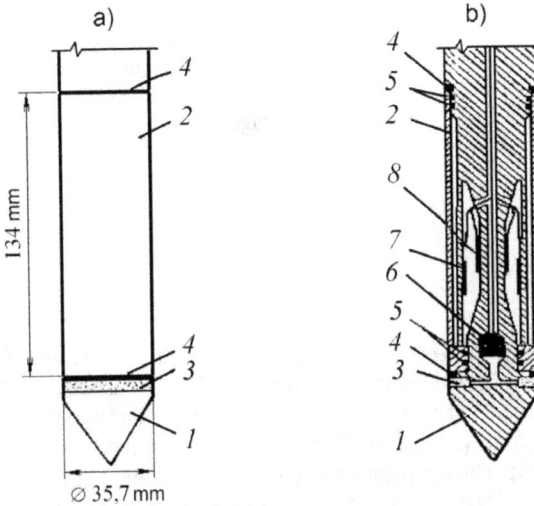

Fig. 1.9 Schematic design of a typical piezocone used in Swedish Standard for cone testing (1992):

a – general view, b – piezocone parts: 1 – cone, 2 – friction sleeve, 3 – filter, 4 – rubber seals to protect piezocone from soil penetration, 5 – rubber ring-seals, 6 – pore pressure sensors signals transformer, 7 – sleeve load cells, 8 – cone load cells

The process of pore pressure measurement with PP-2 is in the operational cone to be penetrated at the required depth and the manometers or vacuum gage readings after 30...40 minutes. The last condition is caused by slow entering of moisture from the soil into the measuring system. If there is a strong sandy layer in the bulk, a borehole is drilled down to a soft stratum; this allows the further penetration to be carried out.

Equipping the penetrometers to carry out **additional mechanical tests of soil** has not been applied in Russia, just only in some isolated cases. The same situation is observed worldwide, though CPT together with mechanical tests is used much more often than in Russia. The Standard GOST 19912-2012 permits special penetrometers application in order to specify some physical and mechanical properties of soil.

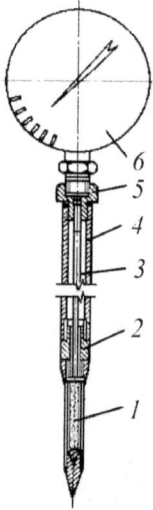

Fig.1.10 Pore piezometer probe PP-2 (USSR – Russia):

1 – ceramic sensor (filter), 2 – body, 3 – polyvinyl chloride tube, 4 – connective push rods, 5 – nut, connecting pressure-and-vacuum gage with a push rod, 6 – pressure-and-vacuum gage

Combination with rotary shear (impeller) has been widely applied in the manual probes. Both removable cone tips and impeller tips are usually applied, i.e. at first cone testing is carried out and then it is replaced with the impeller, i.e. rotary shear testing is carried out (Amaryan 1990). Cones combined with the impeller are occasionally applied. The simplest variant of this device was suggested by Razorenov (1980). He used the cone tip with "blades" in the upper part of it; at first the tip was pushed down and then rotated. The advanced (and, hence, the more complex) penetrometers have been developed aiming at deep CPT with mechanical units. In the former USSR the Research Institute "Energosetproject" developed a rig, thus the idea of combining CPT with rotary shear was realized (fig. 1.11a). The rig VSZ-15 was fixed to the automotive drill unit UGB-50M (Lebedev et al. 1988), i.e. it was the most wide-spread rig in the Soviet Union.

VSZ-15 has a rotary shear block ("rotary actuator') and a CPT block.

The rotary actuator includes a piston hydrocylinder; its rod has angular thread with a large angle of spiral lead. This allows changing rod movement from vertical into rotary one, important for blade system. The amount of rotary shear stress is evaluated by oil pressure in hydrocylinder cavity; the angle of blades' rotation is evaluated by means of special angular gage. Automatic hydraulic recording device is used to register rotary shear measured parameters.

Two variants of the penetrometer are used in the CPT block, namely: mechanical and electric tensometric. In the latter one the penetrometer of NIIOSP (from PIKA-9) incorporating tensometric sensors to evaluate q_c & f_s, the cable located in the tube of push rods and the electronic recording system in the operator's work place are applied.

In the former one special penetrometer of type I interacting with the measurement system by means of rods located inside the tube of outer push rods is applied. Axial thrust is transferred from the cone through the tube of inner push rods onto the measurement device hydrocylinder piston fitted with a manometer. With pressure readings in the manometer cone resistance q_c is evaluated. When rotary shear is carried out the penetrometer is disconnected from the tube of the inner push rods (by means of a jaw coupler) and stays fixed. Side friction Q_s is evaluated after the rotary shear test.

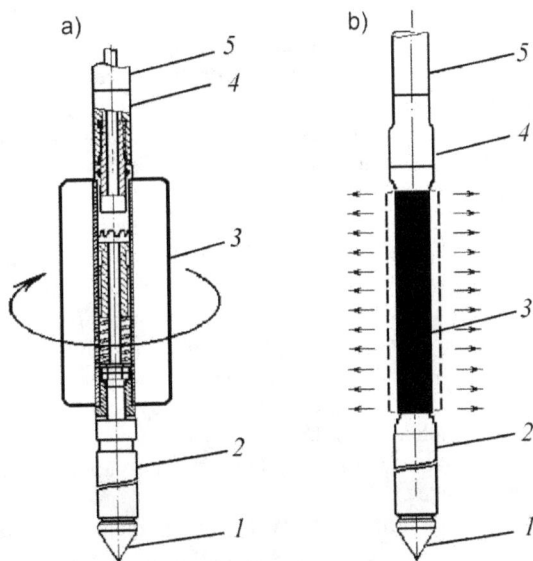

Fig. 1.11 Schematics of penetrometers combining CPT with other mechanical tests of soil:

a – combining CPT with rotary shear (VSZ-15): 1 – cone, 2 – friction sleeve, 3 – impeller (cruciform blade), 4 – coupling arrangement, 5 – standard push rod; b – combining CPT with pressiometry: 1 – cone, 2 – friction sleeve, 3 – pressuremeter (pressuremetric module), 4 – connective zone (coupling arrangement), 5 – standard push rod

Besides full-scale tests, VSZ-15 allows conducting CPT without rotary shear versa.

Combining CPT with pressiometry is more preferable in practice than that with other mechanical tests. Special equipment has been developed in many countries. Lunne et al. (2004) give details of five types of Cone Pressuremeters – CP, designed by different organizations. Dutch specialists Peuchen et al. (1995) underline that annual rise of CP tests is being observed.

Figure 1.11b illustrates the scheme of the most typical penetrometer and pressuremeter location in the non-Russian systems (Peuchen et al. 1995).

Penetrometer pressuremetric module is a cylindrical chamber, the side face of which is composed of a rubber (operational) membrane protected outside with an additional expanding shell (protective membrane). The test consists of radial expansion of the cylindrical chamber, radial stresses and deformations (displacements) being measured. Expansion occurs due to pressure of specified number inside the chamber; gas (nitrogen) or liquid (oil) are delivered into the chamber, thus the specified pressure occurs. Measurements are done with three sensors in three radial directions located 120^0 on-the-mitre relative to each other (view in plan). The diameters of most pressuremeters are more than that of the standard ones – 35.7 mm [pressuremeter LPC – 89 mm, UBC – 44 mm, ISMES – 35.7 mm, Fugro – 43.7 mm (Lunne et al. 2004)]. PC diameter to its length ratio is usually from 1:6 to 1:10.

As a rule, application of pressuremeters – penetrometers is more complicated than that of both "simple" penetrometers and "simple" pressuremeters being used separately. However, there is not any necessity in boreholes apart from "simple" pressuremeters which have to be penetrated into boreholes. Non-Russian specialists usually complicate penetrometers' designs equipping them not only with pressuremeters but pore pressure measurement devices, seismic exploration devices, inclinometers, etc. (Lunne et al. 2004). As a result, the system is no longer a CPT rig as it is, but becomes a complex apparatus to carry out a great number of soil tests. The authors of the book could not find any facts on these apparatuses reliability or their maintenance conveniences.

In Russian practice combining CPT with pressuremeter has not been widely applied, though some special pressuremeters – cone penetrometers PV-60 which are penetrated into the soil but not the borehole – are used (Shvets V.B. et al. 1981). These apparatuses do not refer to the issue since they apply just only penetration principle; cone resistance (q_c) and side friction resistance (f_s or Q_s) are not evaluated.

Combining CPT with geophysical measurements incorporates equipping the penetrometers with additional devices to study soils by means of:
- Seismic methods.
- Electrical exploration methods.
- Radiation methods.

Seismic method of soil study is based on mechanical (seismic) effects of longitudinal and shear waves on the soil as well as their velocity distribution evaluation. As an independent geophysical method it has been applied for over half a

23

century, though together with CPT it has been applied just only during last 15...20 years. When combining seismic method with CPT a seismic receiver is placed inside the penetrometer but the source is behind the penetrometer (as a rule, on the ground surface). Figure 1.12 illustrates principles of the seismic cone survey technique with UBC penetrometer (Great Britain) (Lunne et al. 2004)

Here, the seismic source is located on the ground surface, as shown in Fig. 1.12. Shear waves are generated hitting the beam (plate) ends horizontally with a hammer 4, the beam being loaded statically is equipped with special devices. Shear wave generation is collected by a seismometer.

There is a different way of seismic effects initializing. The source is located not on the ground surface but in the second cone penetrometer being placed apart from the first one (Campanella & Robertson 1986, Baldi et al., 1988 Lunne et al. 2004). Two heavy CPT rigs are used to carry out the tests. The cone penetrometer together with the seismometer has the diameter of 35.7 mm, that with the source – 43.7 mm.

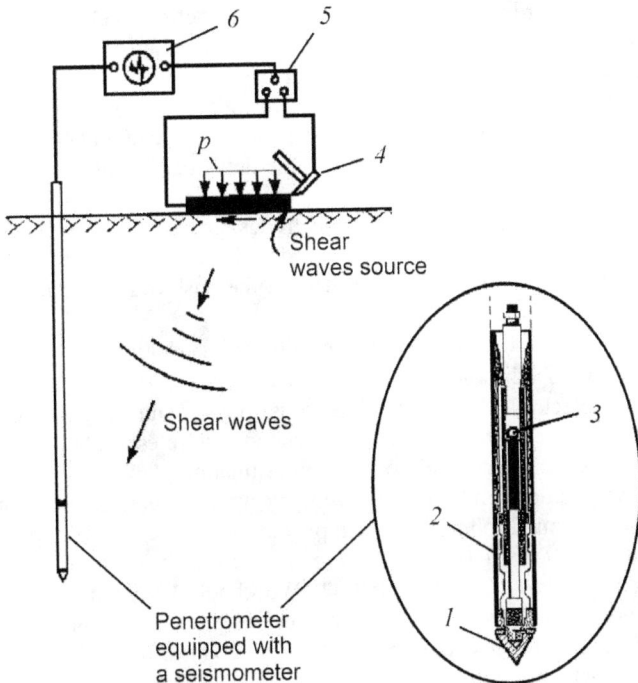

Fig. 1.12 Principles of the seismic cone survey technique with UBC seismic cone:
1 – cone, 2 – friction sleeve, 3 – seismometer, 4 – hammer, 5 – trigger (changer), 6 - oscilloscope (collecting & processing data of seismic oscillation); p – static load

Electrical exploration is based on soil electrical resistance evaluation by which the soil type is defined (each type possesses its own specific electrical resistance, e.g. water-saturated clay – 100...5000 om/cm, wet sand - $2 \cdot 10^5$...$4 \cdot 10^6$ om/cm, etc.). Either seismic or electrical exploration has been used in engineering investigations for half a century but together with CPT it has been applied just only since 1970s.

When combining electrical exploration (seismic exploration) with CPT it is important to use additional equipment placed in the cone tip as well as behind it. Electrodes are referred to this equipment. The first pair of electrodes – measurement electrodes are located in the cone or on the push rod, the other ones – current electrodes are located just on the push rod or on the ground surface either 0.5...1 m apart from the penetration location. When measuring current strength between the current electrodes and the voltage between the measuring ones, specific resistivity of investigated soils is evaluated.

Electrical resistivity of the ground massif depends upon solid particles and pore water resistivity. Pore water resistivity (conductivity) is considered to be the prevailing factor. In order to pump ground water and define its electrical conductivity in lack of solid particles special equipment has been developed. It is also possible to take water samplers, deliver them to the laboratory and define the electrical resistivity with higher accuracy. In practice the extra operations with water resistivity determination are not always carried out as there is equipment in which such procedures are not performed. Nevertheless, the data obtained on water resistivity do increase soil type estimation accuracy.

Figure 1.13 shows the scheme of electrodes' location on the push rod, water resistivity probe and one example of electrodes' location in the cone (Lunne et al. 2004). Two groups of electrodes are located on the push rod (two measurement and two current electrodes):

- Group "A" with a distance between the electrodes – 500 mm.
- Group "B" with a distance between the electrodes – 100 mm.

Group "B" electrodes' zone of influence is spread at much shorter distance than zone "A" but the latter allows to thorough ground texture estimation. Application of both groups ensures the best conditions of soil estimation.

In Russian practice combining CPT with electrical exploration has not been used yet, though in Dynamic Probing Test such penetrometers are widely applied.

Combining CPT with radioactive logging was used in the USSR to a large extent (Ferronsky 1969, GOST 25260-82, Workbook 1988); nowadays it is also applied in Russia but not so much. In non-Russian practice this method is not used very much either (Trofimenkov 1995, Lunne et al. 2004). Lunne et al. (2004) explain this by importance of safety specification adherence, particularly in the situations when the penetrometer saturated with radioactive agents may stay deep in the soil caused by breaking of its connection point. It is evident that the penetrometer must not be left in the soil not only because of economic reasons (the penetrometer is expensive) but because of radiation safety. Extraction of the penetrometer is a costly procedure.

The argument must certainly be paid attention to, but it should not be overestimated. In Russia operations and maintenance of tensometric penetrometers proved to show the following: firstly, situations when the penetrometer "got stuck" in the soil hardly ever occur, secondly, it is easy (within a few hours) to draw the penetrometer from the soil with a sampler of a boring machine.

Radioactive logging is based on radioactive (ionizing) radiation. Up to 1960s it had been applied as a geophysical method apart from CPT.

There exist two methods of radioactive logging: gamma-ray radiation and neutron radiation. In their turn, gamma-ray radiation methods fall into two types (Ferronsky 1969):

- Measurement of gamma-rays intensity (energy spectrum) specified by natural radioactivity of soils – gamma-ray radiation as it is.
- Measurement of secondary gamma-ray radiation occurrence in soil irradiated with gamma-rays source – "gamma-gamma radiation".

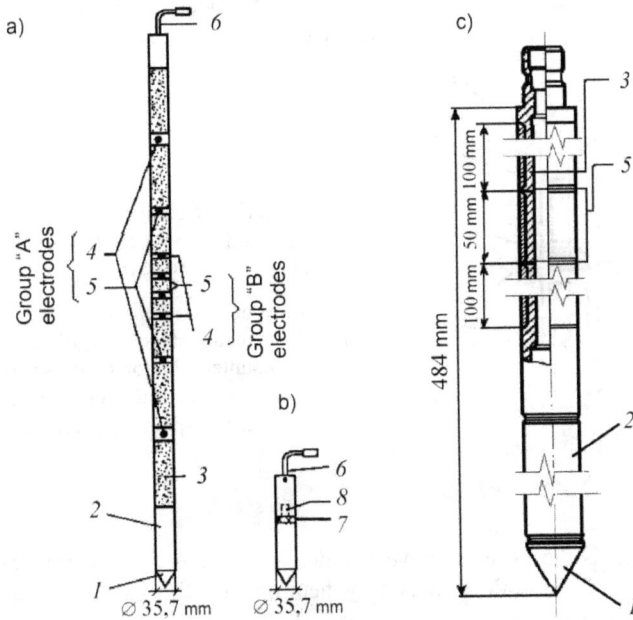

Fig 1.13 Schematic location of extra equipment in combining CPT with electrical exploration:

a – schematic location of electrodes in Dutch probe Graaf et al. (Lunne et al. 2004), *b* – pore water electrical resistivity probe (the same), *c* – location of measurement electrodes after Zuidberg et al. – Holland (Lunne et al. 2004); 1 – cone, 2 – friction sleeve, 3 – insulation, 4 – current electrodes, 5 – measuring electrodes, 6 – cable, 7 – filter, 8 – pore water electrical resistivity measuring cell

26

Since dispersion of gamma-quanta depends upon diffusing medium density, gamma-gamma radiation is used to determine density of soils. Figure 1.14 illustrates the principle of diffused gamma-radiation.

Between a source of radiation 1 and a detector of radiation 2 "measuring & recording" diffused gamma-quanta, special filter is placed – a lead screen 3 preventing direct entering of gamma-rays from the source to the detector (besides lead other materials possessing higher atomic weight can be applied in screens). Detector signals reinforced and formed with a photoelectron multiplier 4 are transferred upward along the cable and then are finally recorded and analyzed.

Neutron radiation is based on the interaction of neutrons with the soil. There are nearly 10 variants of neutron radiation differing in neutron source type, secondary radiation type and the information obtained. When investigating soils the so called neutron-neutron radiation is used; in it the loss of neutrons' energy during their dispersion in the soil is estimated (transformation of the "fast" neutrons into the "heat" ones). Schematic illustration of this process is shown in figure 1.13 taking into account that instead of gamma-ray radiation source neutron source is used, instead of gamma-rays detector – neutron detector. The counterradiation screen is made of lead, paraffin, bismuth, etc.

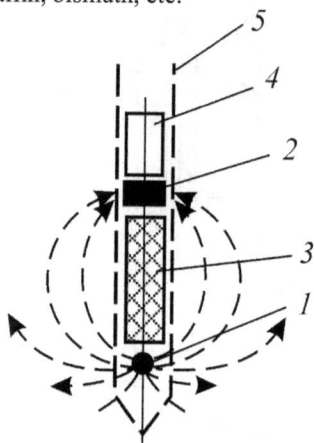

Fig.1.14 Schematic illustration of diffused gamma-radiation (Ferronsky 1969):

1 – source of radiation,
2 – detector(«receiver») of radiation,
3 – counterradiation (lead) screen,
4 – photoelectron multiplier, 5 – cone contour incorporating radioactive device

Hydrogen is the most effective retarder of neutrons; its quantity depends upon water content of the soil. That is why neutron radiation is applied in soil water content estimation.

Figure 1.15 shows two removable penetrometers used in the Soviet penetration-radiation system SPK: the penetrometer combining CPT with gamma-gamma radiation and the penetrometer to perform neutron-neutron radiation (Ferronsky 1969).

As it is shown in figure 1.15 the diameter of the penetrometer (60 mm) exceeds the standard one (35.7 mm).

In penetrometer № 1 (fig. 1.15a) gamma-gamma radiation sensor includes gamma-gamma radiation source – cesium-137 with the activity of 4 *mg-equiv.*

radium. Crystal detector NaJ (T1) with the photoelectron multiplier FEU-35 is located 40 cm apart from it. Two parameters are evaluated when applying penetrometer № 1: cone resistance (q_c) and sleeve friction resistance (f_s); when pulling the penetrometer out gamma-gamma radiation is performed and the third parameter is evaluated – density of soil ρ (or specific weight γ).

In penetrometer № 2 (fig. 1.15b) neutron-neutron radiation sensor is composed of plutonium-beryllium fast neutrons with the activity of $5 \cdot 10^5$ n/s with a filter (screen). Heat neutrons' detector LDNM-2 with the photoelectron multiplier FEU-35 is located 5cm apart from the source. In order to record natural gamma-clutter of soils the additional gamma-radiation sensor incorporating special (scintillation) detector – crystal NaJ with the photoelectron multiplier FEU-35 – is located in penetrometer № 2.This sensor is located 1.1 m apart from the neutron source to exclude the effect of gamma-radiation generating in interaction between neutrons and the environment. Measurements with penetrometer № 2 may be carried out when penetrating and/or extracting the penetrometer.

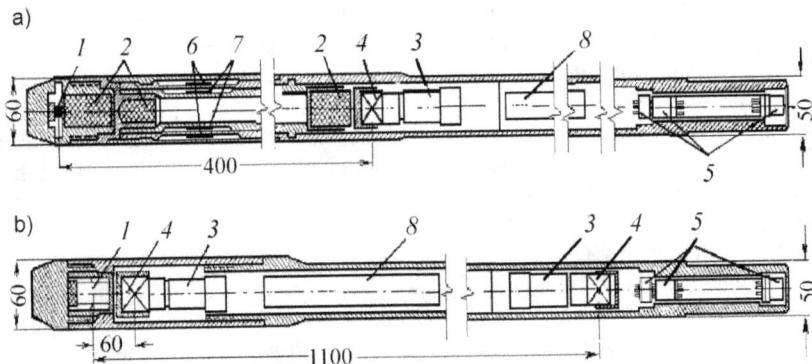

Fig.1.15 Schematics of penetrometers in SPK penetration-radiation system:

a – penetrometer №1 to perform CPT and gamma-gamma radiation; *b* – penetrometer №2 to perform neutron-neutron radiation; 1 – source of radiation, 2 – lead screen, 3 – photoelectron multiplier, 4 – detector of radiation, 5 – pin connector, 6 – cone load cell, 7 – sleeve load cell, 8 – high-voltage power supply with an amplifier

Penetration-radiation tests suggest CPT with penetrometer № 1, cone resistance q_c, sleeve friction resistance f_s and density ρ are evaluated; then penetrometer № 2 is used (in the same hole if possible), water content w is measured.

In contrast to SPK CPT and radioisotope techniques of soil investigation complex Russian system PIKA-15K (Morozov et al. 1999) allows evaluating all the necessary parameters (cone resistance, sleeve friction resistance, natural radioactivity, density of soil and water content) ignoring change of the penetrometer at the measuring location. It is performed due to complex device application made up of two successive penetrometers – the tensometric penetrometer "TK" is located at the foot, the penetrometer "K" for radioisotope measurements is located above; the latter is fixed to the former with a reducer. Unlike SPK the body (tensometric penetrometer included) of this complex device has the standard diameter of 35.7 mm. This considerably facilitates obtained resistance interpretation to CPT. Its total length with electron blocks is 143 cm.

Combining CPT with ecological survey means obtaining information on various contaminants occurrence in the soil (oil products, oils, acids, etc.) alongside with evaluation of q_c & f_s. At present time in non-Russian practice penetrometers with different extra modules are used for this purpose. The well-known among them are (Lunne et al. 2004):

- High-frequency-impedance-measuring penetrometer – HIM-penetrometer.
- Chemipenetrometer – a device for chemical analyses conducting to measure *pH* of pore water, capability of soil to oxidation-reduction reactions, etc.
- Laser-induced fluorescence sensor – LIF.

Dielectric measurement devices are used to reveal non-aqueous-phase-liquid-NAPL in the soil. Electrical resistivity described earlier does not reflect the types of contaminants; it is dielectric constant that can characterize them. Dielectric constant depends on variable electric field frequency, the dependency being quite complicated reflecting a number of factors influence. Dielectric electronics deals with these issues, but specialists on ecology, soil mechanics and geotechnical engineering are not competent here. In order to understand principles of the penetrometers' behavior it is necessary to point out that dielectric measurements need to be done with high-frequency variable electric field (10…500 MHz) ignoring the nature of the applied processes.

Delft Geotechnics (Netherlands) developed equipment to measure dielectric constant of soil in a high-frequency field – HIM-penetrometer. The equipment allows measuring three parameters: dielectric constant, electrical resistivity and *pH*. Schematic illustration of the penetrometer is given in figure 1.16 (Lunne et al. 2004). The cone is penetrated in the soil in a "locked" position when the lower end of the isolator 2 and the outer shell 3 are located on the same level (there is a lack of soil sample camera 4 in this location). When the depth is achieved the outer shell 3 is projected, soil sample camera 4 is formed inside and the soil is pulled into it. High-frequency electric field is generated on the ground surface and transferred to the antenna 5 through the coaxial cable1, thus electric field frequency is measured in the soil. As a result, dielectric constant and specific electric conductivity of the soil are evaluated as functions of this frequency in the interval of 10…500 MHz.

After the measurements soil sample is pulled out of the camera 4, the outer shell 3 is pulled in, the penetrometer "is locked" and then it is penetrated at the next given depth. The cycle is repetitive.

Devices for chemical analyses mounted in the penetrometer are used to measure *pH*, oxidability and chemical composition of both pore water and gas (Lunne et al. 2004). Two approaches are applied:

- Chemical analyses are carried out just in the cone, for this the penetrometer is equipped with special apparatuses.

29

- Chemical analyses are carried out in the laboratory; the penetrometer is used as a sampler of soil, water or gas that is delivered to the laboratory.

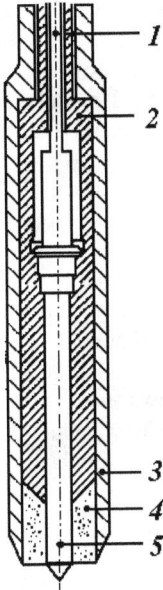

Fig.1.16 Schematic illustration of Delft Geotechnics penetrometer to measure dielectric constant:

1 – coaxial cable, 2 – isolator, 3 – retractable rim receiver in the form of outer cylindrical shell, 4 – soil sample camera, 5 – central (dielectric) antenna

In the first case special analyzers are located above the cone and friction sleeve. By means of this technique *pH* and redox potential are measured.

In the second case the penetrometer possesses mobile parts allowing to cut into the soil as well as to draft water or air. This penetrometer has special containers to place or extract the samplers. The specific equipment is applied, e.g. special filters-screens made of stainless steel are used to draft water, special impermeable injectors, glass or steel vessels are used to select and extract gas samplers, etc.

In general, these technologies greatly differ from CPT as it is. Here, CPT equipment is used in a different way: to select samplers, to penetrate devices performing various functions, etc.

In non-Russian practice such technologies are thought to be an independent trend called "Direct Push Technologies – DPT" but not CPT (Lunne et al. 2004). The techniques discussed in this section (i.e. combining CPT with other tests) may be referred to DPT, especially selection of samplers.

Equipment to apply laser-induced fluorescence – LIF is necessary to reveal oil products concentration in the soil (carbohydrate contaminants). The equipment is considered to be one of the greatest achievements in ecological investigations technique of 1990s. LIF is based on fluorescence spectral analysis of contaminated soil, this fluorescence occurs due to laser irradiation of the soil (in ultraviolet range).

Figure 1.17 shows the scheme of the US penetrometer equipped for LIF (Larsson 1995, Shinn & Bratton 1995).

Ultraviolet radiation passes through the fiber-optical cable 6 onto the optical module 4 and then the sapphire window 5. The soil is irradiated through this window and if oil products (carbohydrates) occur in the soil, it glows. Fluorescence appears due to carbohydrates occurrence in the soil, particularly polyaromatic components. The irradiation driven in the soil is taken by the optical module, then through the second fiber-optical cable it is brought to the surface spectral analysis block located in the laser system 7.

Contamination of soil is estimated by means of spectral analysis results. Concentration of carbohydrates is estimated by fluorescence intensity. Duration of the required data obtaining is 1...2 minutes for every location (depth), thus combining CPT with ecological measurements does not face any difficulties. LIF could be applied to thorough examination of contaminants generating fluorescence. For this it is necessary just to calibrate the results to different conditions when the compositions and concentrations of contaminants are known in detail.

Figure 1.17 shows the penetrometer which allows mechanizing the important ecological operation – tamping of CPT holes. After CPT procedure the lost part of the cone 3 is disconnected; cementing mortar is delivered through the tube 12 and the CPT hole is filled while extracting the penetrometer.

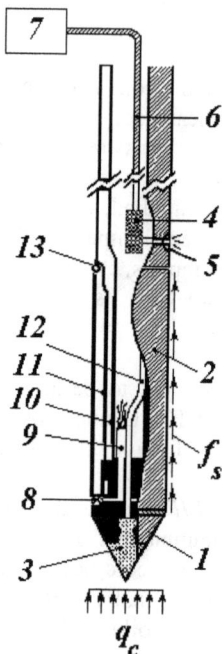

Fig.1.17 Scheme of US penetrometer supplied with equipment for LIF:

1 – cone, 2 – friction sleeve, 3 – lost part of the cone, 4 – optical module (ultraviolet & fluorescence radiation sensor), 5 – sapphire window(through it the soil is irradiated), 6 – fiber-optical cable, 7 – laser system, 8 – Teflon filter, 9 –pore pressure sensor , 10 – cone load cell, 11 – sleeve load cell, 12 –cementing mortar tube, 13 – waterproof seal

31

Temperature measurements of permafrost soils were proved to be very important in Russian construction. In Russia most of the territory is permafrost soils, their mechanical properties are specified not so much by lithological qualities or composition of soils but their temperature either. In fact, in most developed Northern regions of European part of Russia soft frozen soils prevail in permafrost (see section 5.1). Temperature and mechanical properties of the permafrost allow performing CPT, penetration of piles, etc.

Special equipment to measure natural temperature of the soils and specify their mechanical properties is necessary for engineering investigations connected with designing of foundations on soft frozen soils. CPT application is believed to be the most promising trend. Cone Penetration Test with Temperature measurement (CPTT) can considerably raise the efficiency of engineering investigations. CPTT can also be applied as an inspection method in artificial thawing of permafrost soils. Figure 1.18 illustrates the design of Russian penetrometer (BashNIIstroy penetrometer modification) equipped with a temperature sensor (Volkov & Isaev 1985, Isaev 1989).

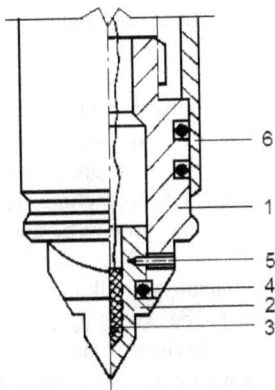

Fig.1.18 Schematic illustration of Russian penetrometer measuring temperature of soils:

1 – main cone, 2 – small cone, 3 – temperature sensor, 4 – seal, 5 – fixing screw, 6 - friction sleeve

The penetrometers equipped with heating elements in addition to temperature sensors are also of great interest (see section 5.1).

Worldwide the issue is paid great attention to either. Some non-Russian as well as Russian professional engineers do not restrict themselves in estimation the permafrost soils only but attempt to use penetrometers equipped with temperature sensors in ecological survey of ordinary not frozen soils. In fact, many organic contaminants are capable to cause exothermic reactions raising the temperature of soil. This gives possibility to study the composition and concentration of contaminants on soils' temperature, thus special penetrometers can be applied. However, the trend is in its infancy, and the principal task is to measure temperature of permafrost soils.

1.3.5 CPT rigs

Penetration is carried out by means of CPT rigs being composed of a cone penetrometer, pushing and extraction device, support-anchor equipment and measuring system. CPT rigs are as follows: self-propelled, i.e. mounted in a truck or a tracked truck or mobile ones, i.e. mounted in a trailer, portable demountable or as pieces attached to drill units. Nowadays, over 50 types of CPT rigs are used worldwide. Great diversity of CPT rigs has been discussed at international conferences time and again. Some of the types being slightly modernized have been successfully applied for several decades; other types have just been developed and they reflect modern tendencies.

Over 10 types of CPT rigs have been applied in Russia, two of them S-832 & SP-59 manufactured in the USSR are the most widely used. As it is shown in section 1.1, depending on thrust necessary to push down the penetrometer CPT rigs are divided into light (up to 50 kN), medium (50...100 kN) and heavy ones (over 100 kN). This affects their sizes: the light CPT rigs are usually of minimal sizes, the heavy CPT rigs – of maximal ones. Reasonability of such a division has been confirmed during last decades. Each type is applied in a particular field, e.g. light CPT rigs are used in the narrow conditions, medium or heavy ones – in pile foundations designing, etc. Let us consider the types separately.

Light CPT rigs are not greatly used in Russian practice but worldwide they are used on a large scale. Fig.1.19 illustrates typical designs of light CPT rigs (Trofimenkov 1995, Ryzhkov 1992). Small handy CPT devices (probes or manual penetrometers) with pushing thrust capacity to be 0.05...1 kN are sometimes referred to a separate group. These rigs seem to be light ones but they require to be applied in a different way as well as in a different field, e.g. CPT on small depth inside the premises, basements, impassible terrains, layerwise control of bulk density, etc. Hanson-5 probe weighing 8 kg (fig. 1.19a), Hanson-10 probe weighing 30 kg (fig. 1.19b) are related to these devices. Productivity of the devices is not very high, although they can be applied in such narrow conditions where any other tests would be impossible. In fact, if a person can find enough room for himself, penetration of the probe is possible. It is evident that these devices are quite cheap.

The depth of penetration, e.g. with Hanson-5 or Hanson-10 is restricted by the weight of the worker. In soils of average density the depth does not exceed 2...3 m, although in weak soils (silt or sapropel) it achieves 7...10 m. In order to increase maximum depth in denser soils, some types of probes are equipped with additional loads placed and fixed on the probe during the test. An old probe (spiral) being applied in Sweden since 1917 is shown in figure 1.20 (Möller et al. 1995).

Manual probes are used to estimate consistency of clayey soils, density of loose sands or bulks. They are not really suitable for soils mechanical properties quantitative estimation, but when used with more accurate tests they appear to be suitable in resolving such problems. Here, calibration of results should be done in accordance with more accurate tests carried out in the same place.

Fig.1.19 Typical light CPT rigs:
a – portable manual probe weighing 8 kg (Hanson-5, a.p. Van den Berg – Holland, Netherlands), *b* – manual probe weighing 30 kg (Hanson-10, the same manufacturer), *c* – light rig weighing nearly 100 kg with 25 kN pushing thrust capacity (Hanson-2.5, the same manufacturer), *d* – light rig weighing 85 kg with 25 kN pushing thrust capacity (Gaudische Machinefabrick – Holland, Netherlands), *e* – light trailer rig with 50 kN pushing thrust capacity (Hyson-5Tf, a.p. Van den Berg – Holland, Netherlands)

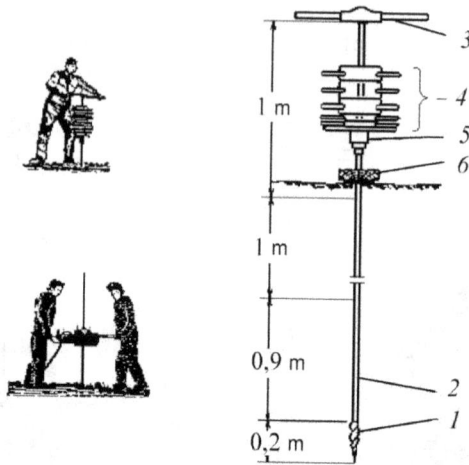

Fig.1.20 Schematic illustration of Swedish manual (screwed) probe penetrated with additional loads:

1 – screwed cone with a diameter of 32.5 mm, 2 – push rod with a diameter of 22 mm, 3 – screwdriving arm, 4 – 10 & 25 kg loads, 5 – 5 *kg* support for loads, 6 – wooden plate with rubber backing

Penetration of probe by hand is typical for the simplest probes. Other CPT systems are supposed to apply mechanical penetration. In demountable systems (see fig.1.19c) penetration is carried out with a worm hand drive mechanism. In more powerful units, even small-sized (see fig.1.19d) penetration is performed by means of a hydraulic actuator. A push rod is clamped by special clamping arrangement and thus, pushing is carried out; this clamping arrangement is firmly fixed to a hydraulic cylinder rod (or two hydraulic cylinders), penetration is performed by the stroke of the cylinder. After that the mechanism unclamps the push rod and while gliding upward along it, the mechanism returns to the primary position. Then, the push rod is clamped again, the cycle is repetitive. The speed of penetration differs (0.5...2 m), but the International Standard (Proposed European Standard 1977) as well as the Standard GOST 19912-2012 limits the speed to 1.2±0.3 m/min. At present the standard speed is taken into account worldwide.

The push rod consists of the same sections – 0.5 m or 1 m in length (the length corresponds to hydraulic cylinder stroke). The sections are joined by means of special end connections or on the thread. Building-up of the push rod, i.e. joining of sections is performed during penetration by hand.

In order to take pushing thrusts anchor devices in the form of boring piles are installed in the rig. Screwdriving of anchor piles is done by hand or with the simplest hand drive mechanisms. In rigs mounted in single-axis trailers (see fig.1.19, *e)* screwdriving is done mechanically without any physical work.

Light mechanized rigs, both demountable and mounted in single-axis trailers (fig.1.19e) are equipped with standard penetrometers, usually mechanical ones of type I. It is evident that CPT soil test tables and formulae given in the Normative Documents are applicable to the results obtained.

Medium CPT rigs, i.e. rigs with 50...100 kN pushing thrust capacity may also be demountable but usually they are mounted in single-axis trailers or they are self-propelled, i.e. mounted in trucks or tracked trucks. Figure 1.21 shows some typical medium CPT rigs (Trofimenkov & Vorobkov 1981, Ryzhkov 1992, Trofimenkov 1995, Lunne 2004).

Medium CPT rigs are widely applied in Russian practice, self-propelled ones prevailing. In last 20...30 years attempts have been taken to exchange rigs mounted in trailers for self-propelled ones mounted in trucks or tracked trucks. It is important to underline that in Russia rigs with pushing thrust capacity of 100...120 kN have predominated for last decades, i.e. CPT rigs between medium and heavy ones. Rigs with larger pushing thrust capacity have not been popular. Thus, division of CPT rigs into medium and heavy ones have always been conditional in Russia.

As a rule, at present, medium CPT rigs are equipped with standard type I or type II penetrometers, tensometric penetrometers being widely used, mechanical penetrometers being not so frequently applied, hydraulic ones being hardly ever used. Russian CPT rigs of Fundamentproject design S-979 and SP-59 shown in figure 1.21 are equipped with type I penetrometers, sometimes being exchanged for type II ones.

The push rod fixing the penetrometer is composed of similar sections 0.5, 1.0, 1.5 or 2 m in length (usually 1 m). Building-up of the push rod is performed during penetration by hand, as in light rigs. The penetration process is analogous to that of light CPT rigs mounted in trailers, i.e. the penetrometer is pushed down by clamping arrangement and a hydraulic cylinder.

Medium CPT rigs are equipped with screw anchor piles but in practice one can manage without these piles application in case heavy truck weight is sufficient to take reactive thrust. In fact it is technologically advantageous since it simplifies and accelerates CPT works.

Productivity of medium CPT rigs is affected by their mechanization, automation and concrete CPT conditions. In practice when applying demountable rigs one can hardly carry out more than three tests 10...11 m in depth taking into account total time costs in eight-hour shift. When applying mechanized self-propelled rigs, one can manage to carry out 3...5 tests. If one can ignore anchorage, productivity may increase up to 4...6 tests per one shift.

Apart from purpose-designed CPT rigs various *attachments (attached implements) to boring machines* are being applied in CPT. Principles of operation of the equipment are just the same as in purpose-designed CPT rigs, and they allow combined CPT to be performed, i.e. alternation of ordinary CPT procedure with main borehole drilling to pass dense soil interlayers and large inclusions.

36

a)

b)

c)

d)

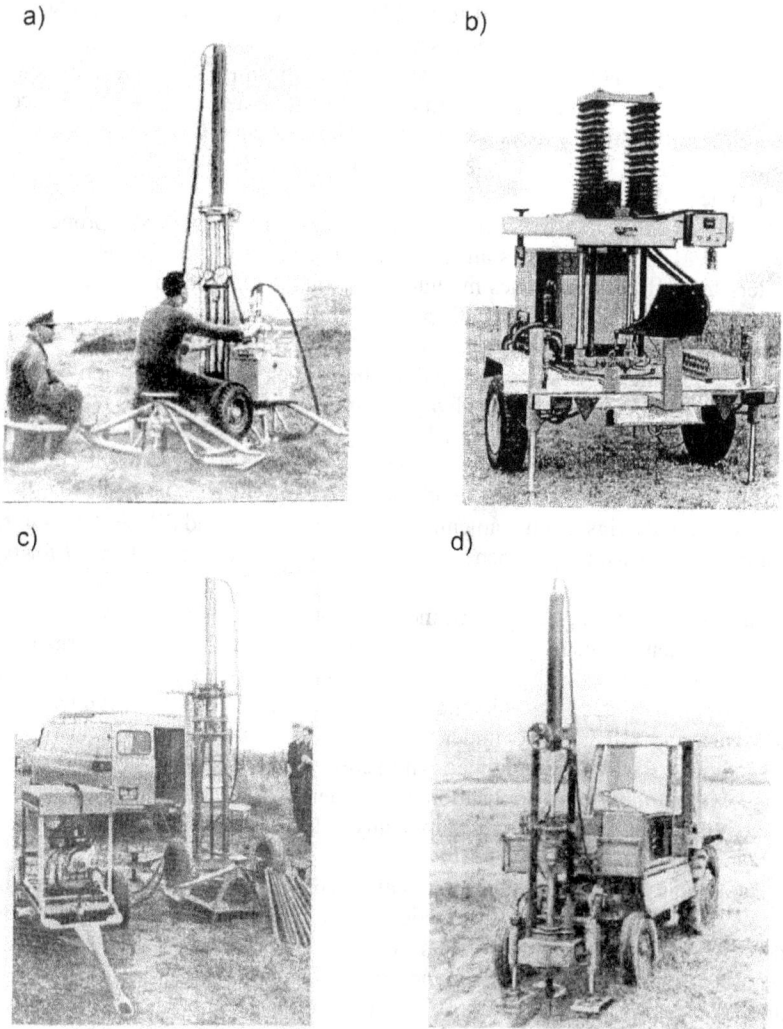

Fig.1.21 Typical medium CPT rigs:
a – rig with pushing thrust capacity up to 100 kN (Gaudische Machinefabrick – Holland, Netherlands), b – single-axis trailer rig with pushing thrust capacity up to 100 kN (Saunda – Italy), c – single-axis trailer rig with pushing thrust capacity up to 100 kN (S-979 – USSR - Russia), d – tracked truck rig with pushing thrust capacity up to 100 kN (SP-59 – USSR - Russia)

Heavy CPT rig with over 100 kN pushing thrust capacity are effective in large amount of work as well as necessity to study soil conditions in considerable depth. Figure 1.22 illustrates some typical heavy CPT rigs.

Thrust machines embrace heavy trucks (fig. 1.22d, e, f), tracked trucks (fig. 1.22b), and combined CPT trucks (fig. 1.22c). It is possible to install equipment in tracked CPT trucks (fig.1.22a).

Power capacity of pushing joints in heavy CPT rigs differ from the analogous ones in medium and light CPT rigs though the principles of operation are the same. The majority of the rigs do not require special anchor equipment due to their sufficient weight capable to take reactive thrust occurring in pushing. In order to ensure reliability of "non-anchor" work, in non-Russian CPT rigs pushing joint is located in the truck's center of gravity.

There exist some variants of CPT rig installation in two trucks, e.g. VSEGINGEO (USSR) – SUGP-10 & SPK penetration radiation units (Ferronsky 1969). The first one applied equipped tracked truck and GAS-63 truck with the enclosure of electronic equipment for data acquisition. The second one applied two automobiles: ZIL-157 truck with the enclosure of pushing equipment and KAvZ-663 bus equipped with measurement devices.

As a rule, heavy CPT rigs are tensometric. Complicated versions to perform additional measurements (see section 1.3.4) are referred to heavy CPT rigs either. Data acquisition is done automatically by means of modern electronic equipment. Software has frequently been applied during last decade; it ensures primary treatment of data obtained in field conditions. This gives added comfort for personnel who work in dustproof and waterproof enclosure with comfortable temperature control, lighting, and all the necessary equipment available at workplace. Heavy CPT rigs ensure the highest productivity of field tests. Taking into account unavoidable noncommercial time costs (travel time to the site, relocation of CPT rigs inside the site, different organizational discrepancies resolving, etc.) heavy CPT rigs can perform over 100 m of penetration in eight-hour shift.

The principal disadvantages of heavy CPT rigs are as follows: their high cost caused by expensive thrust machines applications and large sizes of trucks hampering penetration in narrow conditions.

One of the first heavy CPT rig anticipated a number of modern rigs in its characteristics is thought to have been S-832 CPT rig designed in the USSR (Firestein & Makarov 1964, Ryzhkov & Goncharov 2006). Its series production started in 1963 (fig.1.22f). In spite of rather small thrust capacity (120 kN), just as in medium CPT rigs, it possessed all the advantages of modern heavy CPT rigs: maintainability, high productivity and mobility. The rig was modernized more than once but it did not undergo any significant changes (the developers are: Makarov, Firestein, Kozlovsky et al.). Firstly, ZIL-157 truck, then ZIL-131 truck was used as a thrust machine, pushing and extraction of the penetrometer was carried out by a mandibular pickup device with a hydraulic actuator, and the anchor device presented two screw piles with a hydraulic actuator. As it was mentioned earlier, tensometric penetrometer of type II was applied; data were recorded automatically to chart strips.

a)

b)

c)

d)

e)

f)

Fig. 1.22 Typical heavy CPT rigs:
a – tracked CPT truck (a.p. Van den Berg, Holland, Netherlands), *b* – tracked CPT truck (Fugro Ltd, Holland, Netherlands), *c* – Combi CPT truck (a.p. Van den Berg), *d, e, f* – typical CPT trucks of various designs (Con Tec Ltd; Fugro Ltd, Holland, Netherlands; S-832, USSR – Russia correspondingly)

Despite contradictory primary estimations of some specialists in 1960s, forty years of S-832 operations and maintenance revealed that the design successfully passed ahead of the time. The idea to mount CPT rig in a truck appeared to have been a new one, but promising.

In the following years many manufacturers of CPT systems reequipped trailer CPT rigs into truck ones. A successful decision was to apply *telescopic push rod* (fig.1.23) composed of two sections 11 & 10 m in length instead of compound push rods with sections 1...2 m in length assembled during CPT procedure. This considerably decreased overmaintenance of CPT rig and increased the productivity of CPT procedure up to 70...120 m in a shift (taking into account travel time to the site, replacement of CPT rig inside the site, etc.).

Fig.1.23 General view of S-832 CPT rig in operating position

Tensometric cone penetrometer application (fig.1.6d) and automatic recording of measurements proved to have been a great advance (Ryzhkov & Goncharov 2006). The effectiveness of CPT considerably increased since CPT procedure became a highly mechanized automatic process. Location of recording devices inside the truck's body gave added comfort for an operator. Such working conditions were unusual in the middle of the 20th century, so it was a great advantage either.

CPT rig penetrometer has the parameters which agree with effective Standard GOST 19912-2012 requirements: penetrometer diameter – 35.7 mm, friction sleeve side area – 350 cm^2 (310 mm in length).

Sensor signals come through transmitting transformers on the indicative device located in the truck enclosure. Automatic recorder records the results to two chart strips. The first diagram shows cone resistance q_c, the other one – unit sleeve friction resistance f_s.

Calibration procedure is needed to be done in order to ensure measurement accuracy. For this a dynamometer is loaded through the penetrometer, indicative device readings and those of the dynamometer are compared.

Nowadays, indicative device of CPT rig seems to be obsolescent. Progress in metrology and electronics resulted in displacing automatic recorders with modern electronic devices capable to display information on the monitor and record it to the media to further transmission and processing on PC. At present, CPT data recording is performed applying laptops.

Deep-water CPT is a particular trend in CPT development. Engineering-geological investigations in deep water of seas, reservoirs, lakes or rivers are thought to be rather labor-consuming costly work. Here, traditional techniques of soil study, i.e. drilling and/or monoliths selection become rather complicated. CPT application is believed to be the best solution of the problem. It is evident that deep-water CPT equipment is much more complex than that of on land CPT. It requires to be located on floating vessels – barges, pontoons, lighters, self-propelled cargo vessels. Without special adjustments floating vessel mode for CPT appears to be impossible due to insufficient stability of the push rod which shall behave at lateral bending in large free length. Thus, two approaches are used (fig.1.24):

- The cone penetrometer is pushed down through a conductor pipe or other device excluding loss of push rod's stability (pushing device is placed on the floating vessel).

- The cone penetrometer is pushed down at the reservoir bottom; seabed mode for CPT is applied.

In both cases the penetrometer is pushed down just into the seabed, so non-Russian specialists consider both approaches as a single method – seabed mode (Lunne et al. 2004). The technique is considered to be efficient only in shallow water (up to 30...40 m). In deeper water another CPT technique is recommended: first the hole is drilled, then the cone penetrometer is pushed down-the-hole. Down-the-hole mode allows CPT in very deep water (up to 100 m) (Lunne T. et al. 2004). There are some examples of down-the-hole mode for CPT applications offshore in sea depth of 500 m (Trofimenkov Yu.G. 1995).

There exist CPT rigs with boring casing being pushed down with the cone penetrometer by means of a lifting jack, the cone penetrometer moving before boring casing. In order to take heavy reactive thrust, special loading platform is installed; it is dropped at the bottom being loaded with lead 36 T in weight. Tensometric cone penetrometers are used in deep-water CPT, mechanization and automation of working procedures being of high level. It is obvious that deep-water CPT is rather expensive, but other ways of reservoirs' bottoms studies are even much more expensive and not always affordable.

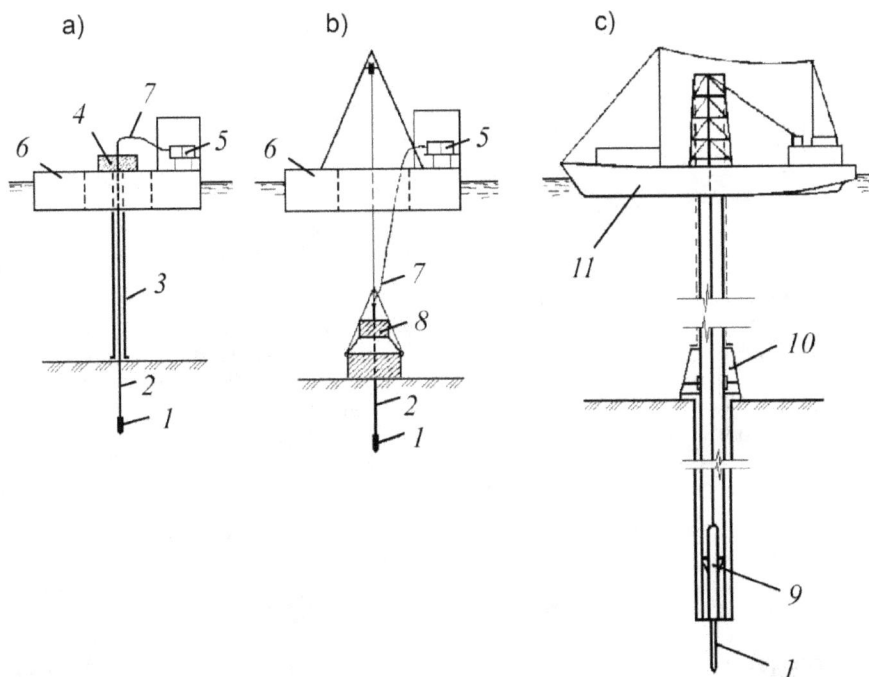

Fig.1.24 Schematics of deep-water CPT equipment (Ferronsky 1969, Trofimenkov 1995, Lunne et al. 2004):
a – seabed mode for CPT, pushing device being located on a floating vessel, b – the same, pushing device being located in underwater unit at the bottom (PSPK VSEGINGEO – USSR), c – down-the-hole mode for CPT (Fugro – Netherland); 1 – cone penetrometer, 2 – push rod, 3 – conductor pipe, 4 – pushing device on a floating vessel, 5 – instrumentation panel with recording equipment, 6 – pontoon (barge), 7 – cable, 8 – underwater unit with pushing device, 9 – borehole anchor, 10 – bottom pickup device, 11 – floating vessel

1.4 BRIEF RETROSPECT OF CPT AND ITS PRINCIPAL DEVELOPMENT TRENDS

CPT as a simplified procedure had been applied for nearly 200 years, though as a method of mechanized soil investigations it dates from the 20th century. "A colonel Volkov" in his "Notes on soil investigation" (Volkov 1836) published in St. Petersburg in 1836 describes manual probes as a thing known to builders of that time. Figure 1.25 illustrates Volkov's manual probes.

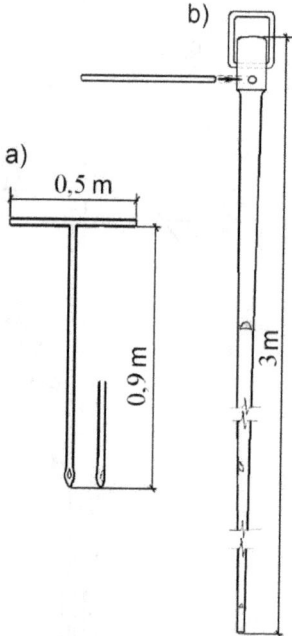

Fig. 1.25 19th century probes (Volkov 1836):

a – manual probe (side view of lower end is shown to the right), *b* – large probe (niche-cuts to collect soil samples are seen on side surface)

The probes were also used in visual estimation of soils by means of small samples extraction. For this, cuts were made on the probe side surface, while turning the probe around longitudinal (vertical) axis the cuts were filled. In small manual probes the only cut was made at the lower end, in large ones several niche-cuts were made (in 0.5...1 m).

In the 19th century probes (both static and dynamic) were probably applied quite often, although there was a lack of publications concerned with engineering investigation issues, and information presented by different authors was contradictory. Cosstay & Sanglerat (1981) believe the probe to have been applied by Collen for clayey soil properties estimation first in France in 1846. Other authors refer to later date of CPT rise, most of them considering the 20th century (Terzaghi & Peck 1958, Bondarick 1964, Trofimenkov & Vorobkov 1981, Lunne et al. 2004).

In 1917 Swedish Ministry of railway transport developed the simplest CPT system (see fig.1.20); it was the modified version of the probes described earlier. The CPT system was different from those shown in figure 1.25 as it had a screwed cone (fig.1.3a) and special loads could be placed on it. The device was applied for clayey soils consistency estimation. The probe was pushed down by operator's physical effort, and then it was loaded step-by-step up to 100 kg. Penetration caused by each loaded step was measured. After that, CPT procedure was continued by means of manual screwdriving, penetration was measured after every 50 turns. The developed CPT system as well as the technique of its application outlived the developers and have been widely used since then.

Manual cone probes (usually without a screw cut) were popular in Denmark, Holland, Switzerland, Norway, etc. (Terzaghi & Peck 1958, Bondarick 1964, Trofimenkov & Vorobkov 1981, Lunne et al. 2004). However, it became evident that the 20th century construction required powerful CPT systems capable to perform deep penetration (not less than 10 m) in any dispersed soils (clayey or sandy) but not manual penetrometers pushed down by physical effort. Here, penetrometer dimensions as well as measurement complex shall guarantee high reliability and accuracy of results. So, from 1930s specialists of a number of countries tried to create just these CPT systems.

Dutch specialists paid great attention to this issue. They designed the first mechanized rigs which predestined CPT equipment development worldwide. The penetrometer was pushed down by mechanical devices – lifting jacks (hydraulic or rack jack) by means of chains, hoists and/or back balance facilities. Figure 1.26 shows CPT procedure on T.K. Huizinga cone penetration system constructed in 1935 (Delft Geotechnics) (Lunne et al. 2004).

Finally, hydraulic lifting jacks happened to have been the most «viable» having excluded other penetration systems. Pushing device in Dutch rigs was placed onto a horizontal platform beam moving up and down along two guide posts. The design presented had become popular worldwide and then was realized in a number of versions (see fig. 1.19 & 1.21). The posts were mounted in a rigid bed that was additionally loaded one or another way. In the first CPT rigs bulks or water tanks were used as the cantledge.

The problem of reactive thrust sensing when pushing down the penetrometer emerged in exchanging shallow CPT for deep one. Screw anchor piles appeared to have been the most effective solution, so all other ways of anchor thrust sensing were excluded then. In the second half of the 20th century rigs mounted in heavy thrust machines became popular, thus the problem was not so urgent any more since the thrust machine weight was sufficient to take reactive thrust without anchorage.

Significant changes in penetrometer designs were made in the second half of the 20th century. In 1929 K. Terzaghi constructed a penetrometer that was pushed down by washing; this allowed significant increase in CPT depth (Terzaghi & Peck 1958). This penetrometer was successfully applied in construction of New-York subway, but nevertheless the idea did not appear to be promising.

Fig.1.26 T.K. Huizinga cone penetration system (Lunne et al. 2004)

In 1936 Barentsen (Holland) constructed the tube-shell penetrometer (see fig.1.2a) that gave possibility to evaluate both cone resistance and side friction resistance (Sanglerat 1971, Cosstay & Sanglerat 1981). The design proved to have been a useful one. It effected CPT rigs development in the 20th century since Barentsen's penetrometers (those measuring two values: cone resistance and side friction resistance) practically excluded the old penetrometers (those measuring only one value: cone resistance). Apart from Barentsen the design was suggested by other specialists as well, but according to Cosstay & Sanglerat (1981) it was Barentsen who constructed this penetrometer.

Vermeiden (1948) & Plantema (1948) greatly improved mechanical cone penetrometers having added a protective mantle to the conical tip; that prevented ring cavity between the outer tube and inner push rod from soil entering (Lunne et al. 2004). Up to present time the mechanical cone penetrometer design has not ever been changed fundamentally. In accordance with the Standard GOST 19912-2012 the design agrees with that of "type I cone penetrometer". Appearance and location of the mantle in modern type I cone penetrometers is shown in figure 1.1 & 1.2.

In Russia type I cone penetrometers have been applied since 1950s in Hydroproject, DIIT, Fundamentprojest, etc. CPT rigs (Bondarick 1964, Trofimenkov & Vorobkov 1981). Nevertheless, the CPT rigs became popular due to GPI Fundamentproject designs that were paid great attention to in 1960s...1980s. At the beginning of 1960s the Institute constructed and practiced type I mechanical CPT rigs: UZK-2 demountable CPT rig and the analogous one mounted in a trailer – UZK-3 manufactured as a stock-produced item under the index S-979. SP-59 self-propelled CPT rig mounted in a wheeled tractor T-16 was designed in 1970s. GPI Fundamentproject CPT rig design is shown in figure 1.2c, S-979 & SP-59 general views are presented in figure 1.21c, d.

Begemann suggested the idea of equipping the penetrometer with a friction sleeve in 1953 (see fig.1.2e). As it was mentioned earlier (see section 1.2.1) Begemann suggested measuring soil resistance in a short length (friction sleeve) located near the cone, but not in a total side surface of the cone penetrometer. The idea proved to have been efficient in the following years when tensometric cone penetrometers appeared.

DEGEBO (Germany) constructed the first tensometric cone penetrometer in 1942 (Bondarick 1964, Lunne et al. 2004). Electric cone penetrometers were of interest in Netherlands, Denmark as well as in other European countries, thus new electric cone penetrometers were designed.

The important step in electric cone penetrometer development is considered to be the tensometric penetrometer of BashNIIstroy constructed in 1963 in the USSR (by Kozlovsky, design engineer) (Ryzhkov & Goncharov 2006). The penetrometer stood the test of time and has effectively been applied. As it was underlined in section 1.3.3 (see fig. 1.6d) the penetrometer is equipped with the friction sleeve and two tensometric sensors to evaluate cone resistance (q_c) and total side friction resistance (f_s). The dimensions of the penetrometer are in accordance with the International Standards and active CIS Standard GOST 19912-2012 as well: cone diameter – 35.7 mm, friction sleeve side area – 350 cm^2 (310 mm in length). The results obtained are recorded automatically: the recorders draw graphs of both cone resistance q_c and side friction resistance f_s in depth variations.

That the penetrometer has widely been applied during several decades reveals its advantages compared to other electric cone penetrometers. This affected the new types of tensometric cone penetrometers designing, e.g. PIKA being well-known in Russia.

The PIKA penetrometer was applied in BashNIIstroy mobile CPT rig S-832 (fig.1.22f, 1.23); its creation in 1963 meant a new level of CPT development: that is automated mechanized CPT rigs of higher productivity and comfort. It became the first CPT rig of series production (1963) in the USSR.

The design features of the CPT rig were discussed in section 1.3.5. It is necessary to point out that the realized design solution happened to have been rather unusual and caused discussions among the specialists of 1960s. The principal objection referred to a builder or surveyor who would need simple but reliable equipment without any electronic or tensometric appliances. However, in practice it was not confirmed. The CPT rig became popular not only in research institutions but in some construction organizations as well. The latter used it to modify pile foundations' design choice; the work was well organized in some organizations of the USSR.

Over 150 S-832 CPT rigs were used in different regions of the USSR in the middle of 1970s. The most effective applications of this CPT rig were when using it in large-scale pile foundation construction and undertime of design and survey works. This situation was typical for Soviet large-scale construction with severe climatic and engineering-geological conditions. In development of Western Siberia territories in 1970s...1980s application of the rig ensured not only foundation designing efficiency but significantly decreased the duration of this development (the towns: Nizhnevartovsk, Raduzhny, Langepas; the large industrial and residential complexes of Tyumen, Tobolsk, etc.). Specific conditions of the region allowed works on pile bearing capacity testing only 3...4 months in a year (the rest of the year the terrain was impassible for transport). In case the builders did not manage to carry out the works in time, they had to finish them in the following year, thus the duration of construction investment cycle was inadmissibly dragged out. Application of the CPT rig totally resolved the problem: amount of CPT in piles decreased several times, works on pile bearing capacity estimation could be performed in the shortest time possible, reliably and economically as well.

Apart from BashNIIstroy CPT cone penetrometer other type II cone penetrometers (with a friction sleeve) were developed and applied in Russian practice. The non-standard penetrometer of SPK VSEGINGEO penetration-radiation rig as well as several modifications of PIKA (NIIOSP) cone penetrometer were discussed in section 1.3.4. By the 21st century type II standard cone penetrometers prevailed both in Russian and non-Russian practice.

A lot of organizations involved in scientific, designing and industrial work took part in creation of effective CPT rigs. Figure 1.27 shows an episode of a comparative test typical in 1960s...1970s. Scientific institutions (Head Institutes – NIIOSP & BashNIIstroy), specialized designing organization Fundamentproject, surveying organization YuzhUralTISIZ took part in this test. Gosstroy of the USSR exercised general management.

Fig.1.27 Comparative tests of S-832, S-979, USZK CPT rigs of different designs held in Chelyabinsk (1969)

(under the management of Gosstroy USSR, members: NIIOSP, Fundamentproject, BashNIIstroy Research Institutes, YuzhuralTISIZ Trust)

Stagnation in scientific-technical work had a negative effect on CPT development in 1990s, i.e. manufacture of CPT rigs stopped, research in CPT applications almost stopped either. Manufacture of CPT equipment was rehabilitated after 2000 (including manufacture of UZS-15/35 – self-propelled CPT rig equipped with type II penetrometers).

When manufactured powerful CPT rigs capable to push down the penetrometer in a significant depth, it became evident that the principal advantages of CPT must be taken into account when resolving the problems of pile foundation construction. Dutch specialists realized this as long ago as 1930s since they had to be engaged in pile applications more than anybody else in Europe due to the soil conditions specific character. They started vast research on full-scale tests of piles parallel with CPT of soils. Later, similar research was put on a large scale in the countries where pile foundations were widely applied. Fundamentproject, BashNIIstroy (NIIpromstroy), NIIOSP, etc. Research Institutes paid great attention to the issue in Russia. Specialists involved in surveying, designing or construction were engaged in the work, analyzed the field tests results.

The problem of pile bearing capacity testing based on CPT data proved to have been a difficult one. Cone resistance and side friction resistance were different form the analogous ones under the foot and on side face of the piles. The difference required to be analyzed carefully due to its complex essence. Any attempts to solve the problems theoretically by means of simplified models of Coulomb's wedge theory, theory of plasticity, etc. were failure. More precise definition of both penetrometers and piles behavior in various conditions was required; it was possible to achieve by thorough experimental research. Here, simulation studies appeared to have been not productive either, as a number of factors of pile behavior were failure when simulated on small trays. That is why the specialists had to refer to field tests. The research was done on a large scale both in Russia and worldwide, great numbers of books on experimental results were published in 1960s…1970s. E.g. in the USSR BashNIIstroy professional engineers saved up 500 results of full-scale tests of piles parallel with CPT of soils in wide range of engineering-geological conditions.

Although the specialists from different organizations suggested their own calculation techniques, often contradictory, knowledge on pile/penetrometer behavior was gradually being specified, accuracy of calculations was being raised. The more accurate the calculations were the less "safety factors" were needed. The most reliable calculation techniques were included in the Normative Documents. In the National Building Codes pile calculation technique based on CPT data was included in 1972 (addition to SNiP II-B.5-67*). The following versions of SNiP included the relevant formulae with different specifications. The issues are discussed in chapter 4 in detail.

There were some difficulties as well in CPT application for soil testing. The issues were of great interest in deep CPT, though research was done in manual probes applications.

First, soil testing based on CPT was done by purely empirical methods. Cone resistances were compared with the traditional soil testing considered as the model.

These empirical dependencies obtained after the statistical treatment were suggested to be applied in practice. However, it soon became clear that it was impossible to obtain any reliable dependencies with a wide range of application. In each lithological variety of soil (even in each genetic variety of the same lithological sampler) different empirical dependencies were obtained. Add to this effect of experiments' techniques, various in different authors.

A number of attempts were taken in 1950s…1960s to resolve the problems theoretically by simplified models of elasticity theory and Coulomb's wedge theory (Berezantsev & Yaroshenko 1962, Yaroshenko 1964, Berezantsev 1966, Ferronsky 1969). When studying deformation properties of soil linear-deformable half-space model was applied, in strengthening properties – Coulomb's loose medium (medium of Coulomb's wedge theory). In the second case "semi-empirical" method was applied when configuration of glide lines was specified by volition (based on experience), and the solution was obtained by Coulomb's wedge theory application. Nevertheless, in both cases the theoretical formulae did not agree with the experimental results.

Applications of elasticity theory models were of greater interest (Reznikov 1961, Vesic 1965, Ryzhkov 1973). The formulae obtained showed dependence of CPT results upon both strengthening and deformation properties of soil. It is evident that the formulae were invalid to be used in practical calculations, though they sorted the variety of known dependencies which could be understood as particular cases of general regularities. Despite the availability of this trend, it was of great interest neither in Russia, nor world wide, though nobody contested the results obtained. The authors of the book believe that this trend deserves consideration to further progress in CPT results comprehension. Chapter 2 discusses the issues in detail.

In last three decades most of Russian and non-Russian professional engineers turned to traditional pattern - empirical dependencies obtaining as applied to each concrete type of soil in terms of large number of tests analyses. In the National Normative Document SP-11-105-97 these dependencies are given. Approximately the same is given in the non-Russian Normative Documents.

New tendencies in complication the penetrometers, the number of measured parameters increase, combining CPT with other tests have been developed since 1980s…1990s. The penetrometers have been equipped with piezocones, inclinometers, thermometers, different adjustments for acoustic, dielectric tests, etc. Wireless connection between the sensors and recording equipment located in the truck enclosure deserves consideration either. The examples of the applications were given in section 1.3.4.

Possibility to carry out various tests as well as simplicity of comparative analyses of the results obtained (without any heterogeneity of soil ill effects) is one of the merits too. However, combining of several tests in one system complicates the problem of the system reliability. It will be shown in practice whether this trend survives. If one can ensure the reliability without excessive costs, multifunctional cone penetrometers will probably be more popular than the traditional single-functional ones.

Attempts to use CPT rigs as the means of pushing down the adjustments not related to CPT are also referred to this tendency. As it was discussed in section 1.3.4 Direct Push Technologies – DPT (Lunne et al. 2004) concept is being developed worldwide; DPT cannot be identified with CPT. Nevertheless, taking into account CPT symposiums of the given period, the specialists' attention is switched over to DPT. One should not conclude that CPT is being excluded by DPT. DPT are directed toward any soil investigation techniques since the technologies allow performing the tests in significant depth. Problems (traditional ones) connected with CPT are not related to this issue.

Nevertheless, in CPT as it is, there exist lots of problems to be resolved or, at least, to be considered. Let us consider three problems.

Firstly, CPT rigs require improvement to ensure their reliability, manufacturability and pushing thrust increase. In practice, difficulties appear due to damage and/or poor behavior of separate CPT rig parts in a large scope of work, those in low-mechanized CPT rigs – overmaintenance. In Russia, experience of the last 30...40 years revealed that in many regions pushing of the cone penetrometer was sufficient 15...20 m in depth. Nevertheless, the depth happens to be accessible just in rather weak soils (soft-plastic and very soft-plastic clays or loams) for Russian CPT rigs. The cone penetrometer is pushed down just in 12...14 m in case pushing thrust capacity is up to 100...120 kN in soils of medium or high density (e.g. bass or loams). Heavier thrust machines application is considered to be the simplest solution since more powerful anchorage decreases CPT productivity.

Nowadays, development of underground in cities or megalopolises requires application of improved heavy CPT rigs with pushing thrust capacity to be 150...200 kN in order to carry out CPT 30...40 m in depth or more. To sum it up, it is also important to develop and improve CPT equipment (support-anchor equipment, pushing & extraction equipment, etc.) as well as the cone penetrometers from the point of view of their capability to work in large pushing thrusts and cone resistances.

Secondly, CPT results need to be computerized. This necessity is due to a great volume of the data obtained (hundreds, sometimes thousands of three-valued or four-valued parameters in every object). It is obvious that the data obtained need to be recorded in a convenient way for OCR (Optical Character Reader). In spite of perfect software for CPT data processing, "manual" input of origins becomes labor-intensive work as well as the weakest link in whole process of data processing. That is why it is necessary to automatize the procedure of OCR as much as possible.

Location of hardware was a subject of discussion some time ago (in the truck enclosure or outside CPT rig, i.e. in cameral conditions). Location of hardware in the CPT rig causes significant costs, as the equipment needs to be adjusted to unfavorable conditions (vibration, variable water content, variable temperature, etc.). Application of laptops proved to be the simplest solution; they can be connected to CPT rig equipment or disconnected when it is necessary as well as delivered to data processing laboratory or one can process data in-situ (in field conditions). Unfortunately, the procedure failed to be applied on a large scale.

Thirdly, the data processing techniques need to be modernized. CPT procedure has not been studied well enough, the applied dependencies are not of high accuracy, sphere of their application is rather limited (in general, the empirical dependencies have been obtained for quaternary soils; loads up to 0.3 MPa), extreme empiricism dominates. Quantitative dependencies given in the Normative Documents sometimes contradict with practice, etc. There are some doubts concerning CPT application in specific soils (collapsible, permafrost, alluvial, technogenic, etc.) or in pile foundation designing, especially bored cast-in-place piles, etc.

In the following chapters the authors attempt to analyze the available information on the issues taking into account our own experience in CPT involving over 40 years of practice. CPT proper is discussed, the penetrometers combining CPT with other tests are not considered.

2 RESEARCH OF CPT PROCESS

Nowadays, principal issues of CPT technology are considered to be more or less examined; thereon, designs and dimensions of the penetrometers as well as CPT technique are standardized both in Russia and worldwide. However, in practice one can face the situations when application of the instructions given in the Normative Documents or results obtained explanation do not agree with the traditional patterns. Here, the specialist shall either know norms and regulations or realize *why* the regulations appeared, as well as understand physical processes standing up for one or another empirical dependency. The chapter presents some results of research, little-known included, ignoring these can result in ineffective, unreliable CPT.

2.1 PHYSICAL AND MECHANICAL PROCESSES OCCURRING IN SOIL DURING CPT

As opposed to traditional methods of mechanical (shear, compressibility, etc.) soil tests where deformation behavior is obvious since the deformations could be observed and measured, CPT procedure is carried out in conditions unavailable to be observed. This caused contradictory ideas of the processes occurring in the soil. A number of computations reflecting soil deformation behavior close to the penetrometer were suggested. The computations were used for great number of rather contradictory formulae connecting different mechanical properties of soil with its resistance to CPT. There exists the prevailing opinion on strong compaction of soil around the penetrometer as well as the opinion on "strongly compacted zone" under the pile foot.

A number of Russian and non-Russian authors carried out the experiments on soil deformation on laboratory trays with a transparent (glass) wall. Another technique when the penetrometer was extracted by cutting of the soil in the tray with a vertical plane was occasionally applied. The experimental results showed that in penetration soil particles shift down and sideways in specific curvilinear paths. Horizontal soil layers "deflect", and vertical ones – "slide apart" sideways.

However, there were some drawbacks in the soil deformation studies. There were some negative attitudes towards stiff (transparent) wall influence on the deformations and penetrometer's extraction (digging out) influence as well. It is evident that these resulted in distrust to the experimental results obtained. Therefore, it is necessary to discuss less-known experiments of BashNIIstroy (NIIpromstroy); here, soil deformation behavior was examined ignoring either transparent walls application or extraction of the penetrometer. Instead, gamma-X-ray introscopy was applied.

The penetrometer was located in the tray's center, but not close to its wall, the pressure (cantledge) imitating upper soil layers weight was imposed on the soil surface –"overburden (natural) stress" (Ryzhkov 1970). Lead balls (pellets) were embedded into the soil to observe soil shifts; the tray was transilluminated with gamma-rays. The tray had the dimensions of 20×20×20 cm, the cantledge – 40 kPa, that corresponded to the depth of nearly 3.5…4 m. The photos were taken after every centimeter of penetration (lead balls were placed in the same plane parallel to film plane and passing through the penetrometer longitudinal axis) so that in overlapping the shots soil shift directions could be recreated. The experiments were carried out in clays with liquidity index $I_L = 0.28$, $I_L = 0.55$, $I_L = 0.6$ (clayey paste with liquidity index close to 1). The experiments did not reveal any differences in various consistency clays' deformation general behavior: just the values of the deformations differed.

Figure 2.1 shows both a shot and a schematic illustration of soil layers shift. As one can see from the figure, soil deformation behavior is almost the same as in the experiments of other authors, i.e. when the penetrometer slides down the tray's transparent wall.

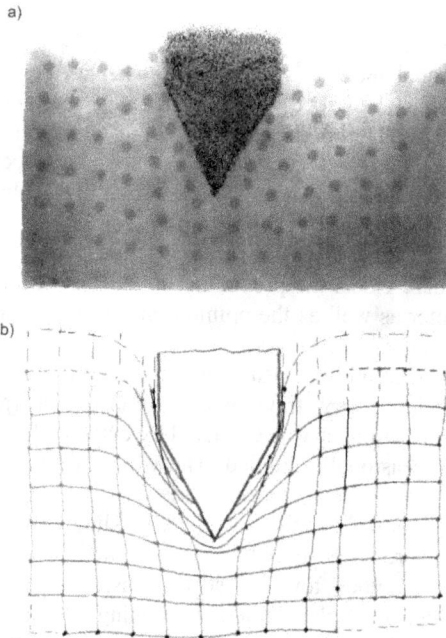

Fig. 2.1 Deformations of clayey soil under the penetrometer: a – X-ray film, b – schematic illustration of soil shift after X-ray film

The analogous conclusion could be made when analyzing schematics of pellets' guidepaths; these were obtained by overlapping the shots corresponding to "adjacent" locations of the penetrometer, i.e. the depths differing in 1 cm. It is necessary to point out that in the experiments vertical component of soil shift prevailed to a considerable degree than in those ones mentioned above, in them the penetration was carried out close to side wall of the tray (paths' deflections from vertical just under the penetrometer were 5...8 %). Thus, motion of pellets (and hence, soil particles) occurs in curvilinear paths "down and sideways", deflection vertical components prevailing.

The doubts that pellets being heavier than soil incorrectly mirror shifts of soil particles were dispelled by the control experiment; in it the tray was turned over, the penetrometer moved from the bottom upwards (i.e. gravity took the opposite direction). Pellets' shift behavior in the given gravity variation did not change at all.

Another series of experiments was concerned with clayey soil compaction studies under penetration. Here, the penetrometer needed to be extracted (dug out) since the accuracy of other possible techniques of soil density-liquidity indexes estimation appeared to have been insufficient for the issue to be resolved. The same demountable tray was used: the tray was filled with clayey paste of various consistencies, the degree of saturation equaled 1; the penetrometer surface was stressed with the same pressure (40 kPa). The cantledge had acted 4 days before the penetration procedure and was kept up to the end of the experiment. The penetration procedure carried out 8...10 cm in depth in the center of the tray was followed by cutting of the soil in the tray, and then soil samples were taken in different lengths from the penetrometer. The speeds of penetration were taken to be 5 mm/min & 5 mm/day. Since the soil possessed the degree of saturation 1, void ratio variations could be estimated in liquidity indexes variation.

The experiments revealed that changes of soil conditions (void – liquidity) are observed just in very small speeds of penetration (Ryzhkov 1971). Figure 2.2a illustrates measured liquidity indexes and void ratios (shown as isolines) of overconsolidated clay around the penetrometer under the speed of penetration 5 mm/day. In the given speed there were not any doubts about soil compaction: liquidity index of clay decreased in 0.04 of decimal fraction, void ratio – 0.08. Nearly the same changes were observed in firm clay ($I_L = 0.63$).

However, the experiments under the speed of penetration 5 mm/min, that is nearly 20 times lower than the standard speed of CPT (1.2±0.3 m/min), did not reveal any changes in void-liquidity in clays of various consistencies – from semi-solid to fluid ($I_L = 0.19...1.2$).

It is evident that in standard speed changing of clayey soil density will be much less probable. Figure 2.2b illustrates the control field experiment results with S-832 CPT rig application. Penetration was carried out with the standard speed of 1 m/min in the alluvial overconsolidated clay with ($I_L = 0.3$) consistency, the depth of penetration being 2 m, the procedure was followed by the extraction of the penetrometer and soil samplers selection as well (penetration was carried out 0.5 m from the edge of the pit vertical slope). As it is shown in figure 2.2b spatial variability of void ratio is of random character, its regular changing is not observed in getting nearer to the penetrometer.

Thus, in CPT clayey soil is hardly ever compacted, though its structure is destructed. Filter consolidation processes are hardly ever revealed in clayey soils either. To sum it up, it is important to draw two practical conclusions concerning CPT in water-saturated clayey soils.

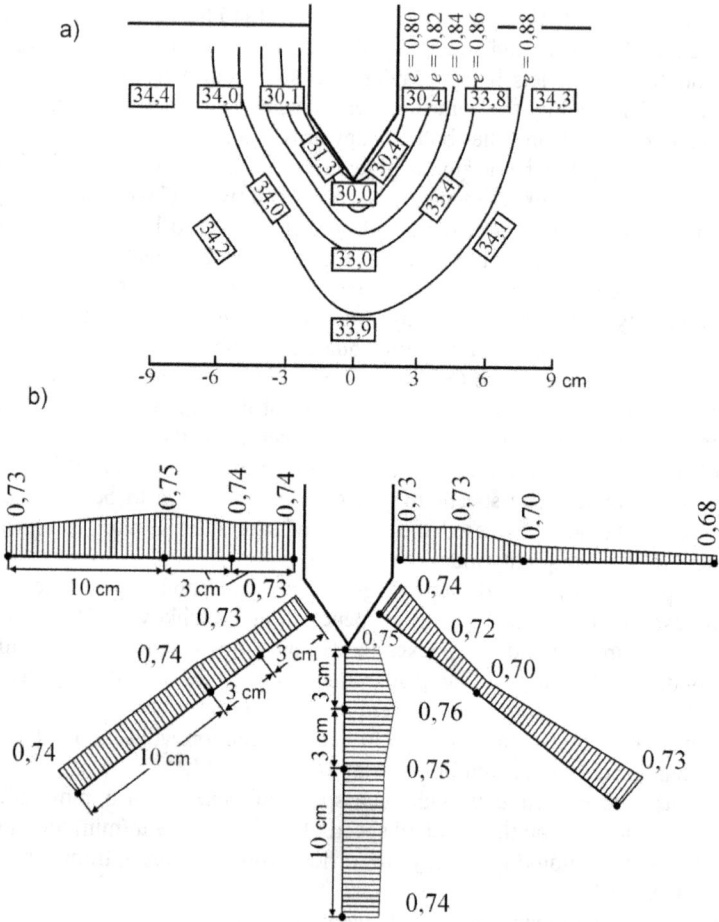

Fig.2.2 Schematics of soil condition change under the penetrometer:

a – laboratory experiment – clay condition change ($I_L = 0,43$) in the tray under the speed of penetration 5 mm/day (numbers in boxes illustrate water content w percentagewise, isolines show void ratios e); b – field experiment – void ratio change e in water-saturated clay under the speed of penetration 1 m/min, 2 m in depth (changes e are shown as diagrams)

Firstly, in CPT the soil acts in absence of drainage ("closed system"), i.e. its destruction agrees with "non-consolidated-non-drainable shear" scheme, and its non-

55

destructive deformations occur without any changes in volume, i.e. in conditions directly opposed to compression.

Secondly, when equipping the penetrometer with additional devices to measure density and/or liquidity (e.g. γ-consolidometer or neutron drymeter), one need to ignore deviation in measurements of clays or loams due to soil compacting. Inaccuracies in measurements will be caused by absolutely different reasons. Some errors may occur in clay sands, though they do not affect the mechanical properties estimation since density and deformability of clay sands depend upon conditions of load action and speed of deformations less than in clays or loams. To a larger extent, it is related to sands capable to be significantly compacted during penetration. However, the fact is of much less importance since the test mode of non-cohesive soils does not particularly affect the results obtained.

2.2 EFFECT OF DESIGN AND DIMENSIONS OF CONE PENETROMETER ON CPT RESULTS

A number of experimental research on **cone penetrometer diameter effect** were carried out by both Russian and non-Russian professional engineers (Terzaghi & Peck 1958, Bondarick 1964, Trofimenkov & Vorobkov 1981, Lunne et al. 2004). The data obtained by different authors are rather contradictory, but in general they reveal rather weak effect of the cone penetrometer diameter on values q_c. In the diameter of 35...80 mm in length values of cone resistance q_c differ just in several percent. Generally, cone resistance q_c decreases with the diameter increase, the decrease being noticeable in dense soils. Numerous field experiments of BashNIIstroy have proved the concepts (Ryzhkov 1971). The issue on rod's diameter effect on cone resistance is paid attention to in the section devoted to pile bearing capacity calculation on CPT data. Nowadays, the problem of cone penetrometer diameter effect appears to be not of great importance, as the dimensions of penetrometers are standardized both in Russia and worldwide, non-standard cone penetrometers being occasionally applied.

The issue on **difference in results obtained by standard cone penetrometers of types I&II** proves to be of greater importance. As it was mentioned earlier, soil resistance related to the mantle in type I cone penetrometer is a part of q_c – cone resistance, but in type II cone penetrometer value q_c indicates just cone resistance. Local unit mantle side resistance in type I cone penetrometers nearly equals f_s – unit sleeve friction resistance in type II cone penetrometers. In a standard cone penetrometer the lateral area of a mantle is 83 cm^2. Taking into account that in clayey soils value f_s is usually 1...3 %, in sands 0.3...1 % from value q_c, it is easy to count that the "addition" of resistance will raise value q_c in 8...25 % – in clayey soils, in 2...8 % – in sands. Thus, cone resistance q_c in type I cone penetrometers shall be larger than in type II ones, approximately in the same percent.

The data were checked experimentally by comparison of results obtained in CPT rigs equipped with types I & II cone penetrometers. In 1969 in Chelyabinsk NIIOSP, Fundamentproject, BashNIIstroy Research Institutes, UralTISIZ Trust carried out comparative tests of CPT rigs equipped with types I & II cone penetrometers: S-979,

USZK and S-832 (fig. 1.27, 2.3 and 2.4) (Kazantsev et al. 1974, Ryzhkov 1992). The works were carried out on clayey and sandy sites (alluvial, dealluvial and eluvial deposits). The results obtained by the following cone penetrometers were compared (fig. 2.3):

- Type I standard mechanical cone penetrometer being applied in S-979 (fig.2.3a) & USZK-3 CPT rigs.
- Type II standard tensometric cone penetrometer being applied in S-832 CPT rig (fig.2.3b).
- Tensometric cone penetrometer of S-832 CPT rig; special tip was screwed on the cone, the dimensions of the tip absolutely agreed with S-979 cone (fig. 2.3c).

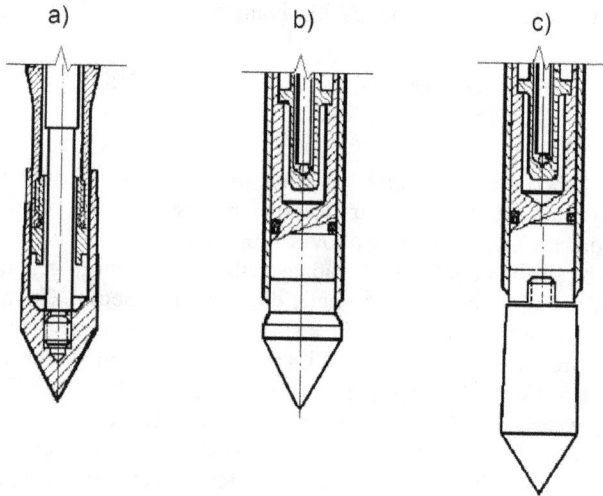

Fig. 2.3 Schematics of penetrometers applied in comparative experiments (Kazantsev et al. 1974):

a – type I mechanical cone penetrometer (with a mantle) of S-979 & USZK-3 CPT rigs, *b* – type II tensometric cone penetrometer (without a mantle, with a friction sleeve) of S-832 CPT rig, *c* – tensometric cone penetrometer of S-832 CPT rig with a special tip, the dimensions absolutely agreed with S-979 cone tip (penetrometer with a mantle and a friction sleeve)

Measurements were done with a high repeatability (fivefold, tenfold), CPT locations with different penetrometers being located randomly(view in plan). *Mean* values of q_c obtained by the illustrated cone penetrometers (see fig.2.3) were compared at every depth. The technique allowed heterogeneity of soil and other random factors to have been significantly decreased.

Figure 2.4 shows q_c comparative results in eluvial loams of very soft and semisolid consistencies. Figure 2.4 also illustrates CPT locations in plan, here, as it was mentioned above, the highest stability of results to random factors effect was ensured.

As it is shown in figure 2.4, the results obtained proved the difference between the resistances q_c, obtained by types I & II cone penetrometers. In some cases the differences appeared to have exceeded 25 %. The greatest attention is paid to the fact that screwdriving of a mantle cone on S-832 CPT rig (friction sleeve was unchanged) excluded the difference, i.e. values q_c, obtained by S-832 CPT rig were practically the same as in S-979 & USZK CPT rigs equipped with type I cone penetrometers. This confirmed the opinion on comparability of CPT results obtained by both mechanical cone penetrometers and tensometric ones; the issue was a subject of discussion in 1960s...1970s. As a matter of fact, the differences between values q_c, obtained by S-832 CPT rig and CPT rigs equipped with mechanical cone penetrometers (S-979, SP-59, etc.) are determined by availability or absence of a mantle in a cone penetrometer but not by measurement techniques.

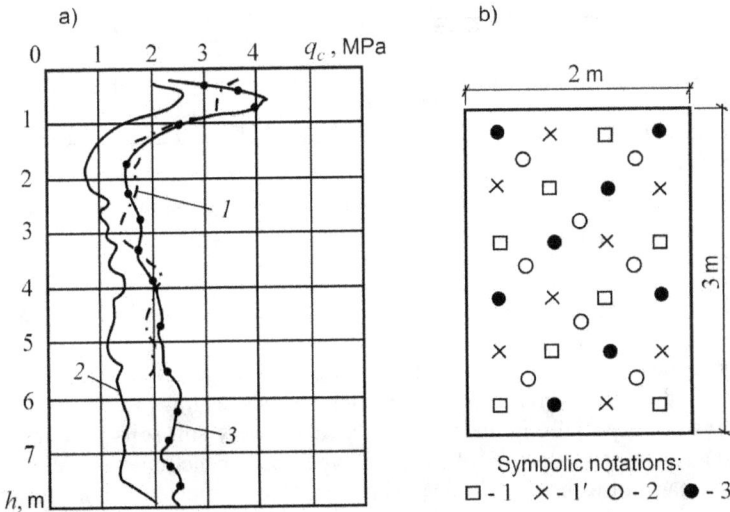

Fig. 2.4 Schematics of comparison in mean values of cone resistance q_c measured by types I&II cone penetrometers (eluvial loams):
a – graphs of q_c changes in depth h: 1 – resistances q_c, measured by type I mechanical cone penetrometer (penetrometer mounted in S-979 CPT rig – see fig.1.21c), 2 – the same with type II tensometric cone penetrometer (penetrometer mounted in S-832 CPT rig – see fig. 1.22f), 3 – the same with tensometric cone penetrometer of S-832 CPT rig equipped with special mantle cone; b – CPT locations view in plan: 1 – CPT with type I cone penetrometer mounted in S-979 CPT rig, 1' – the same with USZK-3 CPT rig (type I cone penetrometer, analogous to S-979, 2 – the same with S-832, 3 – the same with S-832 equipped with mantle cone penetrometer

58

In testing CPT rigs there were some cases when the inner push rod became wedged (rod transferring thrusts on the conical tip) in the outer one of the mechanical cone penetrometer. The results obtained were extremely different from those obtained in the adjacent locations (q_c increased 2...3 times), therefore it was an easy task to find and reject them.

Fig. 2.5 shows the results of the analogous experiment carried out in dealluvial sands of medium fineness as well as in dense and water-saturated ones.

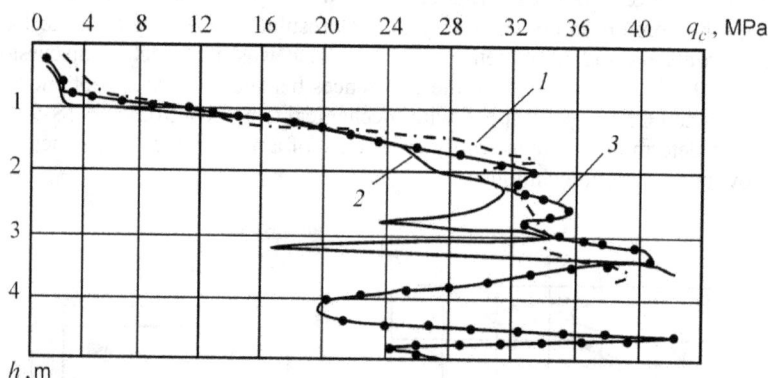

Fig. 2.5 Schematic illustration of comparison in mean values of cone resistance qc measured by types I & II cone penetrometers (dealluvial sands): 1, 2, 3 – see fig. 2.4a

As one can see, the difference between cone resistances q_c in types I & II cone penetrometers is observed, though it is smaller than that in the loams.

Thus, cone resistance q_c values in types I & II cone penetrometers are not equal. It is evident that the difference can be ignored in sands, as it is just 2...8 %, although in clayey soils, especially in underconsolidated and very soft soils it is important to take into account that values q_c, obtained by type I cone penetrometers can be 15...20 % higher than those in type II ones. Therefore, the formulae or tables to evaluate properties of clayey soils, pile bearing capacity in clayey soils on CPT data need to correspond with the concrete type of the penetrometer.

If one knows nothing about the origin of the formulae, value q_c, evaluated by type I cone penetrometer, needs to be reduced in 10...20 % "in stock of reliability". Unfortunately, the ignorance of the fact in the effective Normative Documents (SNiP 2.02.03-85*, SP 11-105-97, SP 47.13330.2012) result in misleading the design engineer and/or unwanted outcomes.

Comparability of results obtained by both mechanical and tensometric cone penetrometers is confirmed by numbers of other experiments carried out by Russian and non-Russian professional engineers, e.g. Yu.G. Trofimenkov (Trofimenkov & Vorobkov 1981) presents results of analogous research conducted by "Fugro". As one can see from the graphs given, in CPT with mechanical and tensometric cone penetrometers the results happen to be the same.

The data given are related to the values of cone resistance q_c, but never to "side friction resistance", i.e. total side friction resistance Q_s and unit sleeve friction resistance f_s. These indexes are incomparable characterizing different subjects of soil resistance measurement – full length of a push rod or a friction sleeve. The issue is discussed in the chapter on pile bearing capacity calculation.

Effect of cone tip angle on the results obtained was thoroughly studied in the middle of the last century. The greatest attention was paid to this issue by the specialists in shallow penetration but not ordinary deep CPT. The depth of penetration did not exceed the height of the cone penetrometer, or the procedure was conducted by means of friction reducer application, its diameter significantly exceeded that of the push rod (see fig. 1.3c,d). Razorenov (1980) gives a number of experimental results obtained with cone tip angles (friction angle) from 20 to 120°. When analyzing hundreds of experimental results obtained from various soils he concludes that "penetration thrusts are invariant in relation to friction angle of the cone tip" or at least hardly depend on it.

Experiments with "deep" rod penetrometers with diameters equal to those of push rods or friction sleeves resulted in the same data in friction angle to have been 30...40°. However, in "sharper" cones effect of friction angle on the results obtained was produced much more. Figure 2.6 shows the experimental results after T. Muromachi (see in Trofimenkov 1995) illustrating that in friction angle α to be more than 40° cone resistance q_c does not depend upon the friction angle, but in smaller angles it significantly increases with the decrease of α.

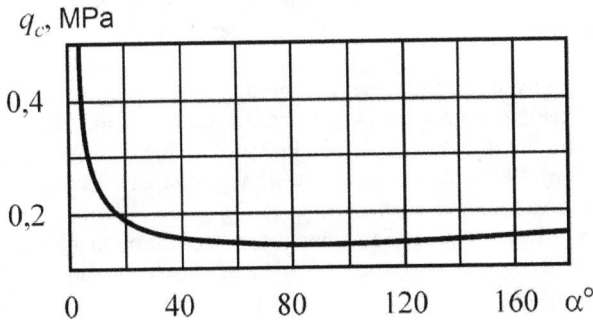

Fig.2.6 Schematic illustration of cone tip angle effect (α) on cone resistance (q_c) in weak clay (author Muromachi, see in Trofimenkov 1995)

Trofimenkov (1995) analyzes the analogous results obtained by other specialists besides the experiments after Muromachi. The differences are affected by the value of the friction angle; here cone resistance does not depend upon the angle. According to different authors the boundary values prove to be the following angles: 30°, 60°, 67°, etc.

In BashNIIstroy the specialists compared the CPT results obtained after standard cone tip application (tip angle – 60^0) and "obtuse" penetrometer without a tip (tip angle – 180^0). S-832 CPT rig equipped with tensometric cone penetrometer was used. The CPT results did not show any significant differences. Under the "obtuse" penetrometer (as under the pile without a tip) soil cone is formed; it moves with the penetrometer (pile) and cuts soil strata just in the same way as the cone penetrometer (pile). Thus, in angles $\alpha > 60°$ it is impossible to reveal any influence of α on cone resistance q_c.

The simplest explanation of cone resistance q_c increase with decrease of cone tip angle α in "sharp" cones ($\alpha < 30°$) is considered to be both increase of cone lateral area and surface friction. In terms of geometrical considerations it is easy to calculate that the lateral surface of "sharp" cones will increase as compared to that of standard ones – $\alpha = 60°$ (base diameter is the same) as follows:

- In $\alpha = 40° – 1.5$ times.
- In $\alpha = 30° – 1.9$ times.
- In $\alpha = 20° – 2.9$ times.
- In $\alpha = 10° – 5.5$ times.

Taking into account that "deep rod penetrometer" having the same diameters of the cone and the push rod (friction sleeve) involves much larger masses of soil than "the shallow penetrometer", amount of friction in cone resistance q_c shall be less in the rod penetrometer than that in the shallow one. Effect of α on q_c in the rod penetrometer is less than that in the shallow one due to the same reason. Taking into account that Razorenov (1980) failed to observe any significant effect of α on q_c in shallow penetrometers (in $\alpha > 20°$), let us consider the effect in rod penetrometers to be rather little.

In general, nowadays, the issue of cone tip angle effect on CPT results as well as the effect of its diameter appears to be irrelevant, since in practice non-standard penetrometers are hardly ever applied. In the International Standards (Proposed European Standard 1977, IRTP 1999) as well as in the CIS Standard GOST 19912-2012 unified tip angle 60^0 is accepted.

Effect of wear on the results obtained was discussed by many specialists in CPT measurement accuracy studies. As a rule, wear of cone (tip) and friction sleeve was estimated. In long-term operation cone tip angle slightly increases (i.e. the cone "becomes obtuse", its height decreases); here the cone surface is planarized due to friction of soil. The surface of the friction sleeve is also planarized and, hence, its diameter decreases.

Figure 2.7 shows the results of research after Schaap & Zuidberg (1982) who studied changes in dimensions of a standard cone penetrometer during its operation.

As it is shown, wear resulted in over 1.5 mm shortening of the cone (1585.2 mcm), damage of the surface occurred unevenly, its shape was slightly crooked, but reduction of the diameter in the cone base was insignificant.

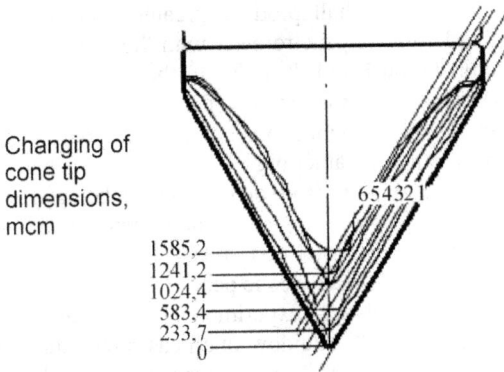

Changing of cone tip dimensions, mcm

1585,2
1241,2
1024,4
583,4
233,7
0

Fig. 2.7 Schematic illustration of changes in dimensions of a standard cone due to wear (Schaap & Zuidberg 1982):
1...6 – contours of cone surface corresponding to different stages of operation (6 stages are given); dimensions are given in microns

To sum it up, one can draw a conclusion that reduction of cone height due to wear is very small and one may ignore it. International requirements on CPT techniques (IRTP 1999) allow rather low accuracy of cone dimensions to be met (fig. 2.8). The CIS Standard GOST 19912-2012 allows reduction of cone height up to 5 mm, its diameter – up to 0.3 mm.

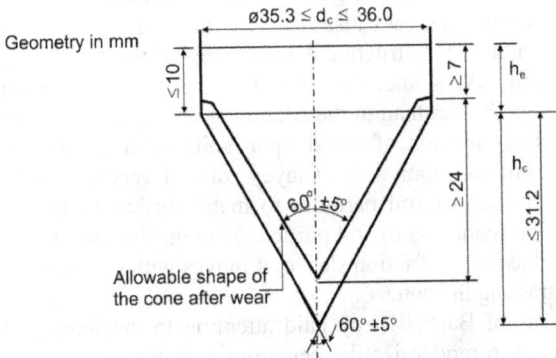

Fig. 2.8 Tolerance requirements for use of cone allowed by the International Recommendations on CPT (IRTP 1999)

Changes of cone surface roughness are much more significant to be taken into account. The issue was discussed in detail when studying shallow penetrometers.

Considering the experiments with shallow penetrometers it is important to underline that cone surface roughness shall produce greater effect on the resistance values obtained than in standard cone penetrometer with the diameter being equal to that of the friction sleeve or the push rod. It is due to the fact mentioned above that in the penetrometer zone of soil deformations is rather small, as the soil is easily projected on the surface without any resistance to penetration. Amount of "friction" in the total soil resistance to penetration is rather big.

The standard penetrometer (rod penetrometer) being pushed down involves much bigger amounts of soil tolerating much larger amounts of soil resistance. Due to this, amount of soil friction in total cone resistance is much lower in the standard penetrometer than that in the shallow cone penetrometers.

Razorenov (1980) analyzes experiments with staged and polished cone penetrometers after Zotsenko & Vagidov. In all cases resistance of soil to penetration of the staged penetrometer proved to be larger than the analogous resistance of the polished one. In coarse sands the difference appeared to be nearly 300 %, in loams with damaged structure – nearly 30 %. Nevertheless, the difference stayed stable in all soil conditions under study (sands and loams with different density were used). This stability significantly simplifies the solution of the emergent problem, since occurrence of cone roughness can always be included in the correction factor. When applying empirical dependencies connecting resistance to penetration with properties of soil, surface condition is included automatically; it is important to take into account that empirical dependencies shall agree with the particular surface condition of the cone.

The analogous situation occurs in wear of friction sleeve that is "polished" by the soil during CPT, particularly by gravel sands or overconsolidated clay. Effect of this wear on the results obtained is much more evident than that of the cone. Unit sleeve friction resistance f_s shows just "friction" of the soil, while cone resistance q_c shows destructive and non-destructive deformations of the significant amount of soil surrounding the cone. As a rule, the value f_s, measured by the friction sleeve with rough surface will be higher than in the friction sleeve with polished surface, since in the polished surface "friction of metal upon soil" is measured, and in the rough surface – "friction of soil upon soil". Clayey soils of very soft of fluid consistency with adhesion (adhesion of soil particles to metal surface of friction sleeve) to be larger than cohesion (cohesion of soil particles) are the exception. Here, a soil film is formed on the surface of the friction sleeve; it moves with the penetrometer and it can be seen when extracting the cone.

The specialists of BashNIIstroy paid attention to the issue in 1960s, thus the attempts were taken to modernize the penetrometers. Friction sleeves in S-832 CPT rig were made with cylindrical grooves (channels) allowing "friction of soil upon soil" in all conditions (fig. 2.9). The issue has been of great importance as "friction of soil upon soil" has always been much more essential than "friction of metal upon soil" for a specialist in foundation construction.

Thus, the penetrometer with increased roughness does ensure "friction of soil upon soil" in all soils (in clays and loams there appeared a soil film on the friction sleeve; it remained after the extraction of the cone), but nevertheless, in practice the solution happened to be ineffective due to the following reasons.

Figure 2.10 illustrates the graphs of both cone resistance q_c and unit sleeve friction resistance f_s changes in depth obtained by CPT in the same soils (sandy and clayey) by the penetrometers with different surfaces. Three penetrometers were used: a new one with rough surface (see fig. 2.9), a smooth one and a rough one with worn-out surface after 5000 m of CPT (3...4 months of work) (Enikeyev et al. 1986).

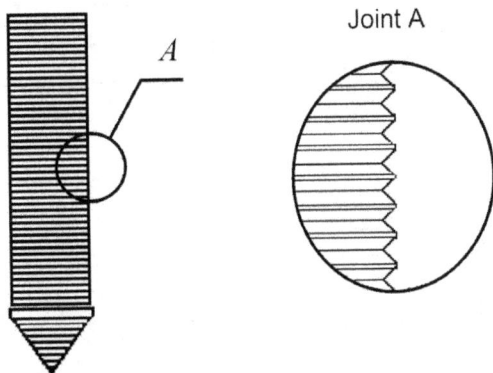

Fig. 2.9 Schematic illustration of the friction sleeve in S-832 CPT rig of 1960s...1970s

As it is shown in figure 2.10, wear of cone had a low effect on the results obtained, i.e. on q_c, though wear of friction sleeve nearly twice reduced f_s.

Abrasive capacity of soil appeared to be larger than it was expected, that is why the grooves were faded rather quickly and the friction sleeve became smooth after two-three months of CPT operation. During this period the degree of wear (smoothing) was changing gradually and it needed to be specially analyzed at every point of time; it was no easy task of itself.

Due to this the version to measure "friction of soil upon soil" was considered to be incorrect. Smooth surfaces were believed to be more preferable; although they showed "friction of metal upon soil", they ensured the stability of measurement conditions. This allowed obtaining relevant information by means of either corrections. The adopted International Recommendations (IRTP 1989, 1999) enabled the professional engineers to apply smooth penetrometers.

However, in practice, smooth friction sleeve is worn-out anyway, its diameter decreases.

a)

b)

Symbolic notations:

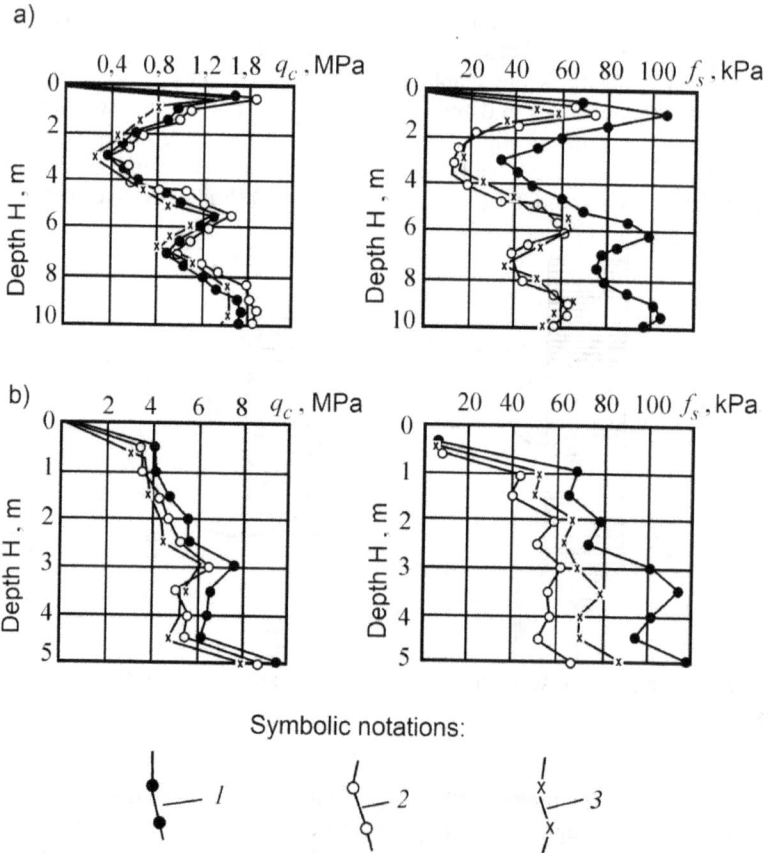

Fig. 2.10 Schematics of q_c & f_s changes in depth of penetration H obtained by S-832 CPT rig with different surfaces of the penetrometers:

a –overconsolidated and semi-solid alluvial clays; b – alluvial-eolian pulverescent sands:
1 – new rough penetrometer (with cylindrical grooves), 2 – smooth penetrometer,
3 – worn-out penetrometer (after 5000 m of CPT)

Figure 2.11 shows the results of research of such changes carried out by Schaap & Zuidberg (1982). There was 1.5 mm decrease in the diameter of the friction sleeve, its surface was curved – this shall enable its contact with soil to be loosened, decrease of radial soil thrust on its surface and decrease of unit sleeve friction resistance f_s. For this reason, the International Recommendations (IRTP 1999) *do not allow* application of friction sleeves with diameters to be less than those of conical tip base d_c. On the other hand, excess of the diameter d_s (friction sleeve) over the diameter d_c (cone base) is restricted in 0.35 mm, this shall exclude any thrust effects on the friction sleeve apart from shear ones. In other words, the diameter of the friction sleeve shall be in the following limits:

$$d_c < d_s < d_c + 0.35 \text{ mm.}$$

Roughness of the friction sleeve surface in the longitudinal direction r shall be in the following limits:

$$0.15 \text{ μm} < r < 0.65 \text{ μm.}$$

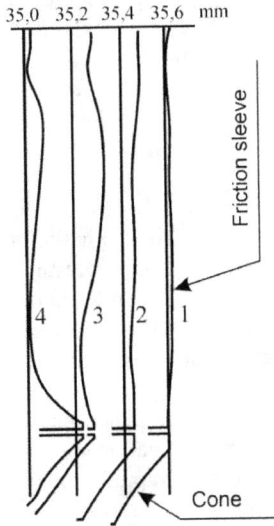

Fig. 2.11 Schematic illustration of changes in dimensions of standard penetrometer friction sleeve due to wear at different stages of maintenance (Schaap & Zuidberg 1982):

1...4 – outlines of friction sleeve surface corresponding to different stages of maintenance (4 stages are given)

Dimensions and conditions of the penetrometers need to be systematically checked and necessary measures need to be taken if required (repair or replacement of the penetrometer).

2.3 EFFECT OF METHODOLOGY AND CONDITIONS ON CPT PROCESS

Besides the described equipment of cone penetrometers, CPT process is also affected by both CPT technique and CPT conditions. The methodological factors are as follows:

- General principle of the resistances to be evaluated, i.e. measurements during uniform penetration, measurements during conditional static equilibrium of the penetrometer or during other penetrometer test modes.
- Speed of penetration (applying the first of the principles mentioned earlier);
- Bending of push rods.
- Temperature in CPT procedure.

- Reliability of measuring devices, especially electronic ones, quality control efficiency, etc.

General principle of CPT procedure became a subject of study due to specialists' desire to minimize the difference in soil behavior under the penetrometer and under the foundation as well. Speeds of soil deformations under the penetrometer are enormously higher than those under the foundations. It is evident that behavior of soil (as any other material) in rapid actions is not identical to its behavior in case it is statically loaded. A surveyor or design engineer having been interested in behavior of soil under the foundation, attempts were taken to modernize CPT procedure by minimizing the mentioned difference. E.g., it was suggested to reduce the speed of penetration up to 2 cm/min or to perform the static tests (to define the "setting-load" ratio) in the given depths instead of measurements. In such an approach speeds of deformations do decrease, however, the principal advantage of CPT is lost – its quickness, possibility to carry out lots of measurements in minimal terms with minimal costs.

For this reason, the approach "in the pure state" was rejected in practice, but the idea to measure static deformations was realized in other approaches. Firstly, it became the basis of a number of special methods of pile bearing capacity testing; here, inventory piles of a small diameter were penetrated and tested, the piles being analogous to penetrometers (piles-penetrometers, reference piles, etc. (Building Code SNiP 2.02.03-85*)). Secondly, there appeared new CPT techniques improving the traditional approach.

Lunne & Keaveny (1995) describe *the dissipation test*; in it the penetrometer equipped with a piezometer is stopped in the given depth, then dissipation of pore pressure around the soil is recorded. In sands the dissipation occurs rapidly, that in clays – in several days (in calculations one needs to consider 50 % dissipation of pore pressure). By means of these tests filtration and consolidation properties of soil are estimated. The authors believe that when performing the tests it is useful to record, besides pore pressure, cone resistance and unit sleeve friction resistance as well. Such tests are applied by many specialists who gained the experience in the procedure.

Relaxation-creep CPT technique of BashNIIstroy is probably considered to be the first step in CPT realization with attained speeds of soil deformations to be close to zero (Firestein & Makarov 1964). Application of this technique (without pore pressure measuring) started in 1962. Here, stabilized deformations corresponding with speeds of deformations tending to zero are evaluated parallel to soil resistance during uniform penetration at separate locations. In these conditions the penetrometer at the given depth is led to conditional limited equilibrium applying a special device in S-832 CPT rig (see fig. 2.12). The pump delivering oil in hydraulic cylinders 2, simultaneously delivers it in a dumper 3, so pressure in the dumper and hydraulic cylinders happens to be the same (as in communicating vessels). When pushing down the penetrometer in a given depth, the pump becomes disconnected; the push rod 5 and the cone 7 are affected by the pressure in the dumper 3. The penetration process retarding, the pressure decreases. Stopping of the penetrometer occurs in attaining "the equilibrium" between pushing thrusts and soil reaction.

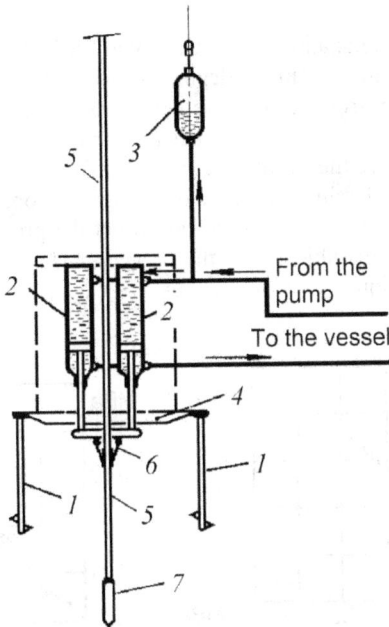

Fig. 2.12 Schematic illustration of push joint equipped with a device to lead the penetrometer to equilibrium in relaxation-creep CPT applying S-832 CPT rig:

1 – anchor (boring) piles, 2 – hydraulic cylinders to penetrate the cone, 3 – air dumper, 4 – frame, 5 – push rod, 6 – hydraulic clamping device to penetrate the push rod and the cone, 7 – cone tip

Relaxation-creep CPT technique is capable to give additional information on type, condition and rheological properties of soil (Isaev et al. 1987, Isaev 1989, Ryzhkov 1992). According to the performed tests (Isaev et al. 1987) the technique can be carried out ignoring dumpers in CPT rigs. There exist some other non-standard techniques as well.

Both measured resistances of soil as they are and those obtained in CPT results (with the speed of penetration – $v = 1$ m/min) are of great interest. One can compare q_c & f_s, obtained in both standard CPT and relaxation-creep CPT at the same location and depth. Figure 2.13 illustrates the comparative results of penetration resistance in standard speed of CPT and relaxation-creep CPT in alluvial clays and sands.

Difference between the "standard" and "equilibrium" values of (q_c & f_s) gives important information about soils under study. It depends upon lithological origin of soil, presence of specific properties in soil, etc. Correlation between "standard" and "equilibrium" values in ordinary lithological soil layers usually preserves stability,

68

but in layers of different origin it significantly varies. Further the issues are discussed in detail.

When performing relaxation-creep CPT with S-832 rig the penetrometer is usually led to equilibrium in 1m of depth, occasionally – in 0.5 m. If required, "equilibrium" of the penetrometer may be carried out selectively at separate depth. Intensive stage of relaxation-creep CPT takes a few minutes (in sands it is faster than in clayey soils). In practice the "equilibrium moment" is determined visually by q_c & f_s curves in chart strips – 1 minute absence of changes is considered to be a criterion. Application of relaxation-creep CPT does not affect the productivity of CPT rigs in case separate locations are subjected to the procedure, but the major part is carried out using standard technique.

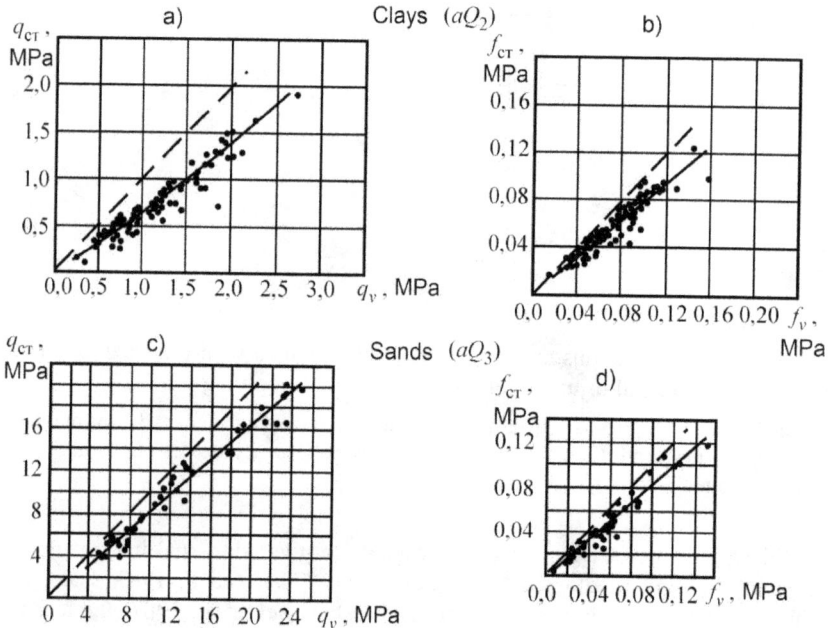

Fig. 2.13 Schematics of comparative results of penetration resistance with the speed of penetration $V = 1$ m/min (for simplicity q_v & f_v are given without the indexes «c» & «s»), in relaxation-creep CPT - equilibrium (q_{cm}, f_{cm}):
a – comparison of cone resistance in clays, b – the same of unit sleeve friction resistance, c – comparison of cone resistance in sands, d – the same of unit sleeve friction resistance (dash lines show absolute conformity of compared values, i.e. $q_{cm} = q_v$ & $f_{cm} = f_v$)

Effect of speed of penetration on the results obtained has been studied by many Russian and non-Russian specialists. Listing the names of the authors who

69

studied the issue and published the results of research in last 40...50 years would take nearly half of a page. Reviews and generalizations of the gained experience were presented once and again, e.g. in Russia the reviews were given in a number of books on CPT published in the second half of the 20th century: Bondarick (1964) published in 1960s, Trofimenkov & Vorobkov (1981) published later, Ferronsky (1969), Trofimenkov (1995) published in 1990s, etc.

Despite some contradictory results the information obtained shows that the speed of penetration has a low effect on CPT results (excluding rheological soils). In the increase of speed from 0.1 m/min to 3 m/min (0.17...5 cm/s) some increase of resistances q_c, f_s, Q_s was observed but after most authors it did not exceed 10...20 %. Here, the results obtained were characterized as unstable with large "uncertainties".

Effect of random factors (heterogeneity of soil, inaccuracy of measuring apparatuses, etc.) happened to be higher than the effect of speed of penetration, thus, most of regularities identified by some authors are hardly ever convincing. In practice, resistances $q_c \& f_s$ in adjacent locations 30...40 cm apart (at the same depth) will probably be 10...20 % different even in homogeneous lithological soils. Taking into account that effect of speed is expressed in changes of $q_c \& f_s$ (or Q_s) just in some percent, it becomes evident that ignoring the necessary replication of measurements the regularities obtained seem to be imaginary, lying in the limiting accuracy of the experiment. Dependence of CPT results upon the speed of penetration may correctly show just *mean* values of resistances q_c, f_s (or Q_s), obtained in several one-type measurements (in several CPT locations).

The results obtained by Kamp (1982) can serve an example of observation the condition. The experiments were carried out in a wide range of conditions (on land, offshore), three CPT procedures being performed in every speed. Changes of resistance to penetration of $q_c \& f_s$ were studied in sands (fine, medium and coarse) in a great variety of speeds 0.2; 1.0; 10; 20; 100 mm/s (0.012; 0.06; 0.6; 1.2; 6.0 m/min). Fugro cone penetrometer was used. CPT was carried out up to 8.5 m in depth with level of ground water to be 1 m in depth, i.e. in water-saturated sands. Table 2.1 shows deviations of resistances obtained in CPT with different speeds (for simplicity $q_v \& f_v$ are given without the indexes «c» & «s»), from analogous resistances $q_c \& f_s$, corresponding to standard speed of 20 mm/s (1.2 m/min). As it is shown in table 2.1., even in considerable change of speed (100/0.2 = 500 times) maximum increase of cone resistance in sands was just 10 %, unit sleeve friction resistance – 29 %.

Great amount of research was carried out in 1960s...1980s. BashNIIstroy studied effect of speed of penetration in clayey soils and silt sands as well. Figure 2.14 illustrates dependencies of resistance to penetration in different soils upon the speed of penetration obtained by Ryzhkov & Enikeyev (1980). The experiments were carried out with fivefold or even tenfold regularity of measurements. In these experiments with different speeds CPT locations were located randomly (view in plan) similar to location of different penetrometers in CPT locations (see fig. 2.4b). Mean values of $q_c \& f_s$ were compared in each speed.

Table 2.1 Deviations of resistance to penetration in sands in speeds of penetration V = 0.012...6.0 m/min from analogous resistances q_c & f_s in standard speed V = 1.2 m/min (Kamp 1982)

Speed of penetration mm/s (m/min)	Deviations	
	Cone resistance $(q_v - q_c)/q_c$ (%)	Unit sleeve friction resistance $(f_v - f_s)/f_s$ (%)
0.2 (0.012)	−10.1	−29.0
1.0 (0.06)	−4.8	−12.1
10 (0.60)	−1.6	+5.0
20 (1.20)	±0.0	±0.0
100 (6.0)	+5.3	+12.8

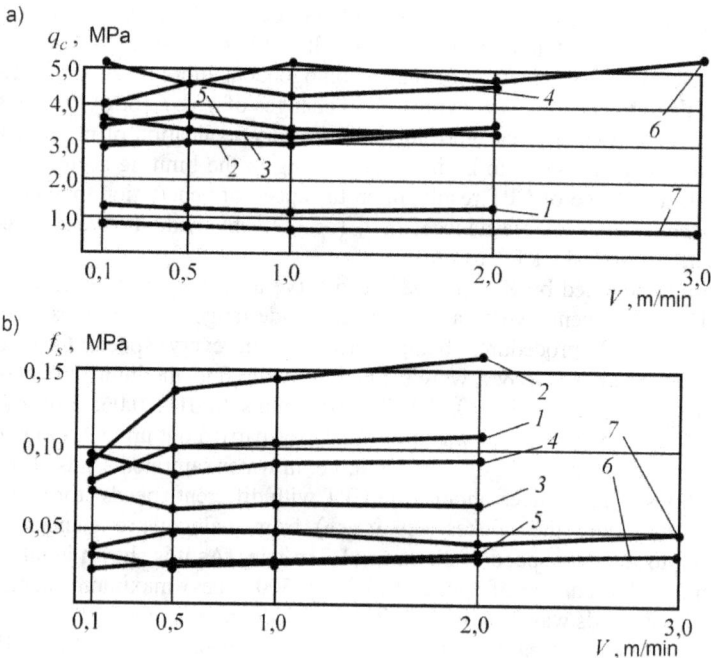

Fig. 2.14 Schematic illustration of cone resistance (q_c) and unit sleeve friction resistance (f_s) dependence upon the speed of penetration (V):

a – experimental results on studies of V on q_c effect; b – the same of V on f_s; 1 – overconsolidated alluvial clays (Ufa), 2 – semi-solid alluvial clays and loams (Ufa), 3 – solid loess-like loams (Lvov), 4 – solid loesses (Herson), 5 & 6 – fine watered alluvial silt sands (Ufa), 7 – soft alluvial-dealluvial loams (Ufa)

As it is shown in figure 2.14, in increase of speed from 0.1 to 2 m/min (i.e. 20 times) resistances q_c either did not increase in most cases or increased in some percent. Unit sleeve friction resistance f_s increased more. A slight increase was observed in unit sleeve friction resistance f_s in semi-solid clays and loams. In both cases the principal increase of f_s was observed in the speed of less than 0.5 m/min. In weak soils the increase of q_c & f_s was not observed. However, it is important to underline that even in significant regularity of changes in the curves given (see fig. 2.14) effect of random factors is present. Perhaps, more frequent regularity is necessary to obtain clear results.

In general, the results on the issue show that it is possible to immerse the penetrometer with rather high speed up to 1.5 m/min without any fear of deviations. In last decades this factor has been recognized both in Russian practice and world wide. In 1950s specialists preferred slow immersion of the penetrometer with the speed of a few cm/min, (e.g. Buisson in 1953 recommended the speed not exceeding 2 cm/min (Bondarick 1964)), but in 1960s…the beginning of 1970s CPT was performed with the speed of 0.5 m/min in most countries. In the National Standard GOST 19912-2012 and International Normative Documents (Proposed European Standard 1977, IRTP 1999) the speed of penetration is specified – 1.2±0.3 m/min, i.e. rather wide range – from 0.9 to 1.5 m/min is allowed.

Taking into account the mentioned experimental results, one can draw a practical conclusion that in a number of cases it is allowable to perform CPT with the speeds exceeding 1.5 m/min. It is obvious that the procedure shall be performed with reasonable precaution taking into account concrete conditions, local experience, and availability of competent specialists.

Another conclusion is theoretical followed by the mentioned experimental results on behavior of soil under the cone. The majority of specialists, particularly non-Russian ones, try to estimate behavior of soil under the penetrometer from the point of view of "pore pressure". According to this theory, increase of pore pressure shall result in decrease of effective stress, and hence, the decrease of soil strength (decrease of its resistance to shear). With speeds of deformations increase pore pressure is to increase either, i.e. more rapid immersion of the penetrometer is to cause higher pore pressure in the soil. The increase of pore pressure is to decrease strength of soil that is to result in decrease of resistance to penetration. In other words, in accordance with pore pressure theory increase in speed of penetration is to decrease both cone resistance and local unit side friction resistance (or push rod).

However, in fact opposite is observed: *increase of q_c & f_s in increase of speed of penetration is a general tendency* being observed even in sands where application of pore pressure theory does not give rise to doubt. Thus, in destruction of soil under the penetrometer the speeds of deformations occurring in it are such that *processes of skeleton viscous resistance* prevail but not the processes of filtration compaction of the soil (and strengthening, correspondingly). From the rheological point of view the soil behaves as "Bingham solid" in which, as it is known, stresses are functionally connected with the speeds of deformations but not the deformations as they are. To sum it up, it is necessary to point out that traditional approaches to the role of pore pressure in destruction of soil are invalid in describing the processes occurring in CPT.

Effect of bending of push rods deserves great attention when applying type I cone penetrometers (see section 1.1). Here, measuring apparatuses (dynamometers) to record both cone resistance q_c and total side friction resistance Q_s are located on the surface. Pushing thrusts being affected by the degree and mode of bending are measured at the zone of penetration of a push rod (inner & outer ones).

Great numbers of specialists worry about bending of the inner push rod transferring thrusts on the cone. Ferronsky (1969) points out that inaccuracies conditioned by reciprocal friction of the inner and outer push rods are unspecified, difficult to be evaluated quantitatively. Resistivity of the inner push rod to lateral bending is not high, the acting thrusts are capable to exceed critical load even in small depths, i.e. cause increase of lateral deformations. The form of bending varies – with one half-wave or several ones, frictions between the inner push rod and the outer one can also be different depending upon the smoothness of their surfaces and the emerging thrusts as well. It is evident that possibility of bending, and hence, inaccuracies in measuring of q_c increases with the increase of effective length of the push rod, i.e. depth of CPT.

To a large extent, the described factors became apparent in comparative tests of CPT rigs in Chelyabinsk (Kazantsev et al. 1974) (see section 2.2). Here, during CPT random character of mentioned inaccuracies was revealed. In one of the CPT procedures carried out by means of type I cone penetrometer in depth of 4 m, cone resistance q_c 2...3 times increased unevenly and stayed stable up to the end of CPT (8 m in depth).This was not observed in adjacent CPT locations though they were located 0.6 m from that one.

Bending of the outer push rod is also undesirable, as it distorts measured total side friction resistance Q_s and allows increasing of bending of the inner push rod, i.e. increases the inaccuracies of q_c evaluation. Besides, bending of the outer push rod distorts the depth of penetration (if one considers the depth in the length of deepened part of the push rod). The more the push rod bends, the less the actual depth of the cone will be than the deepened part of the push rod.

Ferronsky (1969) draws a conclusion on ignoring the penetrometers equipped with measuring apparatuses located on the surface to be applied; he suggests applying the penetrometers equipped with "face space" sensors located inside, i.e. it is reasonable to estimate mechanical properties of soils applying type II cone penetrometers. A number of other specialists are of the same opinion, though there exists another point of view, e.g. Trofimenkov (1995) considers availability of several types of CPT rigs (not less than 5) in the country to be a compulsory condition of effective investigations including type I mechanical cone penetrometers and type II tensometric cone penetrometers as well. Perhaps, this point of view is more correct, but one needs to take into account that in large depths of penetration (more than 10...12 m) it seems to be reasonable to ignore type I cone penetrometers due to increased danger of inadmissible inaccuracies.

When applying type II cone penetrometers the danger of bending of the push rod has the lowest effect on CPT results. Sensors located inside the penetrometer prove to be in the zone of minimum bending moments of the push rod, therefore they shall not depend upon its bends. However, the danger of reducing the depth of penetration as well as the danger of breaking the penetrometer still exists. To a certain extent, the inaccuracies can emerge in anisotropic soils, i.e. in different resistivity of soil in vertical and horizontal directions. However, this occurs just in rather wide deviations not available in depths of 10...15 m[1].

Deviation from the vertical was studied in detail by Graaf & Jekel (1982). In 234 CPT locations at 11 test sites the mean values of deviation were as follows (90 % of reliability):

Depth of penetration, m	10	20	40
Mean deviation, m	0.35	1.2	3.3

As one can see, wider deviations (exceeding 1 m) are typical for significant depths of penetration.

If to estimate the depth of penetration in depth of the deepened part of the penetrometer with the push rod to be bent and deviated from the vertical, then as it was mentioned earlier, errors are possible in estimation of this depth (i.e. its reducing). Figure 2.15 shows the origination of errors (Lunne et al. 2004) that in large depths could be rather significant. Thus, according to figure 2.15, in depth of deepened part of the penetrometer – 35 m, actual depth of penetration happened to be 29 m, i.e. 6 m less.

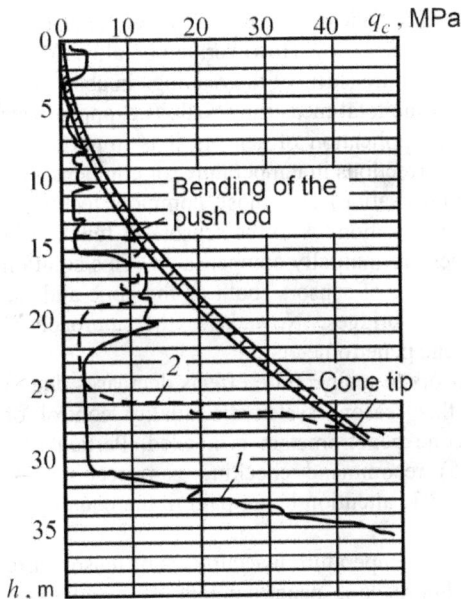

Fig.2.15 Schematic illustration of inaccuracies in depth of penetration estimation [authors Bruzzi and Battaglio (Lunne et al. 2004)]:

q_c – cone resistance, h – depth; 1 – graph of q_c changes in depth regardless of verticality deviation of the penetrometer, 2 – the same with regard to actual depth of penetration (based on recalculation of inclinometer readings)

---

[1] Probability of large deviations in the depth of several meters mentioned in Russian publications of 1980s...1990s is based on a misunderstanding. Once during a meeting of CPT standing committee in 1970, a member of the committee joked about a penetrometer coming out from the ground in CPT procedure with C-832 rig. It is amazing how the joke turned into an objective fact. It was mentioned (in a milder way) as a well-known fact not only in the informal interviews of professional engineers but at the conferences and even in some books of 1990s.

It is obvious that in the large depths the penetrometer must be equipped with an inclinometer to recalculate the depths, but in small depths inaccuracies will be insignificant. It is an easy task to calculate that in the depth of penetration to be 10 m (i.e. in deviation of cone from the vertical − 0.35 m) the actual depth will probably be reduced − 1...2 cm.

It is evident that stiffness of the push rod is also of some importance. In particular, stiffness of continuous push rod (lower and upper sections of the push rod in S-832 CPT rig) is slightly higher than in compound push rods, composed of abutted sections 1...2 m in length. Due to this, deviations of continuous push rods (all other things being equal) shall be lower than those of compound ones.

In general, concrete conditions and required accuracy of results are affected by inclinometer application necessity. T. Lunne et al. (2004) underline that one can ignore application of the inclinometer in case the depth of penetration does not exceed 15 m in depth.

Effect of temperature on CPT results was studied in tensometric cone penetrometers especially in piezometers. With regard to type I mechanical cone penetrometers the issue was not a subject of discussion since dynamometers and manometers applied are usually adapted to operate in a wide range of temperatures. As it was mentioned earlier, other disadvantages cause problems in these penetrometers.

In general, modern sensors in tensometric cone penetrometers show the changes of stresses at different temperatures with high accuracy. However, temperature "drift" of zero readings causes problems, i.e. temperature shifts of the readings in absence of loads. Shifts of zero readings mean shift of the total measuring scale (i.e. appearance of systematic inaccuracy). Due to this, in application of sensors it is important to envisage either automatic introduction of corrections in temperature, or compensation of temperature effects (temperature-compensation).The most common technique appears to be "schematic temperature-compensation" with the sensors to function in the way that their temperature inaccuracies are mutually compensated. This solution is realized in S-832 CPT rig; here, the system of sensors (both of the cone and the friction sleeve) is presented as electronic bridges. Nowadays, the approach is followed by in a number of tensometric cone penetrometers.

As it is known, resistivity of the sensors to temperature effects decreases due to wear of tensometric apparatuses. For this reason, systematic quality control of measuring apparatuses of tensometric cone penetrometers is needed. Particularly, Lunne et al. (Lunne & Keaveny 1995) recommend checking zero readings of tensometric apparatuses before and after CPT, attention being paid to the equality of the outside and inside (the soil) temperatures.

According to Post & Nebbeling (1995) temperature inaccuracies of tensometric cone penetrometers are quite small. In Fugro cone penetrometers ±5° changes of temperature in the adjacent soil caused changes of cone resistance q_c in ±10 kPa. The maximum temperature inaccuracy in the experiments was 130 kPa. For this reason, in sands one should not expect temperature inaccuracies to create significant troubles, as cone resistance q_c of these soils is usually calculated in dozens of mega-Pascal, therefore temperature inaccuracies in them will contain shares of percentage. However, in soils with low resistance to penetration, e.g. in weak clayey soils, effect of temperature deserves attention since it can significantly distort resistances of soil. The authors pay particular attention to possibility of significant inaccuracies occurrence when the penetrometer enters clayey stratum after the soil one. In low-moist sands one can observe more than 30 ^{0}C increase of the temperature of the cone surface (due to friction of sandy particles upon the cone surface). For this reason, the authors recommend stopping the penetrometer before entering the clayey stratum after the soil one, in order to decrease the temperature of the penetrometer to that of the clayey soil.

The problem of protection the piezocones penetrometers from temperature effects are being solved in the same way. Temperature effects in them shall be reduced either by automatic introduction of corrections in temperature or "schematic" temperature-compensation. In the latter pore pressure sensors are installed to ensure compensation of temperature effects, i.e. to ensure reciprocal compensation of temperature inaccuracies of different sensors.

Stability of measuring apparatuses operation deserves the greatest attention with regard to tensometric cone penetrometers. When applying mechanical cone penetrometers inaccuracies occur due to troubles in construction of the penetrometer (bending, wedging of the push rods or other elements) but not to operation of measuring apparatuses (dynamometers, manometers). In tensometric cone penetrometers operation of measuring apparatuses is the principal factor determining the quality of the results obtained.

Stability of measuring apparatuses operation in tensometric cone penetrometers depends upon a number of factors. Besides the discussed temperature effect the results are affected by moisture (in failure of cone proofness), soil particles entering the cone, bending of the cone, different failures of sensors, etc. The manufacturing quality of the penetrometers, maintenance and operation conditions and the degree of wear are of great importance either.

As a rule, in complicated conditions the apparatuses of CPT rigs equipped with tensometric cone penetrometers operate with heavy load. Thus, in spite of the advantages of modern tensometric apparatuses one cannot totally exclude failures in operation. It is evident that manometers and dynamometers in mechanical cone penetrometers and hydraulic ones as well may indicate false readings. For this reason in the National and International Standards (GOST 19912-2012, Proposed European Standard 1977, IRTP 1999) systematic control of accuracy of measuring devices in any types of penetrometers is required.

Manometers and dynamometers of mechanical cone penetrometers shall undergo systematic metrological control. In accordance with the International Normative Documents (Proposed European Standard 1977, IRTP 1989) manometers shall undergo checking not less than once in six months.

For tensometric cone penetrometers this control is realized in calibration of measuring apparatuses. Calibration procedure includes comparison of measuring apparatuses readings of the penetrometer with the "reference" loads on the penetrometer elaborated by pushing equipment of the CPT rig; the loads are measured by dynamometers or other devices *not dependent* on measuring apparatuses of the cone penetrometer.

Calibration procedure can be carried out in-situ or in laboratory conditions. In the first case the load is developed by CPT rig itself, the penetrometer is not immersed in the soil but rests against special adjustment through the dynamometer. During the calibration procedure the load increases or decreases stepwise. Readings of the dynamometer (F) are compared with those of the measuring apparatuses of CPT rig (Π), then the calibration curve is drawn «$\Pi \sim F$», measuring apparatuses are adjusted. The calibration curve is usually a straight line $\Pi = k_a F$, after the adjustment of the apparatuses k_a is equal to1.

In laboratory conditions the penetrometer is checked separately from CPT rig, the load is developed by means of special hydraulic advancing cylinders. Figure 2.16 illustrates general view of a calibration device applied by Dutch firms; the device is adjusted to both in-situ and laboratory conditions (Trofimenkov 1995).

In Russian practice calibration procedure is usually performed in-situ.

The National and International Standards (GOST 19912-2012, Proposed European Standard 1977, IRTP 1989) require calibration procedure to be performed not less than once in three months.

As it was mentioned earlier, the principal problem emerges in shift of "zero readings" causing systematic inaccuracies of measurements. Besides, inaccuracies can also be conditioned by non-linearity between the scale of the device (Π) and actual loads (F), deviation from 1of the angular coefficient k_a of calibration curve $\Pi = k_a F$. The mentioned non-linearity and inaccuracy of calibration curve inclination, i.e. "calibration error" ($k_a \neq 1$), usually occur caused by entering of moisture or soil particles into the cone.

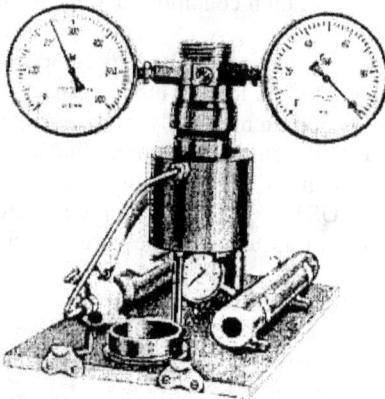

Fig.2.16 Calibration device to check measuring apparatuses of the penetrometers applied by Dutch specialists

Several years of S-832 CPT rig application reveal that failures in strain sensors operation have the specific character (at least with regard to concrete type of strain sensors), e.g. false results of measurements do not occur as single (sprung from somewhere), but they embrace a wide range of depths – the total depths in one or several CPT locations. General pattern of q_c & f_s change in depth usually stays stable.

Schaap & Zuidberg (1982) give results of selective checking of consistency of operation of Fugro tensometric cone penetrometers (6 out of 150 penetrometers were checked). Each penetrometer was checked after 75 CPT procedures 20 m in depth, calibration procedure being performed before checking and after 75 CPT procedures as well. This resulted in inaccuracies to be 2...3 %. Zero readings shift as well as "calibration errors" ($k_a \neq 1$) were considered to be the main reason of inaccuracies. Constant control and cleaning of the cones are required in order to reduce the errors.

In S-832 CPT rig almost the same inaccuracies were found.

However, in practice the actual inaccuracies in CPT are *higher* than those in calibration of measuring apparatuses. Complicated conditions of strain sensors operation in soil do increase the inaccuracies. Thus, in the CIS Standard GOST 19912-2012 differentiated approach is made to identification of inaccuracy in measurement of loading, i.e. both in "ordinary" calibration and in change of soil resistances during CPT. When measuring the load 5 % from it is considered to be an acceptable error, that of soil resistance – 10 % (but not more than 5 % from the maximum parameter). Nevertheless, estimation of measurement errors of soil resistances proves to be a complex technological problem. Given the in-situ heterogeneity of soil the estimation is possible just when comparing mean values of q_c & f_s obtained in repeated CPT of a small site with different penetrometers – controllable and reference ones. Here, CPT locations are to be located randomly (view in plan) slightly distant from each other (see fig. 2.4b). In laboratory conditions the checking is rather hard work as special tray with soil of high homogeneity is to be applied, but it is hardly ever possible. For this reason, in practice control is exercised by "ordinary" calibration procedure.

2.4 THEORETICAL CPT DATA AND SOIL MECHANICAL PROPERTIES DATA INTERRELATION

2.4.1 Empirical and theoretical relationships in CPT practical applications

Most researchers, especially the non-Russian ones believe cone penetration test (CPT) to be a purely "empirical method" and its theoretical interpretation is not very promising. Such an approach seems to be logical, taking into account that during last decades achievements in theoretical study of CPT appear to be quite unassuming compared to the progress in making the equipment for CPT as well as in perfection of the working procedure. However, ignorance of theoretical aspects in CPT application has a negative effect onto its practical efficiency. Poor presenting of physical essence of the processes occurring under the penetrometer makes it difficult to understand the data obtained. It can be obviously seen when applying CPT together with the other methods especially when the data obtained are contradictory. Taking into account purely empirical dependencies creates extra problems as the specialist has to make decisions on one or the other empirical dependency application in particular soils in case CPT has not been used in them before. Considering local experience and "engineers' intuition" tacitly assumes highly qualified specialists to work in the field, but in mass application of CPT it is far from reality.

In researchers' practice it is advisable to use strictly normalized application of any soil test methods with integrated regulations of the procedure, integrated formulae or tables of data obtained interpretation. If there is no way to overcome the diversity of the formulae, there must be, at least, clear and logical system of their application.

With regard to CPT, implementation of these requirements faces considerable difficulties. Empirical dependencies between CPT data and standard properties of soils being obtained in diverse lithological sedimentations (sands, clays, etc.) or even identical ones but of different origin (alluvial, marine, etc.) may differ greatly. Particularly, Solodukhin (1975) in his book published in 1975 gives 24 empiric formulae of Russian authors in which cone resistance (q_c) of various soils and their modulus of deformation (E) are combined. In their article Ziangirov & Kashirsky (2006) give 8 empirical formulae to define the modulus of deformation just only for territory of Moscow.

In the 20th century attempts taken to set approximately integrated theoretical conformities showing common link between CPT data and properties of soils appeared to fail in the sense of their influence on practical implementation of the method. Up to present time empiricism has predominated over in CPT application and it proves to manifest much more compared to any other methods of soil testing. Here, the difficulties emerge not so much because of the variety of empirical formulae but owing to the lack of the system of their implementation, their insufficient adequacy and spheres of their application.

Lunne et al. (2004) try to sort out the data on adequate evaluation of various properties of soil by means of CPT. They present the results on the analysis of a great amount of the actual material relating to evaluating of different properties of soil in accordance with CPT data (12 characteristics). For every property it is suggested an applicability rating of its evaluation according to CPT. The scale embraces 5 levels: 1) high reliability, 2) intermediate reliability – between high and moderate ones, 3) moderate reliability, 4) intermediate reliability – between moderate and low ones, 5) low reliability. Testing of soil strength s_u (shear stress – in Russian publications is usually marked as τ) without dividing the strength into two parameters (φ and c) is considered to be the most reliable; in clays the strength corresponds with the first and the second levels of reliability (high reliability and intermediate one between high and moderate ones). Evaluation of the angle of internal friction (φ) in clays is referred to the third and the fourth levels of reliability (moderate reliability and intermediate one between moderate and low ones), in sands evaluation of φ is referred to the second level. Cohesion (c) is not mentioned. Evaluation of deformation characteristics is believed to be less reliable procedure especially in clays. Here, modulus of deformation (E), modulus of rigidity (G) as well as a number of non-standard deformation characteristics are referred to the fourth and the fifth levels, in sands – to the second and the fourth ones correspondingly. Density of sand layers is referred to the second and the third levels. Permeability coefficient of clays is referred to the second – the fourth levels, the factor of consolidation – to the second and the third ones, thixotropy of clays – to the second and the third ones correspondingly, etc.

The classification given proves to be a conditional one. It is quite easy to give examples not relevant to the scheme. Particularly, in many regions of Russia the characteristics are successfully evaluated by CPT data but following Lunne et al. (2004) they can be hardly evaluated (e.g. φ, c, E of clayey soils). It is also due to the National Normative Document on engineering-geological research SNiP 11-105-97 where one can find the tables compiled on basis of the vast actual material analysis.

In order to systematize the ideas of connection between CPT and the properties of soils it is necessary to make a detailed analysis of the principal ways of theoretical solution of this issue.

2.4.2 CPT typical mathematical models

As it was mentioned above (see section 1.4), theoretical work on CPT has been carried out for a long time, particular interest being special in 1950s ... 1970s. The data obtained by different authors were far from being similar, since they had applied various mathematical models of soil with diverse assumptions and simplifications. Mathematical models of soil were as follows:

- Loose medium being broken according to the Coulomb's law (i.e. $\tau = \sigma \, tg\varphi + c$), behind the destruction zone the loose medium is believed to be rigid (non-deformation).
- Elastic (linear-deformable) medium being deformed according to the Hooke's law (i.e. $\varepsilon = \sigma/E$), the failures do not occur with any loading.
- Elastic-plastic medium being broken according to the Coulomb's law and deformed according to the Hooke's law.

When considering soil as the Coulomb's loose medium (with parameters φ and c) the dependencies $q_c = f(\varphi,c)$ are obtained, in sands $q_c = f(\varphi)$; when applying the model of elastic space (with modulus of deformation E) $q_c = f(E)$, etc. (here and further on under the symbol "f" different functions are meant). In practice CPT can be applied to evaluate both the strength and deformability of soil. Such dual nature is not typical for any other mechanical testing of soil, since every test is expected to evaluate concrete values of strength or deformability of the soil. Even when using the same equipment to study different properties, the techniques prove to be different. It is obvious that this fact does not allow treating CPT as a method of exact evaluation of any soil property. Nevertheless, it demands in-depth understanding of the processes, as the correlation between CPT data and characteristics of soil, i.e. its strength and deformability is a fact of common knowledge.

Application of the elastic-plastic medium removes the contradiction to a certain extent but here another problem arises: the amount of factors effecting on cone resistance q_c considerably increases (shear τ, modulus of deformation E, or even a set of mechanical properties φ, c, E, v). In this case, one and the same value of q_c must correspond with the unlimited amount of mechanical properties combinations φ, c, E, v.

In order to clear the issues it is necessary to examine concrete mathematical models (schemes), showing the process of penetration as it is.

Three models are well-known:

- "Deep foundation" – "classical" axisymmetric problem of limit equilibrium theory of deep foundation bearing capacity.
- Spherical cavity expansion in the soil from radius 0 to r_o (r_o – radius of the cone).
- Cylindrical cavity expansion from radius 0 to r_o.

Figure 2.17 shows the examples of the mentioned above models realization typical in Russian practice.

In non-Russian publications there appear other models of penetration based on both application of non-linear dependency between deformations and stresses and deformation evaluation methods by surfaces of equal stresses, etc. (Lunne et al. 2004) but in Russian they have hardly been applied.

"Classical" model of **"deep foundation" in the Coulomb's loose medium** (as shown in Berezantsev (1965) – fig. 2.17a) assumes loose material destruction zones initiation under the cone (limit equilibrium), they are characterized by a net of crossing sliding surfaces. Figure 2.17, a shows the diagram of a "semi-empirical" approach: the outline of the limit equilibrium zone and sliding surfaces have been accepted in accordance with the experimental results, the further operations have been done analytically by methods of limit equilibrium theory. Ground response influence behind the destruction zones is being modeled by uniformly distributed load. The task given have been solved, dependency of cone resistance q_c upon strengthening characteristics (the angle of internal friction φ and cohesion c), was obtained, deformation characteristics were absent.

Such models are the most suitable when studying the behavior of piles, therefore they are presented in chapter 4 in detail.

The model of **"spherical cavity expansion"** is based on assumption that deformation of soil under the cone is identical to those of spherical cavity expansion from the diameter 0 to r_o, where r_o – radius of the cone. The model was used in different years by a number of non-Russian specialists, e.g. Meyerhof, A.V. Skempton, Gibson, Vesic (Lunne et al. 2004). In non-Russian publications it is usually called SCE (Spherical Cavity Expansion). As a soil model the authors took a deformable elastic-plastic medium with linear breaking as a reference plastic material ($\tau = c$, i.e. in the Coulomb's law $\varphi = 0$).

There exist solutions of this task for a non-breaking linear deformable medium. In Russia the solution was obtained in 1960s by Ferronsky (1969) (fig. 2.17b). It was based on theory of elasticity methods and was the appendix to the well-known

81

solution of Calvin (point force inside the space) to CPT conditions. Cone resistance q_c dependency upon the deformation characteristics of soil (modulus of deformation E and Poisson's ration v), was obtained, strengthening characteristics were absent.

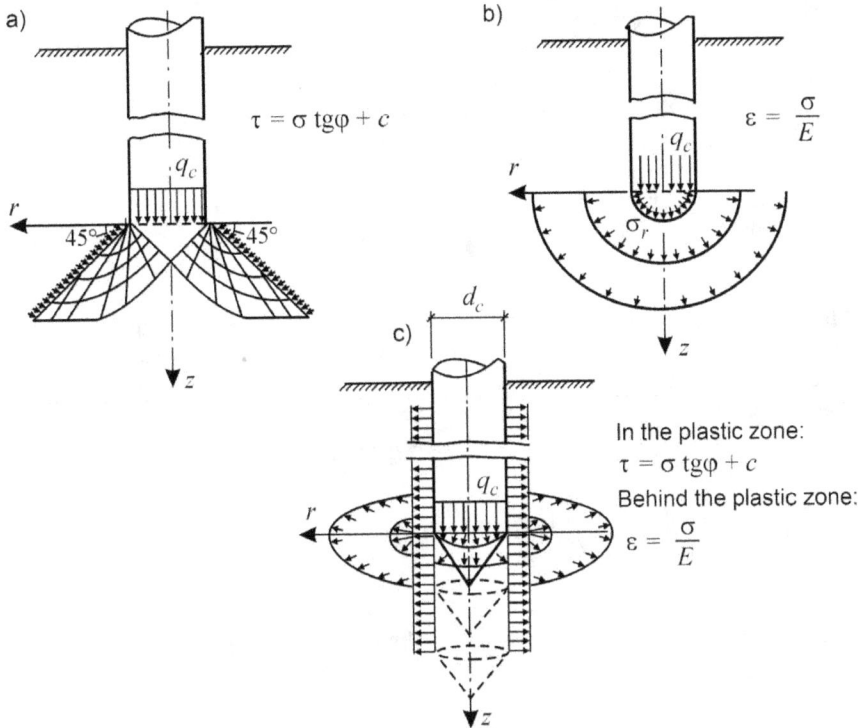

Fig. 2.17 Schematics of design diagrams based on different mathematical models of soil:
a – "deep foundation", soil – the Coulomb's loose medium (Berezantsev's design model, 1965); b – "spherical cavity expansion", soil – linear deformable medium (Ferronsky's design model, 1969); c – "cylindrical cavity expansion" (Terzaghi 1933, Rats 1973, Vesic 1965, Ryzhkov 1973), soil – elastic-plastic medium

When applying the elastic-plastic medium model (in the same task) dependency of q_c upon both strengthening characteristics (limit shear resistance $\tau = c$) and deformation one (modulus of deformation E) is obtained.

In **"cylindrical cavity expansion"** diagram (fig. 2.17c) penetration is considered to expand the cylindrical cavity from radius 0 to r_0, where r_0 – radius of the cone. The model shows soil behavior under CPT embracing the whole process of soil deformation at the same time as the "expanding sphere" shows only a limited part of the deformations, but both solutions are supposed to do quite many assumptions. The principal advantage of both models is the ability to consider soil as an elastic-plastic medium since cylindrical cavity expansion tasks and those of spherical cavity expansion in the elastic-plastic medium are referred to the simplest tasks in theory of plasticity being resolved analytically.

Cylindrical cavity expansion model was firstly suggested by Terzaghi (1933) in 1925, then it was repeatedly modified by a number of authors (Reznikov 1961, Vesic 1965, Ryzhkov 1973). In non-Russian publications it is mentioned as CCE (cylindrical cavity expansion). Its application in CPT is discussed further in detail.

It is necessary to underline that application of both models – spherical cavity expansion and cylindrical cavity expansion – result in the same data. Table 2.2 presents typical formulae obtained on the basis of the models application; the formulae show the dependency under consideration.

Table 2.2 Typical theoretical formulae based on application of spherical cavity expansion and cylindrical cavity expansion models in elastic-plastic medium (Lunne et al. 2004, Reznikov M. 1961)

Spherical cavity		Cylindrical cavity	
Authors	Formulae	Authors	Formulae
Meyerhof (1951)	$q_c = \left[\dfrac{4}{3}\left(1+\ln\dfrac{E}{3\tau}\right)+1\right]\cdot\tau+\sigma$	Hill (1949)	$p_r = \left[1+\ln\dfrac{E}{(5-4v)\tau}\right]\tau+\sigma$ $\left(q_c = \beta\cdot p_r\right)$
Skempton (1951)	$q_c = \left[\dfrac{4}{3}\left(1+\ln\dfrac{E}{\tau}\right)+1\right]\cdot\tau+\sigma$	Menaare (1957)	$p_r = \left[1+\ln\dfrac{E}{(1-v)2\tau}\right]\tau+\sigma$ $\left(q_c = \beta\cdot p_r\right)$
Gibson (1950)	$q_c = \left[\dfrac{4}{3}\left(1+\ln\dfrac{E}{3\tau}\right)+ctg\,\alpha'\right]\cdot\tau+\sigma$	Reznikov (1961)	$p_r = \left[1+2\ln\dfrac{E}{(1-v)^2 4\tau}\right]\tau+\sigma$ $\left(q_c = \beta\cdot p_r\right)$
Vesic (1975)	$q_c = \left[\dfrac{4}{3}\left(1+\ln\dfrac{E}{3\tau}\right)+2,57\right]\cdot\tau+\sigma$	Baligh (1975)	$q_c = \left(1+\ln\dfrac{E}{3\tau}+11\right)\cdot\tau+\sigma$

Footnote: p_r – radial pressure on cavity walls; β – transition coefficient from p_r to cone resistance q_c; E, G – modulus of deformation and modulus of soil shear; τ – shear stress of soil; α' – half of a cone tip angle; σ – pressure in soil before penetration (different in different authors)

In the formulae given (see table 2.2) strength of soil is characterized by one parameter – shear stress τ, corresponding with situation $\varphi = 0°$ when applying Coulomb's law. As computation of foundations includes two parameters of shear to be known – angle of internal friction φ and cohesion c, considering the design diagrams in which strength of elastic-plastic medium is characterized by two parameters – φ & c seems to be more advantageous. Further, the issue will be paid much attention to from this point of view, however, before, it is necessary to analyze the simplified formulae given. The analysis is as follows.

Definition as well as further practical application of theoretical formulae based on models of loose and linear-deformable (elastic) media ($q_c = f(\varphi)$, $q_c = f(E)$) seems to be rather unpromising. Empirical formulae (as well as the corresponding tables) showing the same correlations, i.e. $q_c = f(\varphi)$, $q_c = f(E)$ have been widely used for many decades apart from theoretical research. As it was pointed out earlier, they are different in regions with various engineering-geological conditions, i.e. the parameters reflect specific characters of concrete types of soils. Theoretical consideration of this specific character appears to be unavailable, since the problems of cone penetration are hard to be resolved even for the simplest media which comply with the Coulomb's law and the Hooke's law as well. Consideration of structural features of soils, history of their stress-strain condition, anisotropy and other similar features by means of just mechanical methods would be impossible. The authors (Lunne T. et al. 2004) draw attention to this either. The features shall be allowed for additional empirical coefficients; to determine the coefficients is no easy task either, the same as determination of empirical formulae, e.g. $q_c = f(\varphi)$ or $q_c = f(E)$. For this reason, theoretical formulae reflecting correlation between cone resistance q_c, modulus of deformation E or angle of internal friction of sands φ will not have any advantages in comparison to "standard" empirical formulae.

To analyze formulae based on models of elastic-plastic medium seems to be more efficient. As it is shown in table 2.2, effect of modulus of deformation E on q_c is significantly lower than that of shear stress ($\tau = c$). In all the formulae modulus of deformation is under the logarithm ($\ln\dfrac{E}{\tau}$ or $\ln\dfrac{E}{3\tau}$), while shear stress τ is not only under the logarithm but is an immediate multiplier either. It is important to underline that in formulae given by other authors (Lunne et al. 2004) where complex models were applied, modulus of deformation is also under the logarithm. Thus, taking into account the theory of elasticity one can conclude that to a large extent, CPT results are to indicate strength of soil but not the deformability. This factor agrees with low reliability of evaluation of deformation characteristics of soil in CPT results after Lunne et al. (2004).

However, the conclusion is assumed to be one-sided ignoring correlation between strengthening and deformation properties of real-life soil. For each lithological and genetic variety of soil the corresponding empirical correlation can be determined allowing estimation of modulus of deformation E to shear stress τ (in non-Russian surveys this mode of estimation of modulus of deformation is known to be applied – shear testing of soil). Therefore, one should not believe the conclusion to be automatically valid in real-life soils, as it is related to the reference elastic-plastic medium. Cone resistance q_c reflecting strength of soil inevitably reflects its modulus of deformation; this explains observed correlation between q_c and E in sites with similar soils. The analysis reveals that correlation between q_c and E shall be wider in soils of different lithology and genesis than that between q_c and τ in the same conditions. This does not disagree with practice since empirical formulae to determine modulus of deformation do vary to a large extent.

Another feature of the formulae based on the model of elastic-plastic medium should be taken into account: correlation of modulus of deformation E (shear modulus G) to shear stress τ, i.e. value E/τ or G/τ ($G/\tau \approx E/3\tau$). In concrete soil type this value must be more stable than E and τ separately (due to correlation between E & τ). Here, it is under the logarithm, i.e. its effect on q_c is to be low, and requirements to its accuracy are to be reduced. In these conditions values E/τ or G/τ can be set with regard to some indirect indexes (e.g. physical properties). In non-Russian publications the latter correlation is called a rigidity index marked I_r (Lunne et al. 2004, Vesic 1965).

2.4.3 Application of elastic-plastic model with Coulomb strength and Hooke deformation criterions

In the discussed formulae based on the model of elastic-plastic medium the strength of the medium was characterized by the only parameter – shear stress τ. In order to relate this to the real-life soils and to the real-life investigations as well, it is necessary to find out whether the presented regularities are applicable to elastic-plastic medium with two parameters of strength – φ & c. In other words, it is important to consider the solution of "the mixed problem" of both the theory of elasticity and limit equilibrium theory on penetration in the medium characterized by two parameters of strength – shear φ & c, and compressibility – by modulus of deformation E and Poisson's ratio v. Approximate solution given in (Ryzhkov I.B. 1971,1975,1992),) is considered further. The mentioned "cavity expansion" model is taken on the basis of given analysis of reference diagrams of soil deforming but not as a postulate.

a)   b)   c)

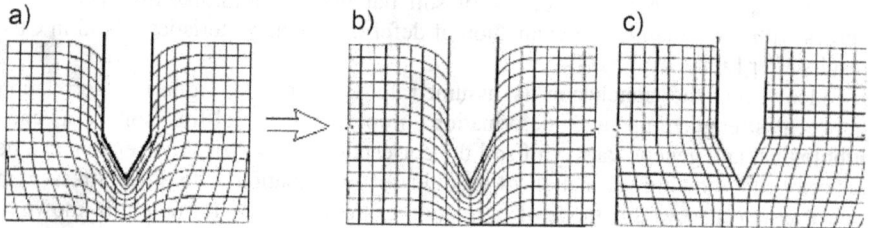

Fig.2.18 Schematics of soil shifts indicating the validity of "cylindrical cavity expansion" model application:

a – real-life penetration due to vertical and horizontal shifts in the medium (soil), b – imaginary penetration due to vertical shifts: resistance to penetration unrestrictedly increases with depth, c – imaginary penetration due to horizontal (radial) shifts: resistance to penetration increases up to a certain limit

Here are some pros of "cavity expansion" model to be applied.

Figure 2.18 illustrates soil shifts during penetration: (a) – real-life case, (b, c) – imaginary ones. In real-life case penetration occurs due to horizontal and vertical shifts in the soil (to be more correct – shifts possessing horizontal and vertical components). In imaginary cases penetration would be realized due to vertical shifts (b) or horizontal ones (c) (see fig. 2.18b,c). In other words, in cases b & c one can suppose that the reference medium is capable to be deformed either vertically or horizontally.

In the first case (b) penetration of the cone would involve the deepened soil layers with pushing thrust to be unrestrictedly increased. In the second case (c) increase of deformations would be restricted by edging of soil particles behind the contour of the cone cross section, due to this pushing thrust would increase up to a certain limit.

As it was said before, both vertical and horizontal shifts occur in real-life soils, i.e. "bending" of soil layers with simultaneous shift of soil particles in the radial direction (see fig. 2.18a). Penetration of the cone due to "bending" of soil layers can last until pushing thrust is sufficient to radial shifts occurrence. Then, penetration will be carried out due to these (radial) shifts; the amount of "bending" will remain constant (in homogeneous soil).

Thus, despite vertical shifts amount of thrust will be the same as in the reference case of absence of vertical shifts, i.e. *penetration of cone is identical to cylindrical cavity expansion* from radius $r = 0$ to r_o, where r_o – radius of the cone.

Due to the importance of the issue the solution of the mentioned problem needs to be considered in detail. The design diagram is shown in figure 2.19. General solution is analogous to the solution after Reznikov (1961), but as opposed to it $\varphi \neq 0$ and $c \neq 0$.

The thrust necessary to push down the penetrometer Q is partly taken to overcome friction between the soil and cone Q_T, the rest of Q_n is taken to deform the adjacent soil (specific values q_T & q_n are shown in figure 2.19b).

$$Q_n = \frac{Q tg\alpha'}{tg(\alpha' + \delta)} - \pi r_o^2 c \cdot tg\alpha',$$ (2.1)

where:

$\alpha' =$ half of a cone tip angle (in standard cone – $\alpha' = 30°$);

$\delta =$ friction angle (for rough cone penetrometer or any other in soils possessing significant adhesion to metal surface of the penetrometer $\delta = \varphi$);

$c =$ cohesion;

$r_o =$ radius of the cone base (see fig. 2.19a).

a)

b)

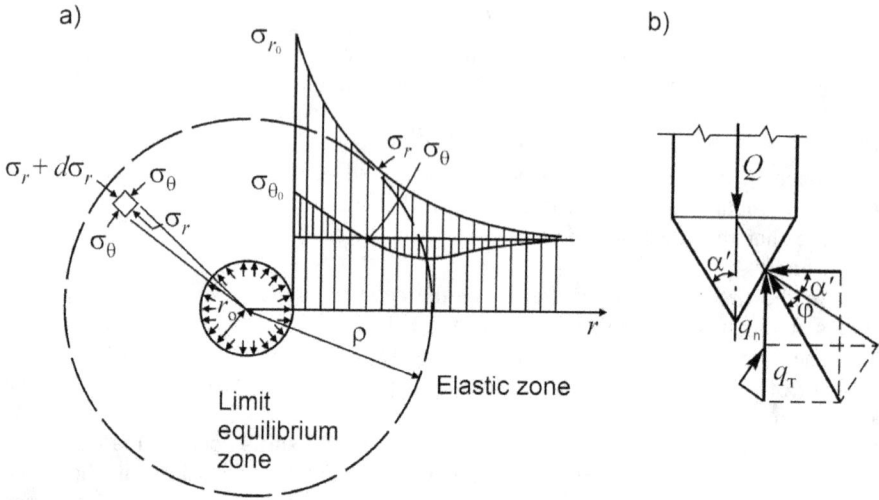

Fig. 2.19 Schematic illustration of the problem on penetration of the cone in elastic-plastic medium after Ryzhkov (1971):

a – horizontal section of the adjacent soil (r_o – radius of the cone, ρ – radius of the limit equilibrium zone); b – forces acting on the cone (q_τ & q_n – parts of total unit soil resistance reflecting cone surface friction and deformations of the adjacent soil)

In Terzaghi (1933) as well as a number of other authors dealing with the theory of plasticity (see Reznikov 1961) it is confirmed that radial stresses in cavity expansion may be considered not dependent upon the cavity diameter and, hence, radial stresses on the cone σ_{ro} are stable as well. The force Q allowing the penetration works at each optional section s; the work equals to that of the radial pressure σ_{ro}, necessary to displace soil particles behind the cone cross section:

$$Qs = \sigma_{ro}\, s \int_0^{2\pi} \int_0^{r_0} r\, dr\, d\theta + Q_T s\, \pi r^2 = \sigma_{ro} s \pi r_0^2 + Q_T s , \qquad (2.2)$$

$$\sigma_{ro} = \frac{Q - Q_T}{\pi r_0^2} , \qquad (2.3)$$

where:

    r & $\theta =$ polar coordinates of the points under study;

    $r_o =$ the same as in the formula (2.1);

    $Q_T =$ part of the force Q, taken to overcome friction between the soil and cone surface.

87

Stresses around the cone in the limit equilibrium zone are expressed as follows:

$$\sigma_r = \left(\sigma_{ro} + \sigma_c\right)\left(\frac{r}{r_o}\right)^{\xi-1} - \sigma_c, \tag{2.4}$$

$$\sigma_\theta = \xi\left(\sigma_{ro} + \sigma_c\right)\left(\frac{r}{r_o}\right)^{\xi-1} - \sigma_c, \tag{2.5}$$

in the zone of linear deformations:

$$\sigma_r = -\frac{A}{r^2} + \sigma_g, \tag{2.6}$$

$$\sigma_\theta = \frac{A}{r^2} + \sigma_g, \tag{2.7}$$

$$\xi = tg^2\left(45 - \frac{\varphi}{2}\right), \tag{2.8}$$

$$\sigma_c = \frac{c}{tg\varphi}, \tag{2.9}$$

$$A = \rho(\sigma_g + \sigma_c)\, sin\varphi, \tag{2.10}$$

where:

$\sigma_g =$ natural pressure (overburden stress), that is pressure in the soil at the considered depth before penetration;

$\rho =$ radius of the limit equilibrium zone:

$$\rho = r_o\left[\frac{2\left(\sigma_g + \sigma_c\right)}{(1+\xi)\left(\sigma_{r0} + \sigma_c\right)}\right]^{\frac{1}{\xi-1}}, \tag{2.11}$$

$r_o =$ the same as in the formula (2.1).

From the assumption of identity between penetration of the cone and cylindrical cavity expansion follows

$$r_0 = \int_0^\rho \varepsilon dr + \int_\rho^\infty \varepsilon dr, \tag{2.12}$$

where the summand shows deformations of soil in the limit equilibrium zone, and the addend – in the zone of linear deformations (ε – relative deformations).

88

Having done the necessary transformations and simplifications, cone resistance may be presented as follows:

$$q_c = \beta \psi \tau_g, \qquad (2.13)$$

where:

q_c = cone resistance of soil (of elastic-plastic medium).

$$\beta = \frac{tg(\alpha' + \varphi)}{tg\alpha'}, \qquad (2.14)$$

$$\psi = \left\{ \left(\frac{E}{\tau_g} \right)^{1-\xi} \frac{2}{1-\xi} \left[\frac{(1+\xi)^2 \xi}{2\sqrt{\xi}(1+2\xi)\left(2-\xi-\xi^2 \dfrac{c}{\tau_g}\right)} \right]^{1-\xi} - 1 \right\} \frac{1}{tg\varphi}, \qquad (2.15)$$

$$\tau_g = \sigma_g tg\varphi + c, \qquad (2.16)$$

φ, c, E = angle of internal friction, cohesion, modulus of deformation correspondingly;

σ_g = the same as in the formulae (2.6, 2.10);

α' = the same as in the formula (2.1);

ξ = see the formula (2.8).

Figure 2.20 shows graphs of dependency of the value ψ upon the ratio E/τ_g & φ.

Other authors obtained the similar solutions as well. In 1960s Vesic (1965) when studying cylindrical cavity expansion in elastic-plastic medium obtained the formula to determine pressure upon the cavity walls σ_r, similar to the formula (2.13) but given as a binomial and applied in the solution presented Ryzhkov (1971). Numerical values of the principal parameters are slightly higher in Vesic than those in the formula (2.13), as radial pressure σ_r is to be lower than q_c in β times (formula (2.14)). Besides, Vesic (1965) applied G/τ_g (G – shear modulus equal to $E/(1+\nu)$, where ν – Poisson's ration) but not E/τ_g.

It is necessary to point out that when studying CPT Vesic turned to spherical cavity expansion but not the cylindrical one. Table 2.2 presents the formula based on spherical cavity expansion.

Value β (formula (2.14)) characterizes the transition from stresses upon walls of imaginary cylindrical cavity to cone resistance. It shows the behavior of soil in the zone of *maximum normal stresses* (in contact with the cone surface). In this zone normal stresses are close to cone resistance q_c in their values, i.e. one-two degrees higher than pressures of soil shears in standard laboratory tests. Clayey soil in high normal stresses $\sigma_\varphi = 0.7...1.2$ MPa (fig. 2.21) acquires special properties (Baldi et al.

1988, Peuchen et al. 1995) – it behaves as a medium with $\varphi \approx 0°$ (in high cohesion c), that corresponds to $\beta \approx 1$.

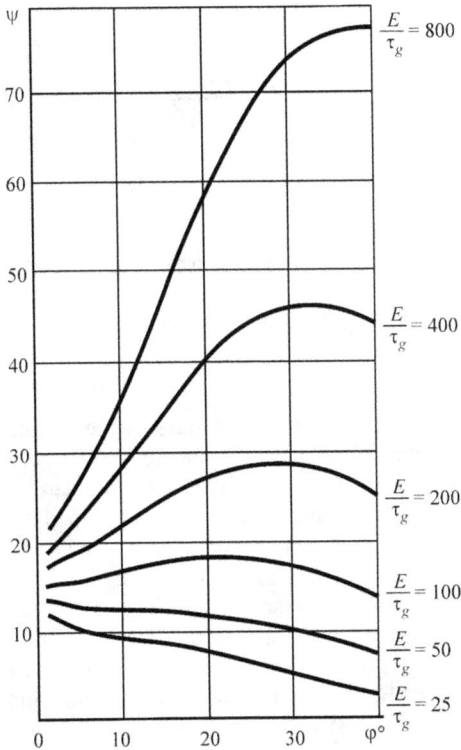

Fig.2.20 Graph of dependency of ψ upon φ and E/τ_g in the formula (2.13)

The research of BashNIIstroy revealed that in sands the deviations from the Coulomb's law are found out if the normal stresses prove to be very high ($\sigma_\varphi > 10$ MPa), i.e. for most sands β is not equal to1, but is in the limits of 2.6...4.3. This significantly increases the differences between the results obtained in sands and clayey soils even if angles of internal friction do not significantly differ in them (e.g. $\varphi = 26...29°$ is possible both in sands and clayey soils).

Formula (2.13) is slightly different from those given in the table 2.2 but in $\varphi = 0$ it can be transformed to the same ones (value E/τ_g will be under the logarithm after elimination of the uncertainties, etc.). Here, the principal features of the formula (2.13) are analogous to those given in the table 2.2.

Firstly, according to the formula (2.13) effect of modulus of deformation E on q_c as in the formulae of table 2.2 happens to be lower than the effect of strength τ_g. Value E is in power of $(1 - \xi)$ being < 1 in all $\varphi > 0°$. In change of the angle of internal friction in the range of $\varphi = 0... 35°$ the value $(1 - \xi)$ changes in the range of 0...0.72, the most typical being 0.2...0.3 in clayey soils. Thus, even in $\varphi \neq 0$

correlations between q_c & E are more likely to be explained by specific character of soil as a material with the strength approximately reflecting its deformability than by mechanical processes under the cone penetrometer.

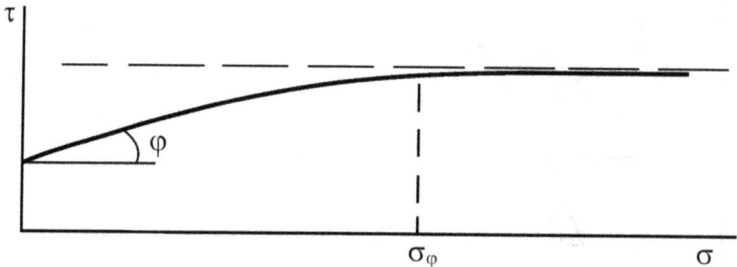

Fig. 2.21 Schematic illustration of dependency of limit shear stress τ upon normal stresses σ in wide range of their change

Secondly, in the formula (2.13) as well as in table 2.2 ratio of the modulus of deformation to shear stress is the important determinative parameter ("rigidity index" E/τ_g). Here, shear stress attains necessary definition: instead of "non-standard" value of shear stress τ, not anchored to any normal pressures, there appears definitely determined value τ_g agreeable with the natural pressure (overburden stress) σ_g, i.e. the pressure in the soil before penetration of the cone (see formula (2.16)).

At the same time, the formula (2.13) reveals some new features of relation between resistance to penetration and properties of soil. Thus, effect of modulus of deformation E is dependent upon the angle of internal friction φ. When φ increases the effect increases either, e.g. if $\varphi = 10°$, tenfold increase of the modulus of deformation increases cone resistance q_c just 2.45 times by the formula (2.13), in $\varphi = 35° - 5.25$ times. Besides, as followed from the graph (see fig. 2.19) in soils with $\varphi = 15...25°$ & $E/\tau_g = 50...150$ (e.g. in overconsolidated or underconsolidated clays) cone resistance q_c is to be evaluated by shear stress τ_g, angle of internal friction does not particularly effect on q_c (values of ψ in this zone are affected by φ mot much, i.e. zones of curves $\psi = f(\varphi, E/\tau_g)$ in it slightly incline to horizontal).

To sum it up, it is important to underline that the mathematical models of soil behavior under the cone must be paid greater attention to, than attempts to obtain "more accurate" dependencies capable to replace the existing empirical formulae. Nowadays, CPT is a sphere in which practice leaves theory behind, and numerous, often contradictory, empirical dependencies need to be systematized and grasped in order to guarantee further development. Nevertheless, the reason is conditioned by conformability between general theoretical dependency and practical results obtained, i.e. evaluated values of q_c shall not significantly differ from the measured ones.

The simplest and the most reliable way of checking the validity of the formula (2.13) "in general", i.e. ignoring its "inaccuracies" in special cases, seems to be comparison of results evaluated by this formula accompanied with generalized empirical material – data of the tables 1...5 given in appendix "И" of the Normative

Document SP 11-105-97 (SP 47.13330.2012). These tables are the result of statistical treatment of a great number of experimental data related to comparison of cone resistance q_c with mechanical properties of soil φ, c and E.

Tables 2.3 & 2.4 present both data for clays and soils taken from the appendix «I» of SP 11-105-97 (SP 47.13330.2012) and the results of cone resistance q_c evaluated by the formula (2.13) for corresponding correlations of φ, c and E (evaluated values of q_c are marked as q_c^{theor}). The depth was considered to be equal to 5 m, i.e. $\sigma_g \approx 90$ kPa, value β equaled 1 for clays; for sands it was evaluated for each value of φ in $\alpha' = 30^0$.

Table 2.3 Comparison of generalized experimental data for clays given in SP 11-105-97 accompanied with results of q_c evaluated by the formula (2.13)

Empirical data (table. 5, appendix "И" SP 11-105-97)				Theoretical evaluation by the formula (2.13) for E, φ and c given in the previous columns
q_c (MPa)	E (MPa)	c (MPa)	$\varphi^{\,o}$	q_c^{theor} (MPa)
0.5	3.5	0.025	14	0.7
1	7	0.030	17	1.3
2	14	0.035	18	2.1
3	21	0.040	20	2.7
4	28	0.045	22	3.5
5	35	0.050	24	4.0

Table 2.4 Comparison of generalized experimental data for sands given in SP 11-105-97 accompanied with results of q_c evaluated by the formula (2.13)

Empirical data (tables 2&3, appendix "И" SP 11-105-97)			Theoretical evaluation by the formula (2.13) for E & φ given in the previous columns
q_c (MPa)	$\varphi^{\,o}$	E (MPa)	q_c^{theor} (MPa)
2	26	6 (17)	2.4 (4.9)
3	28	9 (18.5)	3.7 (6.0)
5	30	15 (21)	5.4 (7.0)
8	32	24 (25)	8.1 (8.2)
12	34	36 (30)	12.7 (11.9)
18	36	54 (38)	19.8 (14.4)

Footnote: Values for alluvial and fluvioglacial sands are given in brackets, for other ones – behind the brackets

As it is seen from the tables 2.3 & 2.4 agreement between theoretical and averaged experimental values of q_c seems to be quite satisfactory apart from weak soils – clays in $q_c < 1$ MPa and in particular, alluvial sands in $q_c < 5$ MPa. Perhaps, fine and pulverescent alluvial sands of loose and normal structure are affected by rather indefinite factors ignored by the adopted model of soil. Perhaps, it is connected with the decreased value of σ_φ (see fig. 2.20) in these soils, or the decrease of soils in volume during their destruction (low parameter of "critical porosity" after Casagrande). Besides, it is important to take into account that in CPT, deformations of soil occur in considerably insignificant changes of porosity; they need to be characterized by the higher modulus of deformation than in standard tests of soil with an odometer or stamp. It is obvious that these are different in different soils.

2.4.4 Principal application trends of the discussed theoretical dependencies

When studying the possibility of the practical application for theoretical solutionmentioned above, it is important to consider the issue on reciprocal correlation of mechanical properties of soil in detail. As it was mentioned in the previous section, cone resistance q_c reflects both the strength and deformability of this soil. Thus, it is to be evaluated by the complex of mechanical properties – angle of internal friction φ, cohesion c and modulus of deformation E, but not just by a single property. This means that even in ideal conditions when the soil is destructed in accordance with the Coulomb's law and deformed (before destruction) in accordance with the Hooke's law it is *not* possible to obtain either the formula $\varphi = f(q_c)$ for elastic-plastic medium, or the formula $c = f(q_c)$ or the formula $E = f(q_c)$, since in fact, q_c must be a function of several variables, i.e. $q_c = f(\varphi, c, E)$. Each value of q_c may correspond with any amount of φ, c and E combinations.

As it was noted when considering reference soils (in $\varphi = 0$), this bottleneck can be overcome if the solution is not restricted with mechanics only, but special character of the soil as a geological thing is taken into account. Strengthening and deformation properties of any soil are determined by one and the same factors – porosity, water content, structure, mineral composition, etc. These factors show the history and conditions of the soil generating. For this reason, in the concrete genetic type of soil there will always be correlation links between its mechanical properties.

Figure 2.22 shows the schematics of dissipation characterizing reciprocal correlation of mechanical properties of alluvial and dealuvial clayey deposits in Ufa. Figure 2.22 shows that the link between compared characteristics is rather weak (especially between φ & c, as clays and loams were studied as an integrated manifold). The correlation coefficients are in the following limits: $r = 0.35...0.5$. However, given such a large number of compared values (105 locations), to validate the correlation (i.e. conditions $r > 0$ with confidence probability – 0.995) it is sufficient of the correlation coefficient to be obtained $r > 0.25$ (Gmurman 2000), i.e. schematics of dissipation shown in fig. 2.22 do confirm the correlation.

Closeness of the correlation links is usually not high, but the closer the range of the soils, the higher it will be. For example, in considering just only loams the correlations will be closer than in joint consideration of loams and clays; in considering just only alluvial deposits the correlations will be closer than in joint consideration of quaternary alluvial or, e.g. Neogene deposits, etc. The form of the links is also determined by the conditions of the soil formation, i.e. its genesis and successive history. For example, the link between strengthening characteristics and deformation ones both in alluvial and moraine loams will vary: given the same strength the deformational characteristics in moraine loams will be higher, etc.

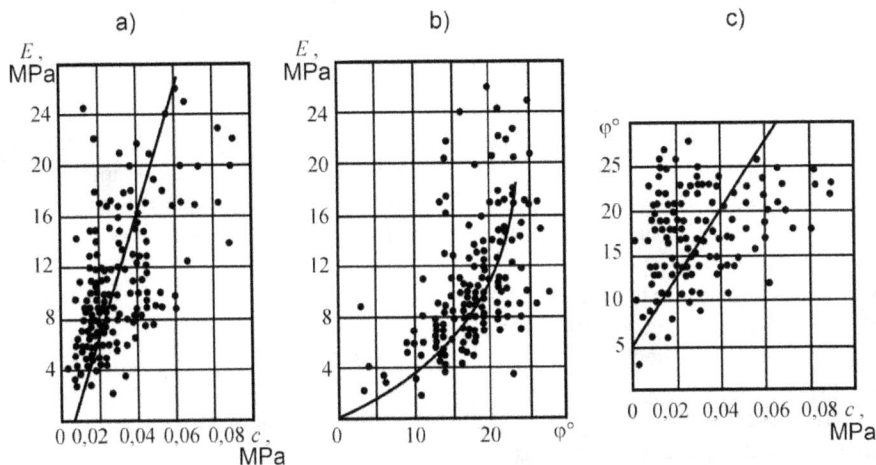

Fig.2.22 Schematics of correlation links between different characteristics of mechanical properties in soils of limited genesis (alluvial and dealuvial clays and loams in Ufa):

a – correlation link between the cohesion (c) and modulus of deformation (E), b – the same between the angle of internal friction (φ) and modulus of deformation (E), c – the same between the cohesion (c) and angle of internal friction (φ)

Thus, any characteristics of mechanical properties of soil may approximately be expressed through any other one; the closer the range of the soils, the more accurate it will be. Therefore, in case the dependency $q_c = f(\varphi, c, E)$ is valid, the approximate dependencies $q_c \approx f(\varphi)$, $q_c \approx f(c)$, $q_c \approx f(E)$ are assumed to be valid either. This fact totally changes the situation of possible theoretical expression of the link between CPT data and characteristics of soil. Instead of accurate dependency as a function of three variables $q_c = f(\varphi, c, E)$, one can use three approximate dependencies mentioned earlier: $q_c \approx f(E)$, $q_c \approx f(\varphi)$, $q_c \approx f(c)$; their form will depend on the type of correlation link between the indexes of the soil properties.

With regard to stated ratios between characteristics of concrete type of soil and the formula (2.13) as well, one can obtain approximate formulae or tables linking cone resistance q_c with each mechanical characteristics, i.e. $E \approx f_E(q)$, $\varphi \approx f_\varphi(q)$, $c \approx f_c(q)$. Depending upon the ratios mentioned which are characterized as E/τ_g in the adopted model – these formulae can characterize types of soil not presented in tables

94

2.3 & 2.4. For example, the formula (2.13) may be transformed on the assumption of the value $a = E/\tau_g$ to be known, as follows:

$$q_c = \beta\psi\tau_g = \beta\Psi E/a.$$

Then

$$E = (a/\beta\Psi)\, q_c = k\, q_c, \qquad (2.17)$$

where value k shall depend upon a, i.e. upon E/τ_g and φ.

Value E/τ_g in one and the same lithological and genetic type of soil is always more stable than its components E & τ_g separately. Vesic (1965) applies the analogous value – "rigidity index" $I_r = G/\tau_g$, where $G = E/2(1+v)$ – shear modulus and gives the table of the "rigidity index" typical values for different soils:

- Rocks $\qquad\qquad I_r = 250...700$
- Sands from loose to dense $\qquad\qquad I_r = 100...500$
- Water-saturated clays from underconsolidated
  to overconsolidated $\qquad\qquad I_r = 20...600$
- Mica clay sands $\qquad\qquad I_r = 20...50$

If one considers rigidity index as E/τ_g, then its values will be 2.5...3 times higher than in Vesic since $G = E/(2(1+v))$. Thus, in sands they must be in the interval of 250...1200, in clays – 50...1500. According to the data of investigations carried out in Bashkortostan E/τ_g ratios for alluvial clays and loams are usually in the limits of 100...300, for sands - 200...800 (3...10 m in depth).

If we substitute the most typical values of c & φ in the formula (2.13), we obtain the simplified formulae being agreeable with the known empirical dependencies (see chapter 3). Thus, for sands the formula (2.13) is as follows:

$$E = (2.5...3.1)\, q_c, \qquad (2.18)$$

for clays and alluvial loams

$$E = (5.6...7.4)\, q_c, \qquad (2.19)$$

for moraine loams

$$E = (6.5...8.5)\, q_c \qquad (2.20)$$

etc.

It is evident that we can obtain the simplified dependencies for φ & c from the formula (2.13) in the same way. The following transformation of the formula (2.13) seems to be reasonable:

$$q_c/\tau_g = \beta\Psi. \qquad (2.21)$$

When applying the graphs in fig. 2.19 as well as the fact that in sands $\beta =$ 2.6...4.3, and in clays $- \beta \approx 1$, it is easy to calculate multiplication of $\beta \Psi$ typical for both sands and clays. In sands and gravel soils values $\beta \Psi$ are to be 75...300, in clays and loams $- 15...70$, i.e. much lower.

Thus, value q_c / τ_g can characterize belonging of soil to one or another lithological type (sands, clays, etc.). It is of practical importance on condition that instead of τ_g, parameter f_s $-$ unit sleeve friction resistance is used $-$ close to τ_g in order of values. Figure 2.23 shows the relevant schematics of f_s & τ_g dissipation in alluvial, dealuvial clayey and fluvioglacial sandy soils.

Figure 2.23 shows that one can observe both correlation and analogy of numerical values between the compared parameters, particularly in clayey soils. Disagreements between τ_g & f_s for clays and loams are in the limits of 20...30 %. In sands one can also observe the same analogy, though separate values of f_s happen to be significantly lower than shear stress τ_g.

One can explain the analogy between τ_g & f_s by radial stresses that effect on the friction sleeve and exceed the natural pressure σ_g as well are compensated by the defects of contacts between the friction sleeve and the soil, and in clays and loams it is due to the defects of their structure. In sands separate small values of f_s are probably connected with the effect of friction sleeve wear, being stronger in sands than in clays.

In general, one can believe that $q/f_s \approx \beta \Psi$.

These agree with years of practice which reveals that in clayey soils the values of unit sleeve friction resistance f_s prove to be the same as in sandy ones, while values of cone resistance q_c in sands are usually several times higher than in clays or loams. In other words, q_c / f_s in sands and, hence, in gravels prove to be higher than in clayey soils. Thus, from 1960s q_c / f_s is used in investigations to identify clayey and sandy soils. As it was noted in chapter 1, the approach was suggested by Begemann (1965) after purely empirical research. At present, identification of sandy and clayey soils applying q_c / f_s is widely used both in Russia and worldwide. The International Recommendations (IRTP 1999) require two indexes to be considered:

- *"friction index"* $I_f = q_c / f_s$;
- *"friction ratio"* being often translated as "parameter of friction" (Trofimenkov 1995), $R_f = f_s / q_c$, i.e. parameter opposite to friction index.

Friction ratio R_f is used by non-Russian specialists to a larger extent than friction index. It is usually expressed in percentage. The issue is discussed in chapter 3 in detail.

Thus, application of soil model as elastic-plastic medium reveals the variety and principal features of known empirical dependencies connecting CPT data with mechanical properties of soil. To a certain extent these dependencies become special cases in the general integrated system.

a)   b)

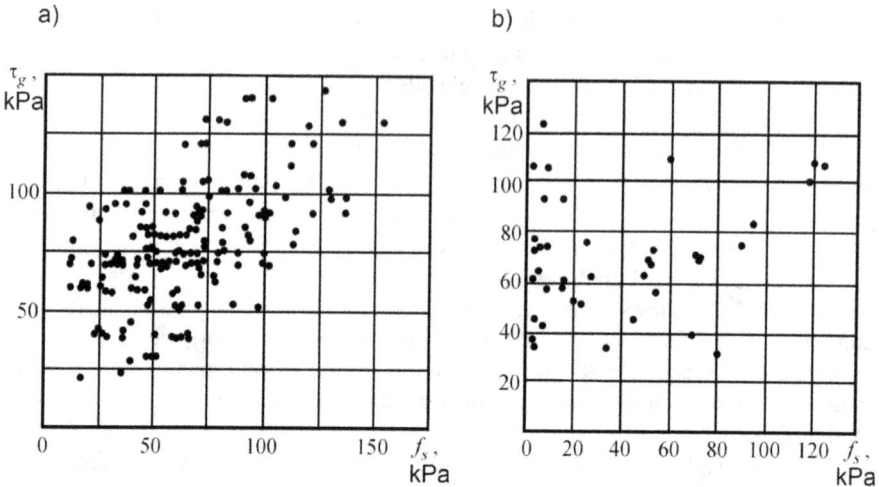

Fig.2.23 Schematics of dissipation showing connection between unit sleeve friction resistance f_s and values of shear stress in natural pressure (in overburden stress) τ_g:

a – clays and loams (aQ, dQ), b – sands (f-$g$$l$Q)

It is evident that the complexity of processes in CPT does not allow them to be modeled with accuracy capable to force out the applied in practice empirical dependencies by theoretical formulae. However, deep understanding of their physics, rational estimation of the processes occurring in the soil allow the engineer to achieve success in comprehension of the variety of known empirical dependencies, the features of local conditions and effective solutions of a number of complex problems which one cannot always foresee and fit to any known patterns.

2.5 THEORETICAL EVALUATION OF ACCURACY AND NUMBER OF MEASUREMENTS IN CPT

The principal factor lowering trust to CPT (as well as to other express-methods) is fear that the data obtained (standard properties of soil, pile bearing capacity, etc.) are approximate, not sufficient of the reliable designs to be guaranteed. The opinion is based on one-sided understanding of possible errors in estimation of foundations which refer to the accuracy of test results. In fact, soils at any site are heterogeneous; the results, no matter how accurate they are do not ensure obtaining the same results at the other sites as well. Apart from accuracy of the data obtained, their sufficient amount is also needed, i.e. maximum possible completeness of conditions of the investigated site. Few accurate tests are usually insufficient. As a rule, this drawback is hidden: having gained the number of tests, i.e. the number of investigated locations at the site one can find it out.

Therefore, when estimating tests in two criteria – accuracy and maximum possible completeness at the investigated site, one needs to draw a conclusion that express-methods being a failure in accuracy as compared to traditional ones, win in the second criteria – maximum possible completeness at the investigated site. This totally refers to CPT allowing the site to be estimated in numbers of locations, this being unavailable in more accurate tests (more expensive and laborious correspondingly).

From the theoretical point of view one needs to consider in detail whether inaccuracy of tests is compensated by their completeness.

Figure 2.24a illustrates a site (of a random form) divided into n sectors. The numbers of sectors as well as the overall dimensions are not of principal importance; but for easiness-to-use let us consider the sectors to be 3×3 m, their number being several hundreds.

Let us suppose that a property of soil marked F is estimated at the site (or pile bearing capacity of given dimensions). In every i-sector in the given depth the value of required characteristics is considered to be constant, equal to one of m possible values (j-value), i.e. equal to a special value F_{ij}. The number of values (m) reflects the expected accuracy of the required index F. For example, if the required index – modulus of deformation with its expected value to be in the interval from 1 to 50 MPa, and the expected accuracy of estimation to be ±1 MPa, then $m = 50$. If the accuracy is ±0.01 MPa, then $m = 5000$, etc. The size of the interval is not of principal importance as it is seen from the following considerations.

a)   b)

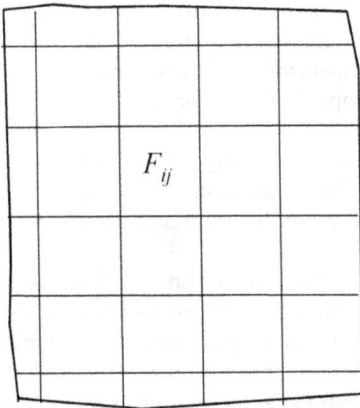
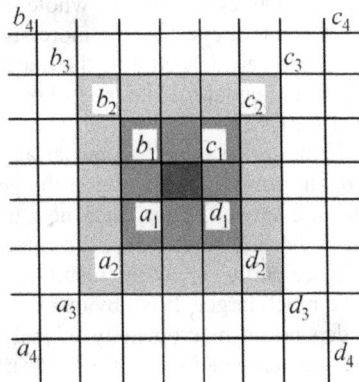

Fig.2.24 Schematic illustrations of estimation technique of required index F changeability at the site:

a – division of the site into sectors; b – zones of the results obtained spreading: $a_1b_1c_1d_1$ – maximum heterogeneity of soil, $a_2b_2c_2d_2$ – high heterogeneity, $a_3b_3c_3d_3$ – medium heterogeneity, $a_4b_4c_4d_4$ – homogeneous soil

Before starting the tests, let us suppose complete uncertainty to be taken, i.e. values F_{ij} are equiprobable. Here, the number of possible cartograms, i.e. dimensional locations of values F_{ij} on n sectors is mn, probability of each variant (ij-variant) before the tests will be the same, equal to $p_{ij}=1/(mn)$. If to take the uncertainty of the situation as entropy H, then according to information theory (in bits) it is as follows (Yaglom & Yaglom 1973):

$$H = -\sum_{ij=1} P_{ij} \, log_2 P_{ij},$$ (2.22)

where p_{ij} = probability of ij-variant, i.e. j-value of F_{ij} in i-sector (F_{ij}).
In equiprobable variants (total uncertainty) the entropy equals to

$$H = n \, log_2 m .$$ (2.23)

The uncertainty will decrease after the first test in any sector. Let us suppose that in k-sector the test has been carried out with the following result: F_{kt} (i.e. $i = k$, $j=t$), then the equality of probabilities is disturbed. The value of required index corresponding with the result F_{kt}, will attain the maximum probability, the closest possible values $F_{k,t-1}$ и $F_{k,t+1}$ – lesser probability, the further ones ($F_{k,t-2}$ и $F_{k,t+2}$) – much lesser, etc. Spreading F_{ij} will depend upon the accuracy of the performed test. It must correspond with possible inaccuracies spreading, being known in this mode (from statistical data analysis of past experience). Changes of probabilities at k-sector will decrease the entropy H; this will effect on the total entropy of the site, since the entropy of values F along the whole site is a sum of entropies of separate sectors. The more accurate the test is, the more significant decrease of H will be. For example, if the test had zero inaccuracy, the uncertainty in k-sector would disappear: probability of F_{kt} would equal to1, i.e. $p (F_{kt}) = 1$, and entropy H in k-sector would equal to zero ($H = 1 \cdot log1 = 0$).

If the test was carried out at any other site, the entropy would decrease in the analogous way depending upon the accuracy. This would decrease the total entropy of the site as well. Thus, each new test will decrease the uncertainty concerning F_{ij} value at the site, i.e. decrease the entropy.

If we partly spread the result obtained to the adjacent sectors, the decrease of H will be much larger. It is obvious that the possibility of such extrapolation of results will depend on heterogeneity of soils at the site. Let us point out four conditional cases of heterogeneity (see fig. 2.24b):

• Maximum heterogeneity not allowing any extrapolation of results behind the test sector.

• High heterogeneity allowing extrapolation of results just in a row, i.e. in the nearest to the test sector adjacent sectors.

• Medium heterogeneity allowing extrapolation in two rows in all directions.

• Homogeneous soil allowing extrapolation in 3 rows in every direction.

If the dimensions of the site were taken to be 3×3 m, the admissible extrapolation in the homogeneous soil would be 10.5 m in every direction, that in medium heterogeneity – 4.5 m, in maximum one – 1.5 m (i.e. inside the site).

The more homogeneous the soil is, the more significantly the total entropy H decreases. Difference in entropies before and after the test characterizes the quantity of information according to Shennon (see in Yaglom & Yaglom 1973):

$$I = H_0 - H_1,$$ (2.24)

where H_0 & H_1 = entropies before and after the test correspondingly.

The second and further tests at other sectors of the site will result in the analogous decrease of entropy H, thus H_2, H_3, H_4 ..., etc. will be obtained. In every test the difference between the previous and following values of entropy will give the quantity of information about the site (in bits). This technique allows estimating the total quantity of information contained in any group of results of any accuracy and any heterogeneity of soil.

Numerical analysis of different situations revealed the paradoxical result at first sight: the quantity of information obtained due to numbers of approximate tests can exceed the quantity of information obtained from a small number of accurate tests because of heterogeneity of soil. Thus, for example, calculations done for reference conditions show that if the site is divided into 100 sectors in homogeneous soils (according to the criteria adopted), 20 approximate tests with inaccuracy of a single test to be ±30 % characterize the site as one "accurate" test with inaccuracy to be ±5 %. In medium heterogeneity of soil the same number of approximate tests is equivalent to two accurate ones, in high heterogeneity of soil two accurate tests are equivalent to 6...8 approximate ones, in maximum heterogeneity of soil – only 5. The book (Terzaghi & Peck 1958) presents the quantity of information on the number of tests dependency diagrams with regard to the sites of various heterogeneity; the diagrams give a detailed analysis of the described regularities.

Analogous conclusions can be made in case the mean values of required characteristics F are considered. Here, it is easier to apply the value of standard deviation of results (σ) or the relative value – coefficient of variation $(\upsilon = \sigma/F)$. Consideration of mean characteristics deserves attention either, as averaging of test results is one of the stages of estimated characteristics determination (according to the International Standard ISO-2394 and the National Standards of many countries, Russia included).

Admitting the validity of normal distribution law, one can use the facts that in increase of results N standard deviation and coefficient of variation of the mean value decrease inversely to N. The value of coefficient of variation υ is to be $\upsilon = \sqrt{\upsilon_1^2 + \upsilon_2^2}$, where υ_1 & υ_2 – coefficients of variation, conditioned by inaccuracy of the test and heterogeneity of soil correspondingly. Coefficient of variation υ_2 for homogeneous soils may be admitted to be 0.025, in medium heterogeneity – 0.1, in high heterogeneity – 0.2. Inaccuracy of tests may be characterized by coefficients of variation to be 0.2 (approximate test) and 0.025 (accurate one). Relative inaccuracies of tests are to be equal in accordance with theory of errors rules:

$$\Delta = t_\alpha \upsilon, \tag{2.25}$$

where t_α = Student's coefficient, depending upon admitted confidence probability α and number of test results.

The calculations of inaccuracies of average results done in tests of different accuracy revealed the prevailing effect of heterogeneity of soil. For example, inaccuracy of the average result from two accurate tests happens to be equivalent to the average one from 15 approximate – in homogeneous soils, six approximate tests – in soils of medium heterogeneity and only three approximate tests – in soils of high heterogeneity. It is evident that these refer to the site as a whole as well as to any sector of it, to any separate geological element.

Thus, heterogeneity of soil significantly reduces the quality of accurate tests. It makes the results, so to say, less impressive. For this reason, *in a number of cases insufficient quantity of tests proves to be more dangerous than insufficient accuracy.* At the same time, accurate tests are costly, complicated and time-consuming. In fact, in soil testing with a stamp or full-scale pile testing with a static load the procedures usually take several days and they are much more expensive than any other express-method. These complicated costly methods are impossible to be applied in a scope allowing factual heterogeneity of soil to be taken into account, while application of CPT, being a simple technique as it is, solves the problem.

However, it is important to underline that in given arguments there is one weak point. Apart from random inaccuracies the results of any tests may contain systematic inaccuracy which cannot be excluded by increase of a number of tests. This error may be excluded by means of comparison of approximate results with the accurate ones, i.e. reference ones. On one hand, the problem may be resolved by parallel application of independent results of several approximate tests. On the other hand, it is evident that these tests cannot give systematic inaccuracy just in one direction. In the following chapters one can find out that occurrence of systematic inaccuracy in estimation of soil properties or pile bearing capacity in CPT data has its deep-rooted causes.

Thus, significant number of CPT locations at the site is not the single condition of this technique to be efficiently applied. Its application with other techniques adjusting CPT results is also very important. It is obvious that the number of results in accurate tests may be rather insignificant; it is CPT that being a cost-effective solution allows optimally reducing it. In a number of cases the data of previous investigations can be admitted as adjusting information, including adjacent territories similar in engineering-geological conditions.

To sum it up it is important to underline that CPT is to be considered as a necessary element of investigations completing more accurate tests provided soil conditions are allowable. Few accurate tests, however good they may be, without CPT (or other express-method capable to "feel" the whole site) characterize just those sectors where they were carried out.

2.6 SOME GENERAL PRACTICAL GENERALIZATIONS

The results of research given in this chapter allow the following conclusions to be made:

1. CPT is a mechanical test of soil reflecting both its strength and deformability. Cone resistance q_c generally characterizes a set of mechanical properties of soil, i.e. from mathematical point of view it is a function of several variables; separate estimation of soil is possible if one applies additional information. Nevertheless, in real-life soil one can always observe mutual correlation of mechanical characteristics (φ, c, E), allowing approximate expressing of each property through other characteristics. Although closeness of the correlation is quite low, and the form of link appears to be different in soils of different lithological and genetic types, the occurrence of the closeness explains approximate empirical dependencies like $q_c \approx f(E)$, $q_c \approx f(\varphi)$, $q_c \approx f(c)$, characteristic for each concrete type of soil.

2. Empirical dependencies like $q_c \approx f(E)$, $q_c \approx f(\varphi)$, $q_c \approx f(c)$ can be considered as particular cases of general theoretical dependency $q_c = f(\varphi, c, E)$ as applied to concrete types of soil. Each type is generally characterized by its own range of "rigidity index" (E/τ_g) values, its ranges of mechanical characteristics (φ, c, E) as well as a number of other features. As a rule, these soil types correspond with both deposits of particular lithological type (sands, clays, etc.) and of particular genesis (alluvial, moraine, etc.).

The mentioned simplest regional dependencies may be obtained by statistical treatment of archival data containing CPT results carried out parallel with other more accurate methods of soil tests. Determination of empirical dependencies (formulae or tables) must be done differentially as applied to concrete soil types. For example, sands are to be considered separately from clayey soils, soils of different genesis are to be treated separately either, etc. The smaller the range of soils, the higher the reliability of these dependencies will be. Choice of formulae for modulus of deformation which according to theoretical aspects are to be greatly diversified proves to be the most differentiated.

3. Conditions of soil behavior under the cone penetrometer are characterized by significant speeds of deformations with processes of skeleton viscous resistance prevailing over filtration compaction. The following prove the fact:

- Extremely weak compaction of clayey soil under the penetrometer.
- Tendency of cone resistance increase in rise of speed of penetration.
- Connected with the previous one, cone resistance decrease in transition of the moving penetrometer in conditional limit equilibrium ("relaxation-creep" CPT – see section 2.3).

Thus, cone resistance reflects the properties of soil corresponding to rapid mechanical effects, i.e. strengthening properties correspond to *non-drainable shear*, deformational ones – to *unstabilized deformations*. Estimation of standard properties of soil corresponding to stabilized condition (finished compaction) is "indirect". It happens to be possible due to the fact that the characteristics determined in rapid loading (in fast speeds of deformations) are correlated with characteristics corresponding to slow loading allowing complete compaction to be ensured.

4. In soil, under the cone there occur high normal stresses significantly exceeding pressures characteristic for foundation bases on natural bed. In pressures given, the Coulomb law in clayey soils is broken: the graph «$\tau \sim \sigma$» is distorted turning asymptotically to straight line «$\tau = const$» (equivalent to $\varphi = 0$ in high value of «c»). The process will unlikely occur in disconnected (sandy, gravel) soils, since distortion of the graph «$\tau \sim \sigma$» corresponds with extremely high normal stresses which hardly ever occur in CPT. This fact significantly increases disagreement between the discussed theoretical dependencies and real behavior of soil under the cone. Firstly, mechanical characteristics of soil corresponding to "ordinary" conditions of its behavior in the bases of shallow foundations are evaluated by CPT just approximately. Secondly, it explains the principal reason of significant difference in empirical formulae $q_c \approx f(E)$, $q_c \approx f(\varphi)$, $q_c \approx f(c)$ in sands and clayey soils.

5. Designs of penetrometers (type I or type II – in accordance with the National Standard GOST 19912-2012) effect on the results obtained. Mantle cone penetrometers of type I (see fig. 1.1) increase cone resistance q_c, i.e. in one and the same soils it is slightly higher when applying type I cone penetrometer than type II one without a mantle. It can clearly be seen in clays and loams in which the mentioned increase of q_c may attain 10...20 %. In sands the increase does not usually exceed 2...3 %, so it can be ignored in practical applications. Total side friction resistance of type I cone penetrometers (Q_s) and unit sleeve friction resistance of type II cone penetrometers (f_s) represent absolutely different values (the first one – force, the second one – stress) relating to different zones of soil. One can fail to compare them.

6. Insignificant deviations from standard dimensions of the cone penetrometer (diameter, cone tip angle) have a low effect on the results obtained.

Nevertheless, application of non-standard cone penetrometers is believed to be undesirable since the uncertainty of "inaccuracies", possibility of their summarizing with other errors (methodical) can significantly decrease the accuracy of the results obtained. Providing non-standard cone penetrometers are to be applied, it is useful to make some corrections in the results obtained. The corrections are to be determined for every particular case when comparing the results of standard and non-standard CPT.

As a rule, changes in dimensions of the penetrometer due to wear do not effect on the results obtained as the decrease does not exceed segments of millimeters. The International Norms (IRTP 1999) restrict the decrease of the cone diameter in 0.4 mm, its height – in 7 mm (in relation to standard dimensions – see fig. 2.8). The National Standard GOST 19912-2012 determines stricter limitations: the cone diameter – up to 0.3 mm, the decrease of height – up to 5 mm. In the International Norms the greatest attention is paid to the cone diameter to exceed that of friction sleeve in type II cone penetrometer or the adjacent zone of the push rod in type I cone penetrometer upwards of 0.35 mm. The stricter requirements of the National Standard GOST 19912-2012 automatically agree with the conditions given.

In man-made roughness of the friction sleeve and the cone as well the following conditions occur: "friction of soil upon soil" is ensured, however the surface is rapidly worn-out turning into the smooth one in 1...2 months. Since the changes in roughness of the friction sleeve greatly effect on the results obtained, the maximum stability and definition of measurements is achieved in smooth surface of the cone as it is admitted in the International Norms.

7. The speed of penetration has a low effect on the results obtained. Despite some contradictory experimental data in different authors as well as "data spread", the tendency of small increase of cone resistance in increase of the speed of penetration (in the interval of speeds to be 0.1...5 m/min one can observe the change of cone resistance in average 20...30 %) does not give rise to doubt. In the admissible by the Standards interval of speeds to be 1.2±0.3 m/min (20 mm/s) one can ignore the effect, i.e. the results obtained may be considered to be independent upon the speed of penetration.

8. Measurement of cone resistance in transition of the penetrometer to limit equilibrium ("relaxation-creep" CPT) allow obtaining additional information concerning the type, condition and rheological properties of soil characterizing its behavior in decrease of speeds of deformations approaching to zero. Gradual decrease of resistances q_c & f_s in time characteristic for most cases of "relaxation-creep" CPT ("relaxation-creep" tests) as well as the increase of these resistances in increase of the speed of penetration in traditional CPT reveals predominance of the processes of skeleton viscous resistance of soil over the processes of filtration compaction.

9. Deviation of the penetrometer from the vertical in significant depths of penetration may cause great distortions of the actual depth of penetration, thus reducing it. The error can be avoided when applying the penetrometers equipped with inclinometers. Ignorance of inclinometers is possible in the depths up to 15 m, as deviations in the depths given have a low effect on the depth of penetration measurement.

10. Theoretical analysis shows that approximate in-situ tests can ensure a lot more information than few accurate ones. It is due to the fact that accuracy of the particular test does not guarantee the typicalness of the result obtained (i.e. its typical nature for the given site). The developed mathematical modes allow in determining the required characteristics in the expected inaccuracies numerically estimating the total information content of approximate tests stating the proportion between the set of approximate tests and the number of accurate ones. Generally, this means that to a certain extent, inaccuracy of estimations can be compensated by their number. For this reason, the pros of CPT are to be seen in preforming a great number of CPT procedures being possible in high productivity of the equipment applied. Owing to the fact that in increase of measurements systematic inaccuracies (errors) do not decrease, application of CPT in combination with few accurate tests is believed to be optimal; this allows corrections of CPT results to be made, thus reducing systematic (for particular site!) inaccuracies up to minimum.

3. PRACTICAL METHODS OF CPT APPLICATION IN SITE SURVEY

3.1 INVESTIGATION OF SOIL STRATIFICATION FEATURES (PLOTS OF LITHOLOGICAL SECTIONS)

3.1.1 Identification of soil lithological varieties

Study on soil stratification features as well as a plot of lithological sections is the most important task in any site surveys. The solution of the problem is complicated by the fact that "accurate" methods of identification of soil lithological varieties (drilling, pitting, etc.) give information just on separate site locations spaced in significant distance from each other. According to the Recommendations SP 11-105-97 (SP 47.13339.2012) even in the most detailed surveys (to work out work paper) the distances are in the limits of 20...100 m. One can only imagine what it is located between the boreholes or pits. On the other hand, generalized studies of large massifs of soil (geophysical, etc.) do not guarantee the desired accuracy of the identification. For many years the Russian and non-Russian specialists have studied whether CPT is capable to resolve the problems given, i.e. to determine lithological origin of deposits being cut by the penetrometer. The issue is of great importance since CPT allows investigating of the site in "a denser grid" than that by means of traditional drilling techniques (in analogous material and time expenses).

In section 2.4.3 it was underlined that in increase of the angle of internal friction φ and rigidity index E/τ_g as well, cone resistance q_c increases more intensively than unit sleeve friction resistance f_s. Thus, sands possessing significantly higher values of φ & E/τ_g are to have higher values of q_c in the same values of f_s. Besides, decrease of φ in high normal pressures (fig. 2.21) in sands corresponds with many times higher pressures of σ_φ than in clayey soils. For this reason, in the adjacent zone of the cone clayey soil behaves as plastic material with $\varphi \approx 0$, while sand keeps its previous value of φ in this zone. This additionally increases the difference in values of q_c typical for sands and clayey soils (in insignificantly different values of f_s).To a certain extent, it is shown in the formula (2.21) where curvilinearity of the relationship «$\tau \sim \sigma$» was reflected by β, and more intensive increase of ψ with the increase of both the angle of internal friction φ and E/τ_g can be seen in fig. 2.20. Unit sleeve friction resistance f_s differs insignificantly in sandy and clayey soils as it is seen in figure 2.23. Lower values of the angle of internal friction φ in clayey soils are compensated by cohesion "c" being significantly higher than in sands.

As it was mentioned above, it was Begemann (1965) who discovered the possibility of identification of sands and clayey soils by CPT data; earlier he suggested the cone equipped with a friction sleeve. He stated empirically that relationships between cone resistance q_c and unit sleeve friction resistance f_s in sands possess much higher values than those in clays. Figure 3.1 shows classification of soils in accordance with the relationship between q_c & f_s exactly corresponding to that published in 1965.

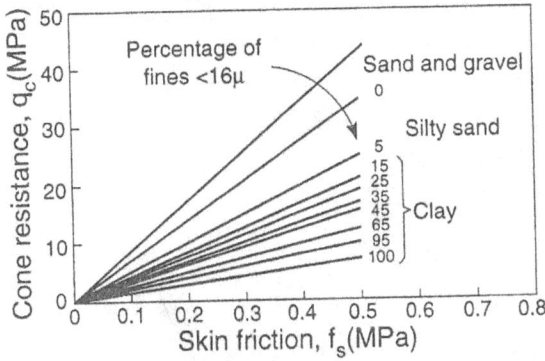

Fig. 3.1 Classification of soils in relationships of cone resistance (q_c) and unit sleeve friction resistance (f_s), suggested by Begemann (1965)

After two years, in 1967 the American specialist Schmertmann (1967) published his results of analogous research. Taking the value f_s/q_c he managed to determine the principal groups of soils under classification in Florida, though not in detail as in Begemann:

$f_s/q_c < 0.5$ %   – soft rocks, shell rock;
$f_s/q_c = 0.5...2$ %   – sands, pulverescent soils;
$f_s/q_c = 2...5$ %   – clay sands;
$f_s/q_c > 5$ %   – clays.

In 1960s...1970s in the former USSR the analogous research was carried out as well (Belyaev 1970, Ryzhkov 1970) with the results not significantly different from the given above. In particular, in the Normative Document of the USSR SN-448-72 regulating CPT application, the instructions on identification of sands and clays by this technique were given.

Then, the issue was studied by a number of specialists who suggested one or the other adjustments. It soon became clear that the accuracy of the technique was not high, i.e. one failed to determine the percentage of clayey particles. It was noticed that the ratio q_c/f_s named "*friction index*" (see section 2.5) or the reciprocal, i.e. f_s/q_c – *friction ratio* failed to definitely characterize lithological type of soil: it was necessary to take into account the values q_c & f_s as they were (particularly q_c). For this reason, the charts have become popular; in them in the coordinates $q_c \sim f_s$, $q_c \sim f_s/q_c$ or $q_c \sim q_c/f_s$ the zones corresponding to one or the other groups of soils under classification are defined. At present, this approach has become the most popular. Figure 3.2 illustrates the chart of Douglas & Olsen (1981) applied in practice; the chart was made on the basis of data treatment of surveys in the western regions of the

USA. Figure 3.3 shows the analogous chart of Robertson et al. (1986) made after the data of survey in Canada (the penetrometers were equipped with piezometers, thus instead of q_c the authors applied corrected cone resistance q_t, with regard to the features of the concrete piezometer – see section 3.2.3. for details).

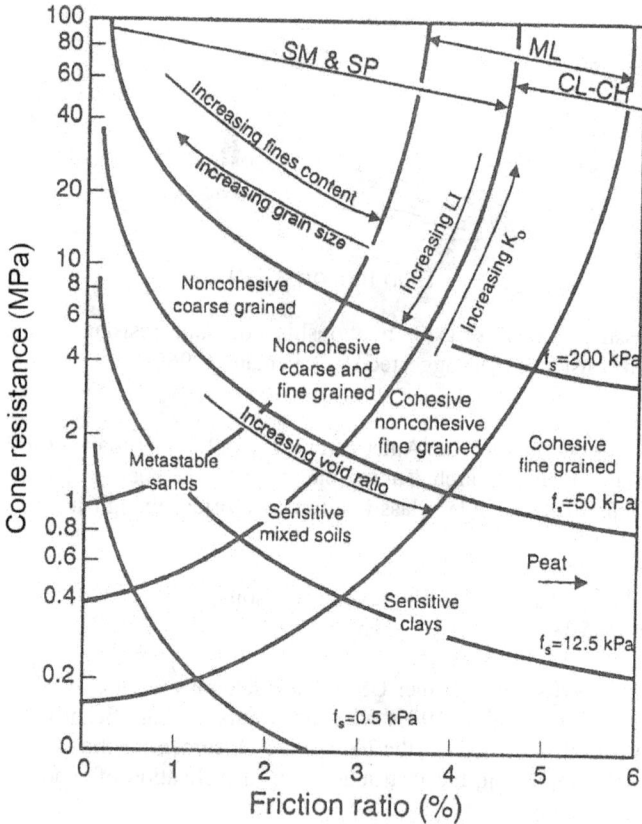

Fig. 3.2 Chart to define types of soil after Douglas & Olsen (1981):
(I_L – liquidity index, K_o – coefficient of earth pressure at rest)

As it is shown in fig. 3.2 and fig. 3.3 the charts do not greatly differ in content despite the fact that they were obtained in different regions. The authors use different terminology and $q_t \neq q_c$ however, ignoring the slight difference between q_t & q_c one can compare the results of identification of soil corresponding with each chart. For example, in $q_c = 1$ MPa, $f_s = 50$ kPa friction ratio will be $R_f = f_s/q_c = 50/1000 = 0.05 = 5 \%$. According to Douglas et al. (fig. 3.2) q_c & R_f correspond with "cohesive finely dispersed soil", i.e. in the terminology of the National Standards of Russia – "clay" or "loam". According to Robertson et al. it is "clay" that corresponds with this combination (zone 3 in fig. 3.3).

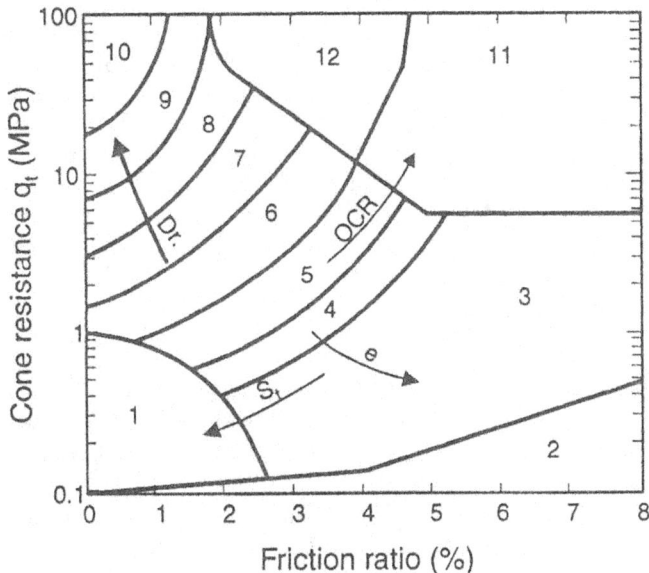

Fig.3.3 Chart to define types of soil after Robertson et al. (1986):

e – void ratio, S_t – sensitivity, OCR – overconsolidation ratio, I_D – density index, q_t – corrected cone resistance $q_t = q_c + (1 - a) u_2$, see in section 3.2.3 for details.
Types of soils: 1 – sensitive fine grained, 2 – organic material, 3 – clay, 4 – silty clay to clay, 5 – clayey silt to silty clay, 6 – sandy silt to clayey silt, 7 – silty sand to sandy silt, 8 – sand to silty sand, 9 – sands, 10 – gravely sand to sand, 11 – very stiff fine grained (overconsolidated or cemented), 12 – send to clayey sand (overconsolidated or cemented)

In $q_c = 1$ MPa, $f_s = 15$ kPa, $R_f = 1.5$ % is obtained; this corresponds to sensitive mixed soils after Douglas et al. It is clayey silt or silty clay after Roberson et al.

In $q_c = 5$ MPa, $f_s = 100$ kPa, $R_f = 2$ % is obtained. This corresponds to noncohesive coarse grained soils after B. Douglas et al., i.e. according to the National Terminology – to "sands". It is silty sand or sandy silt after Robertson et al.

Although there are some differences, but they are not significant.

It is important to underline that absolutely different soils correspond to one and the same ratio R_f. For example, figure 3.3 shows the vertical $R_f = 2$ % which "cuts" almost all highlighted zones – from sensitive clays to cemented sands. The same refers to cone resistance q_c: both sandy and clayey soils correspond to one and the same value of q_c especially in the zone $q_c = 1...10$ MPa; this can be seen in fig. 3.2 and fig. 3.3.

Nevertheless, one should pay special attention to the fact that different soils correspond to one and the same values of q_c & R_f at sites of different regions (sometimes of one and the same region). The reasons are connected with specific engineering-geological conditions of every region, i.e. with different origin and the following geological history of soils under study. Lithological similarity of soils does not exclude the differences in their structure and texture which effect on the values q_c & R_f. Besides, numerous random factors greatly effect on R_f. Lunne et al. (2004) inform that specialists in CPT are of the opinion that measurements of f_s, i.e. unit sleeve friction resistance are less accurate and less reliable as well compared to measurements of q_c, i.e. cone resistance. When applying various non-standard penetrometers it is the most visible as the dimensions of the friction sleeve influence the results obtained more than the dimensions of the cone. Besides, it is important to note that f_s/q_c, as the quotient of two random values, is less stable than f_s & q_c separately, i.e. possesses higher dispersion and contains higher inaccuracies as well.

Accuracy of measurements is affected by one feature of stress-strain condition of soil under the cone. The measurement results of q_c always show mechanical properties of soil both in the adjacent zone and distant one below the cone. The cone approaching the denser layer or weaker one "feels" it, i.e. cone resistance q_c increases or decreases gradually when approaching the roof of a new layer. This feature shows itself in a different way in soils of different density and compressibility. Lunne et al. (2004) point out that in soft materials "sphere of influence" of the cone does not exceed two-three diameters, i.e. in standard cone penetrometer – 7...10 cm. In stiff materials the "sphere of influence" increases up to 10...20 diameters of the cone, i.e. up to 35...70 cm. This means that in dense soils the cone penetrometer does not allow the borders of thin alien layers to be determined. It is possible in thickness of layers to be 70...75 cm, values of cone resistance q_c at the contact zone of layers will inevitably be distorted. On the contrary, in weak soils one can accurately discover the location of alien layers 10 cm in thickness (Lunne et al. 2004). It is obvious that in thickness of 3...5 cm (strip clays) one can fail to determine the borders of such layers with high accuracy.

In general, the factors mentioned above, cannot be strictly regulated. When solving the concrete engineering-geological problems one has to take into account the experience of surveys in the region given and to use the techniques not mentioned in the general Normative Documents yet. Here, the most useful is the study of experience concerning identification of soils by CPT in various regional conditions, since any situation may have precedents occurred and studied in other regions. Some typical examples of generalized local experience of CPT application for identification of soil types are given below.

Figure 3.4 illustrates the chart for identification of sandy (a) and clayey (b) soils in the city of Moscow worked out by Ziangirov & Kashirsky (2006); the chart is based on engineering investigation data acquisition of Mosgorgeotrust. Compared to the discussed charts this one differs in details, as 18 types of soils are given here. For

109

this reason, the authors preferred not to unite sandy and clayey soils into one chart to avoid the complexity.

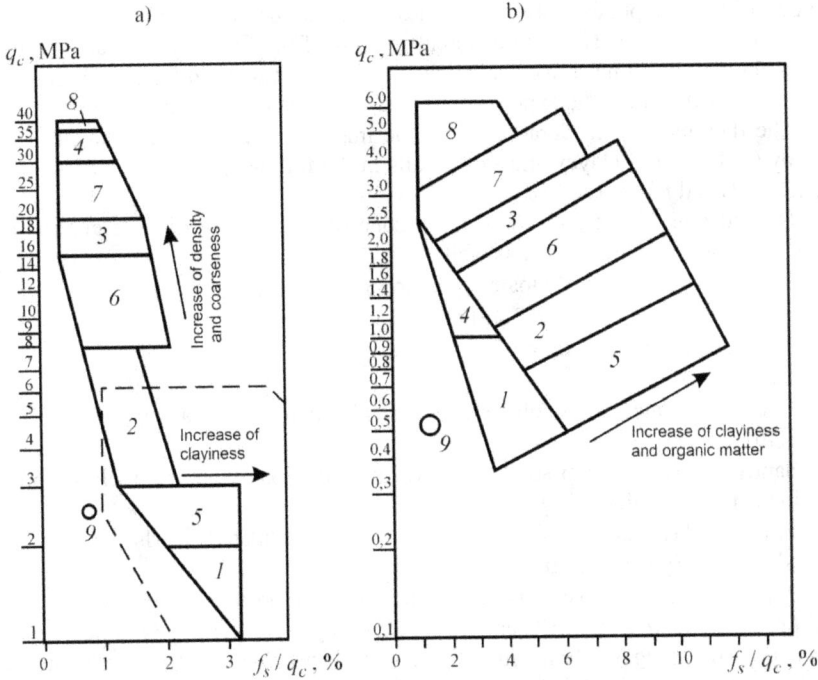

Fig. 3.4 Charts for identification of soils on the territory of Moscow (Ziangirov & Kashirsky 2006):

a – sandy soils (dashed line shows the zone of possible location of clayey soils in accordance with chart b): 1 – pulverescent loose sandy loams, 2 – fine sands, 3 – gravel-pebble soil, 4 – gravel sands and gravel-pebble soil, 5 – pulverescent sands, 6 – sands of medium fineness, 7 – coarse sands, 8 – fine and pulverescent sands of Cretaceous period, 9 – loose sands of medium fineness ($e = 0.75$) – single test;

b – clayey soils:1 – clayey soils of soft and very soft consistency, 2 – loams, clays, alluvial sandy loams, lacustrine marsh soft marls with organic admixtures, 3 – Jurassic clays and lacustrine marsh overconsolidated marls with organic admixtures, 4 – pulverescent sandy loams inclined to liquefaction, 5 – organic-mineral soils and highly moisturized peats, 6 – lacustrine marsh overconsolidated marls, 7 – overconsolidated semi-solid moraine sandy loams and loams, 8 – sandy loams and semi-solid and solid moraine loams with gravel and pebble inclusions, 9 – sand of medium fineness saturated with oil products (carbohydrates)

Engineering-geological conditions of Moscow differ in exceptional diversity, thus, the study is useful for researchers of Moscow and other regions as well.

It is important to note that according to the data (Ziangirov & Kashirsky 2006) values of q_c in clayey soils exceeded 6 MPa nowhere. This gives rise to some doubts on reliability of table 4 in the appendix "И" SP 11-105-97 where solid consistency of clayey soils corresponds with the values of q_c to be up to 20 MPa. Years of application of S-832 CPT rig in different regions of the former USSR also revealed that in clayey soils cone resistance values hardly ever exceeded 5 MPa. Perhaps, the table was based on both the experimental data and certain extrapolation of these data.

The data under consideration show the impossibility of accurate identification (just by CPT data) of clayey and sandy soils in the following range of CPT data – q_c = 1...6 MPa and f_s/q_c = 1...3 %.

Dashed line (figure 3.4a) shows the zone of possible location of clayey soils. Here, sandy soils can be located as well:

• Pulverescent loose sandy loams (position 1, fig. 3.4a).

• Fine sands (position 2, fig. 3.4a).

• Pulverescent sands (position 5, fig.3.4a).

And clayey soils:

• Overconsolidated semi-solid moraine sandy loams and loams (position 7, fig. 3.4b).

• Sandy loams and semi-solid and solid moraine loams with gravel and pebble inclusions (position 8, fig. 3.4b).

• Jurassic clays and lacustrine marsh overconsolidated marls with organic admixtures (position 3, fig. 3.4b).

• Pulverescent sandy loams inclined to liquefaction (position 4, fig.3.4b).

For example, q_c = 2.2 MPa and f_s/q_c = 3 % may correspond to both pulverescent sands (position 5, fig.3.4a) and Jurassic clays or lacustrine marsh marls (position 3, fig. 3.4b). q_c = 3.2 MPa and f_s/q_c = 2 % may correspond to both fine sands (position 2, fig. 3.4a) and moraine loams or sandy loams (position 7, fig. 3.4b).

One should draw an important conclusion from the generalized data shown in figure 3.4 – soils can be related to clays in f_s/q_c > 3.2 %, to sands – in q_c > 6 MPa. In other cases one must take decisions in additional information to be available: results of control drilling, results of surveys at the adjacent territories or, at least, availability of local experience in the given region.

Nevertheless, the conclusion has to be restricted by regional conditions. It can be seen from comparison of generalizations shown in fig. 3.4 with the charts of American and Canadian specialists (fig. 3.2 & fig. 3.3). One can notice that despite different geological conditions of Eastern Europe (Moscow) and Northern American territories the results of applications prove to be the analogical, though there exist some differences as well. According to both the chart of Robertson et al. (fig. 3.3) and the chart of Douglas et al. (fig. 3.2) q_c > 6 MPa generally corresponds to sandy soils. However, position 11 (fig. 3.3) is related to very dense finely dispersed clayey soils. According to the authors, f_s/q_c > 3.5 % corresponds to clayey soils in general, however, even in this case position 12 (fig. 3.3) does not correspond to this, as it is related to cemented sands.

The analysis of the results of analogous research carried out in other regions does not significantly change the given concepts. Small friction ratios f_s/q_c and high values of cone resistance q_c correspond to sands in all the authors. Vice versa, large values of f_s/q_c are typical for clayey soils. There occur situations when any single results do not agree with any rules stated earlier (see position 9, fig. 3.4a), as it is seen in all the authors. Unfortunately, some specialists hide these situations.

The majority of regional charts are worked out in a simplified form ignoring any details, but they are often supplied with the tables giving one or the other definitions.

Figure 3.5 illustrates the simplified chart worked out by Belyaev (2005); the chart is based on the analysis of the results obtained of surveys carried out in the city of Samara and Samara region as well (location of axes q_c & f_s/q_c in fig. 3.5 is changed in relation to the original for simplicity of comparison with the charts given above).

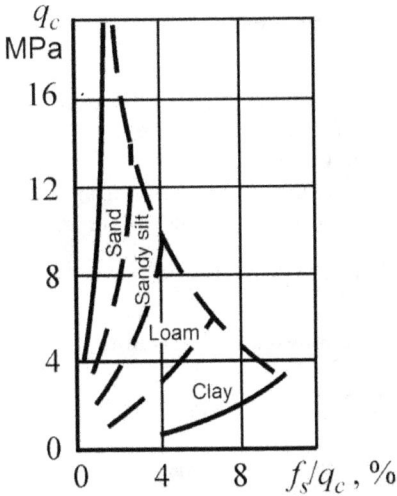

Fig. 3.5 Chart to define types of soil in Samara and Samara region after Belyaev (1970)

V.P. Belyaev suggests special table for detailed identification of soils (see table 3.1).

If to compare the recommendations given by Belyaev with ones given by the authors mentioned above, one needs to point out the following:

- In sands of Samara region friction ratios f_s/q_c are in the same range as in Moscow region (in table 3.1 $f_s/q_c = 0.5...2.3$). The same situation can be found in clayey soils, but the lower limit of f_s/q_c is a little bit higher than that in Moscow: in table 3.1 the ratio f_s/q_c is 2 % higher while in Moscow it is nearly 1%.

- Apart from the data obtained for the territory of Moscow, Belyaev found very dense clayey soils in Samara region with $q_c > 6$ MPa: they were some types of loams (with q_c up to 9 MPa) and some types of sandy loams (with q_c up to 13 MPa).

112

Table 3.1 Typical values of friction ratio f_s/q_c of soils in Samara region (Belyaev 2005)

Age and genesis of layers	Geomorphological element	Fineness of sand or plasticity index of clayey soil	Friction ratio f_s/q_c (%)
$aQ_{II, hz}$	III terrace of Volga river II terrace of Samara river	Fine and pulverescent sand	0.5...2.3
$aQ_{II\text{-}III, hv, hz}$	II & III terraces of Volga river	4...6	2.0...4.0
		8...12	2.7...5.2
— » —	I & II terraces of Samara river	13...17	3.0...6.0
$aQ_{III, hv}$	II terrace of Volga river	13...17	2.3...5.8
$aQ_{II, hz}$	I & III terraces of Volga river	13...17	2.5...5.5
		18...22	3.0...6.1
dQ	Dealluvial slopes	15...25	3.0...8.0
N_2, P_2^t	Watershed & slopes	17...30	6.0...10.0

Table 3.2 shows criteria of identification of soils worked out by Norshayan et al. for Bashkortostan (Norshayan & Ryzhkov 1998). Table 3.2 shows the more careful approach to application of friction ratio; its function is limited by relation of soil to clays, loams or sands (there is a lack of sandy loams in Bashkir region near Urals), here, the border between clays and loams is not clearly expressed. More detailed division is made in cone resistance q_c which becomes the principal criterion of identification of soils. In sands the highest value of friction ratio f_s/q_c is 2.5 %.

Table 3.2 Criteria of identification of soils in Bashkortostan (Norshayan & Ryzhkov 1998)

Type of soil	Cone resistance q_c (MPa)	Friction index $I_f = q_c/f_s$ (friction ratio $R_f = f_s/q_c$, %)
Soft and very soft loam	< 1.0	10...50 (2...10)
Overconsolidated loam	1.0...2.5	10...50 (2...10)
Semi-solid loam	2.5...4.0	10...50 (2...10)
Soild loam	>4.0	10...50 (2...10)
Soft and very soft clay	< 1.0	< 40 (>2.5)
Overconsolidated clay	1.0...2.5	< 40 (>2.5)
Semi-solid clay	2.5...4.0	< 40 (>2.5)
Solid clay	> 4.0	< 40 (> 2.5)
Pulverescent sand	< 1.5	> 50 (< 2.0)
Fine sand	1.5...2.5	> 50 (< 2.0)
Sand of medium fineness	2.5...5.0	> 50 (< 2.0)
Coarse sand	5.0...7.0	> 50 (< 2.0)
Gravel sand	7.0...12.0	> 50 (< 2.0)
Gravel	> 12.0	> 50 (< 2.0)

The charts given above allow the following conclusion to be made: it is impossible to reliably define the type of soil just only by means of CPT data. Almost all specialists both in Russia and worldwide agree on necessity of additional information. It is obvious that the results of drilling with obtained lithological columns with description of soil types are the most reliable additional information. It is evident that the number of boreholes in presence of CPT can be significantly reduced. Nevertheless, rather successful attempts have been taken in non-Russian practice in application of one more kind of additional information. These are the data obtained by piezocones, i.e. the cones equipped with piezometers. Although in Russian practice piezocones are rarely applied, the techniques of their application in identification of soils do deserve attention.

Based on pore pressure measurements, one or the other auxiliary parameter is determined; the parameter is applied as the criterion of identification of soil type. "Pore pressure parameter" Bq is usually applied as that (Lunne et al. 2004):

$$B_q = \frac{u_2 - u_0}{q_c - \sigma_{vo}} , \tag{3.1}$$

where:

$u_2 =$ pore pressure measured behind the cone;

$u_0 =$ in situ pore pressure ("stabilized" pore pressure before penetration);

$q_c =$ cone resistance (corrected cone resistanceв q_t to piezocones geometry);

$\sigma_{vo} =$ total overburden stress (natural stress) in soil in given depth.

The necessity of q_c to be corrected emerges from difference between the outer diameter of the cone base and the diameter of its inner element where the thrusts are measured (if only it is not shown automatically on the measurement scale).

The authors of the chart (fig.3.6) suppose the chart to be applied together with that one (fig.3.3).

As it is shown in figure 3.6, pore pressure parameter B_q, showing relative value of excessive pore pressure tends to increasing with decreasing of soil particles in size. It is generally higher in clayey soils than in sandy ones, and in fine sands it is generally higher than in sands of medium fineness, etc. Nevertheless, the tendency is not considered to be a general rule: zones characterizing each type of soil embrace wide ranges of B_q values including negative ones. For example, in clays (position 3) B_q values may vary from -0.1 to 1.3, in sands of medium fineness (position 9) – from -0.05 to 0.1. Zero values of B_q showing the situation when pore pressure does not increase are possible in all types of soils: vertical $B_q = 0$ crosses every zone except 1 & 2. One can conclude from fig.3.6 that in $B_q > 0.2$ soil is related to clayey one (i.e. is not a sand). However, one cannot absolutely trust in this, since other authors suggest a different border. Figure 3.7 illustrates the diagram $q_c \sim B_q$ showing the results of earlier experimental research on the issue (probably, the first research on the issue – 1984) of Senneset & Yanbu (see Trofimenkov 1995). As one can see from figure 3.7, the border of sands is not $B_q \le 0.2$, but $B_q \le 0.4$. There is a tendency of B_q increase due to increase of liquidity index (consistency) but not only to decrease of

114

soil particles. According to Senneset & Yanbu, soils with $q_c \leq 2$ MPa and $B_q \geq 0.4$ can be referred to clayey ones.

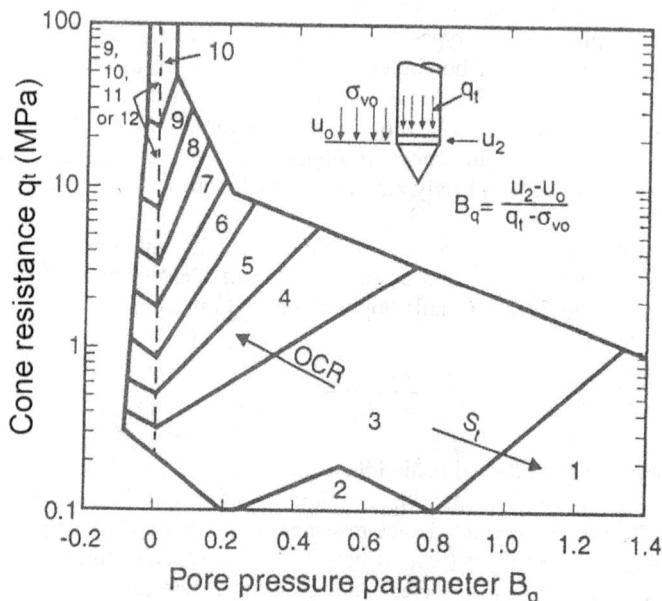

Fig. 3.6 Chart to define types of soil in data obtained by piezocones after Robertson et al. (1986):

The indexes are the same as in fig. 3.3: S_t – sensitivity, OCR – overconsolidation ratio, q_t – corrected cone resistance;

Types of soils: 1 – sensitive fine grained, 2 – organic material, 3 – clay,4 – silty clay to clay, 5 – clayey silt to silty clay, 6 – sandy silt to clayey silt, 7 – silty sand to sandy silt, 8 – sand to silty sand, 9 – sands, 10 – gravely sand to sand, 11 – very stiff fine grained (overconsolidated or cemented), 12 – send to clayey sand (overconsolidated or cemented)

There exist the ideas of using other data as the additional information in identification of soils. In 1970s in BashNIIstroy Enikeyev (1980) found out that in "relaxation-creep" CPT (see section 2.3) soil resistances in transition to equilibrium condition is not equal in different soils. In majority of clayey soils the transition occurs by gentle decrease of resistances q_c & f_s, while in majority of sands the transition occurs by gentle increase of f_s. However, Enikeyev found out the same (and even clearer) increase of resistances in pulverescent clayey soils possessing collapsible properties. His further research was based on the study of collapsible soils by "relaxation-creep" CPT data, as it was of practical importance. The issue is discussed in chapter 5 in detail.

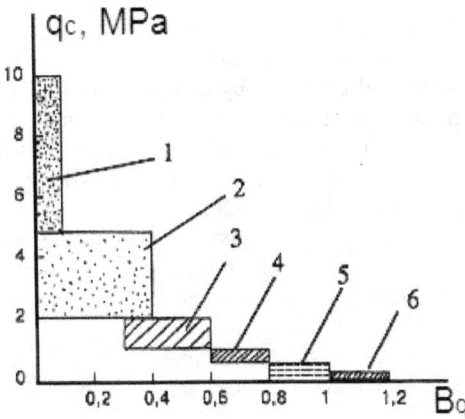

Fig. 3.7 Diagram of Senneset and Yanbu (see Trofimenkov 1995) to define the type of soil by results of CPT with a piezocone:

q_c & B_q – the same as in fig. 3.6;
1 – dense sand, 2 – sand of medium fineness, 3 – clay, 4 – overconsolidated clay, 5 – underconsolidated clay, 6 – very soft clay

Thus, the additional "indirect information" does increase the probability of correct identification of soils, although it does not exclude the errors in a number of particular cases. One cannot rely upon the given identification in case important decisions are to be taken.

One can rely upon "the direct information", i.e. the results of drilling. This does not depreciate the recommendations given (charts, tables), as joint application of CPT and drilling gives possibility to significantly decrease expensive drilling operation if replaced by quicker and cheaper CPT. Here, drilling plays the role of the *correcting* factor to specify decisions preliminary taken after CPT results.

3.1.2. Lithological profiles plots

When making lithological profiles, the important merit of CPT application is the possibility to apply more convenient *quantitative* information (q_c & f_s/q_c) instead of *descriptive* information of lithological columns; the former one allows any interpolation to be easily performed by means of software. There appears the possibility to state fair borders between different soil layers taking into account interpolation of q_c & f_s/q_c values in the zone between the adjacent "CPT locations" (measurements from one position of CPT rig) instead of random schematic

116

presentation of the situation. Schematic illustration of the interpolation is presented in figure 3.8.

The borders between different soil layers may be given in isolines of definite (boundary) values of q_c & f_s/q_c. The values are to be chosen in accordance with one or the other chart (as charts given above) and corrected by comparison of CPT results with drilling ones at the site under study or at the adjacent site. In simple cases (in availability of two-three engineering-geological elements) one can restrict himself with application of just one parameter – friction ratio f_s/q_c (or friction index q_c/f_s).

In the approach given the general principle of lithological profiles patterning agrees with the principle of horizontal patterning in given elevations being widely applied in geodesy. The specific feature is that pivotal values (corresponding to elevations in geodetic surveying) are located strictly in vertical rows, since these are CPT results.

CPT locations chosen for a lithological profile are located in a given order, and then every section between the adjacent CPT locations is examined step by step.

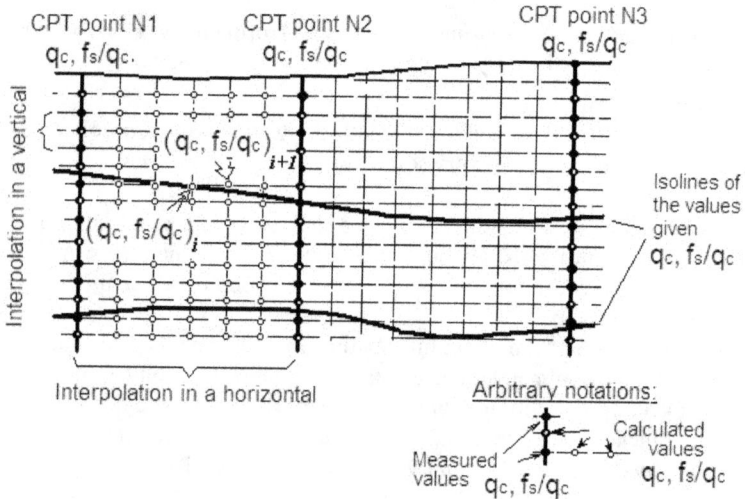

Fig. 3.8 Positioning of isolines of given values of q_c & f_s/q_c by interpolation of parameters obtained in CPT locations

The space between each pair of CPT locations is divided into tiny cells of a random form (usually 0.2 m in height, 1 m in length) located in horizontal rows. Particular values of required parameters, i.e. q_c & f_s/q_c (or just f_s/q_c) are evaluated in each cell. These values are the result of measurements in the edge cells, as they correspond to one or the other CPT location. In obtained edge values of $q_c, f_s/q_c$ (or just f_s/q_c) intermediate values of these parameters are calculated in each cell. For this, linear interpolation of values of the applied parameter is performed in each horizontal row of cells, so that concrete value of this parameter, i.e. $(q_c, f_s/q_c)_i$ corresponds with each i-cell. Isolines are to pass through those cells where values of $(q_c, f_s/q_c)_i$ will "cross" the defined boundary values of q_c, f_s/q_c. In order to ensure the smoothness of isolines the crossed cells are to be divided into smaller sections evaluating q_c, f_s/q_c in each section then.

The described method is realized in a number of BashNIIstroy software and was successfully applied even in 1980s.

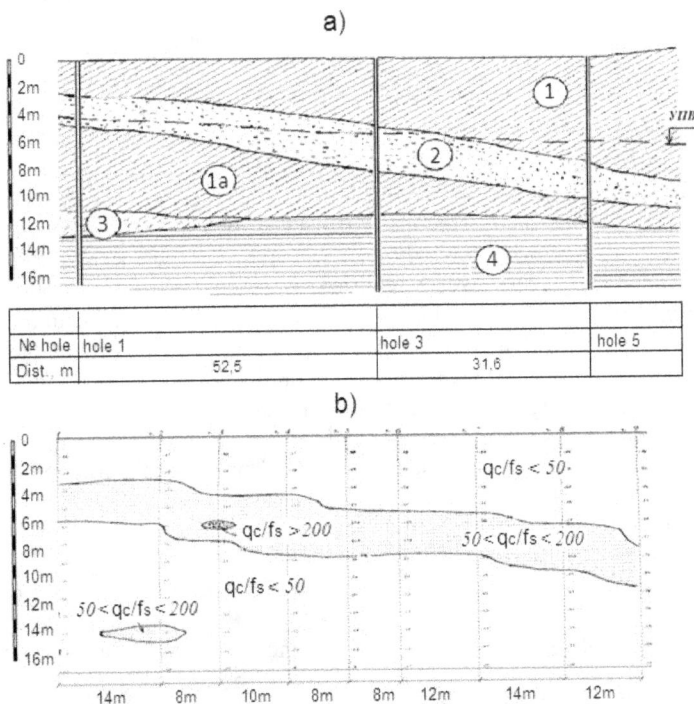

a)

b)

Fig. 3.9 Schematics of soil stratification features estimation by traditional method (drilling) and CPT data:

a – lithological section made by hand, obtained after data of drilling: 1 – underconsolidated loam aQ_{II-III}, 1a – the same, overconsolidated, 2 – fine sand aQ_{II-III}, 3 – plastic sandy loam aQ_{II}, 4 – semi-solid clay N_2;
b – section made by computer, obtained after CPT data, reflects the ratio q_c/f_s; zones q_c/f_s ≤ 50, 50 < q_c/f_s < 200 and q_c/f_s ≥ 200 are pointed

118

Unfortunately, the presented principle of lithological sections patterning as isolines has some drawbacks. Transition from one lithological variety of soil to the other is not always gentle. As a rule, the transition happens to be sharp reflecting rapid changes of sedimentation conditions typical for geological history of continental deposits, alternation of accumulative and erosion processes, etc. They are the techniques based on interpolation of results that ensure gentle transition from one type of soil to the other. Nevertheless, this drawback is not a significant one in case CPT locations are densely located. It seems to be more reliable than identification of borders between the layers "by eye", applying descriptive data given in lithological columns. Here, it does not exclude the following correction based on any additional information.

Figure 3.9 illustrates two lithological sections: the first one is made by hand after the data of drilling, the second one – by computer after CPT data applying just one parameter – friction index q_c/f_s. CPT results obtained at one of the sites in Ufa were taken as origins.

As one can observe from fig. 3.9 the location of sandy stratum (position 2) after both CPT data and drilling as well, are not significantly different. Here, according to CPT results the roof of sandy stratum is less even than after the data of drilling. According to the data of drilling, sandy loam (position 3) occurs as a layer, but after CPT data – as a lentil. The taken simplified approach giving three zones in one criterion ($q_c/f_s \leq 50$, $50 < q_c/f_s < 200$ & $q_c/f_s \geq 200$), makes it impossible to differ sand (position 2) from sandy loam (position 3), however, in control boreholes to be available it does not cause any problems. When applying CPT, reliability and details of the obtained section significantly increase. In traditional approach information on soil types was obtained in three locations only (three holes), while in the discussed approach nine CPT locations were added. It is evident that the danger of "a weak" alien lentil to be missed greatly decreased.

Therefore, if there is the possibility of correcting the results after control drilling data, the efficiency of CPT in patterning the lithological sections goes without saying even in the simplest case when applying only one criterion of identification of soils - (q_c/f_s).

In application of several criteria, general principles of patterning are the same. One or the other chart to define the soil type is to be taken. The algorithm shall determine the relation of the pointed cell (fig.3.8) to one or the other zone on the chart, e.g. when applying the chart (fig.3.4a) zone 5 will be characterized by the following conditions to be observed:

1) $2\text{ MPa} < q_c < 3\text{ MPa}$.

2) $(3.6 - 0.8q_c) < f_s/q_c\,100\% < 3.2$.

The rest of the zones will be characterized by their own conditions.

One must correct the relation of the cells to one or the other zone at the sites in case the holes and CPT locations are located close to each other.

119

3.2 DETERMINATION OF SOIL PROPERTIES

3.2.1 Typical issues of soil investigations

Apart from the majority of standard mechanical soil tests (in compressibility, shear, etc.), capable to evaluate one or the other property of soil (modulus of deformation, shear parameters, etc.), CPT results generally reflect a complicated set of soil properties which one can fail to determine *accurately* by standard properties. As it was mentioned in chapter 2, conditions of soil behavior under the cone are significantly different from those typical for bases of most foundations, as they correspond to extremely high speeds of deformations and extremely high normal stresses as well. For this reason, soil behavior in CPT cannot be understood by standard properties applied in calculations of bases, i.e. in other conditions to be needed. Besides, ignorance of the fact does not give possibility to rely upon homogeneous values of cone resistance q_c to correspond with a property of soil.

In section 2.4 it was underlined that any dependency linking cone resistance with a property of this soil, i.e. $\varphi = f(q_c)$, $c = f(q_c)$ or $E = f(q_c)$ may just be *approximate*, as it is a particular case of a more generalized dependency $q_c = f(\varphi, c, E)$. The existence of these simple dependencies is conditioned by mutual correlation of strengthening and deformation properties of soil as well as unequal effect of properties reflected by these characteristics on cone resistance. This makes it difficult to study connection between CPT and properties of soil experimentally, since the specialist looking for the simplest and convenient formulae or tables has to deal with dependencies hardly ever expressed in general case. These dependencies are clearly manifested just in limited ranges of engineering-geological conditions. For these reasons, any empirical generalizations are assumed to be made based on a number of initial data, each type of geological deposit being studied separately.

Another problem is concerned with special character of in-situ experiments, i.e. with comparison of CPT results and other techniques of soil testing. The comparisons have to be made in strong influence of random factors, first – influence of heterogeneity of soils. Even in homogeneous lithological soil the parameters of its mechanical properties obtained in excavations 1...2 m apart from each other, may differ in 10...20 %. Absolute conformity of CPT location and any other soil test location (or a borehole with selection of monoliths) is not possible. Preliminary CPT procedure exactly in place of impending test is usually not appropriate, as it can effect on the results of such a test. Combining of the CPT location with the location of the borehole is impossible in all the cases. For these reasons one has to compare the results of measurements related to the locations spaced (in horizontal) 1...2 m apart from each other, when there is not any danger of mutual influence of the performed tests.

When showing the results of such a comparison as a diagram of dissipation, one has to face with the effect of heterogeneity of soil, causing the increase of "spreading", i.e. with the decrease of coefficient of correlation. In these conditions, estimation of this "spreading" proves to be of great interest: the spreading would be studied in *a reference case* if the connection between the property of soil and cone resistance was functional (without any inaccuracies) but not correlated, but between the CPT location and the location of the testing (or selection of monoliths) the distance was 1...2 m. At first sight this variant seems to be unreal, but it can be easily performed when comparing CPT results by one and the same penetrometer in locations located 1...2 m apart from each other. Such a research was carried out in BashNIIstroy in 1970s: CPT results were compared in locations spaced apart from each other (1...2 m) with comparison of values of cone resistance in the equal depths (Ryzhkov I.B. 1973, 1992). 20 pairs of CPT locations were studied; the research was carried out 10... 15 m in depth by S-832 CPT rig at five sites in the cities of Ufa and Tolyatti (426 pairs of q_c values were compared). Figure 3.10 illustrates two diagrams of dissipation: the first one shows the results of comparison of cone resistance q_c, the second one – values of friction index q_c/f_s. In each pair of CPT locations the compared values referred to equal depths and were located in a distance of 1 m in depth.

As it is shown in fig. 3.10, the spreading of values happens to be significantly larger than it would be anticipated ignoring the experiments. To a greater extent, it refers to the values of friction index q_c/f_s, which being the function of two random variables, is extremely sensitive to the influence of random factors. Nevertheless, the stronger connection "$q_c \sim q'_c$" is characterized by lower coefficient of correlation - 0.76.

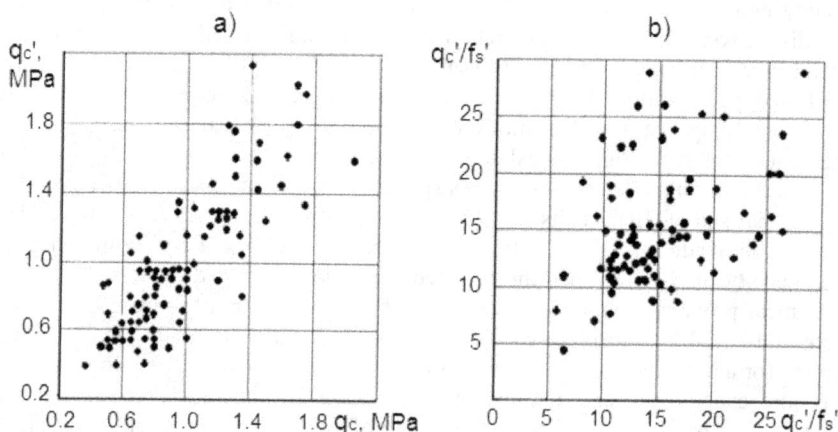

Fig. 3.10 Schematics of dissipation obtained in comparison of CPT results in adjacent locations spaced 1...2 m apart from each other in equal depths:
a – in comparison of cone resistances q_c (q_c & q_c' – the results of measurements in two adjacent CPT locations in equal depth), b – in comparison of friction indexes q_c/f_s (stroke means the same).

Table 3.3 gives coefficients of correlation of the analogous connection "$q_c \sim q'_c$" for five sites applying S-832 CPT rig.

Table 3.3 Closeness of connection between cone resistances q_c & q'_c, measured at the same depth in the distance (in horizontal) of 1...2 m

Site	Soil conditions	Number of pairs $q_c \sim q_c'$	Coefficient of correlation
3d residential community, Ufa (fig. 3.10)	Alluvial clays and loams	95	0.76
Quarters 90-97, Ufa	Lacustrine alluvial clays	66	0.78
UHZ plant	Fluvioglacial clays and loams	53	0.76
UZSS plant (at present - USK plant)	Alluvial clays	45	0.79
Industrial community of Tolyatti	Dealluvial loams	164	0.79

It is necessary to point out that the analysis of the results of CPT rigs testing carried out in Chelyabinsk (see section 2.2) revealed that the spreading of values showed heterogeneity of soil and was not dependent upon the geometry of the penetrometer. In CPT with mechanical cone penetrometers disagreement between values of q_c in the distance of 1...2 m happened to be almost the same as in tensometric cone penetrometers.

The coefficients of correlation given in table 3.3 allow the important practical conclusion to be made: in in-situ experiments connected with comparison of mechanical properties of soil and CPT results at locations spaced 1...2 m apart from each other, *coefficient of correlation cannot exceed* 0.8...0.85 due to heterogeneity of soil. This should be taken into account in revealing any local dependencies, in testing and anchoring general dependencies to local conditions. For example, in comparison of cone resistances q_c with modulus of deformation E, angle of internal friction φ, cohesion c or other parameters, one can expect coefficients of correlation to be in the limits of 0.7...0.8, and in majority of cases – 0.6...0.7. Some publications describe cases when coefficients of correlation were obtained to be 0.95...0.98; these results show either random situations possible in small number of results or situations when a specialist excludes unsuitable results because of his burning ambition to obtain narrow correlation or the lack of his own scientific knowledge.

When considering limitations of coefficients of correlation "from above", caused by heterogeneity of soil, it is necessary to take into account their formal-mathematical limitations "from below". In case the obtained coefficient of correlation r happens (in absolute value) to be lower than a value $|r|_{min}$, the occurrence of correlation ($r \neq 0$) is not proven, hence, obtaining of any empirical dependencies is inadmissible. In small number of results obtained, these limitations ("from above" &"from below") will probably be close. Values of $|r|_{min}$, calculated for confidence probability $\alpha = 0.95$ and $\alpha = 0.995$ (in brackets) are given below:

- in 10 locations on the diagram of dissipation $|r|_{min} = 0.55\ (0.77)$;
- in 20 ones   —— » —— $|r|_{min} = 0.38\ (0.56)$;
- in 30 ones   —— » —— $|r|_{min} = 0.31\ (0.46)$;
- in 40 ones   —— » —— $|r|_{min} = 0.26\ (0.40)$;
- in 50 ones   —— » —— $|r|_{min} = 0.23\ (0.36)$.

Thus, e.g. in 10 locations on the diagram of dissipation the coefficient of correlation equal to $r = 0.54$ means (in $\alpha = 0.95$) the absence of correlation, and in 40 locations $r = 0.27$ proves its occurrence (in the same confidence probability given).

These data confirm the given condition that obtaining of any empirical dependencies showing the connection between CPT results and physical-mechanical properties of soils is to be based on large amount of initial data (at least, 30...40 results are desirable).

3.2.2 General concepts for practical evaluation of soil properties

As it was mentioned in sections 1.4 and 2.4.1, issues on evaluation of soil properties by CPT data have been studied for decades since the beginning of CPT application. At first, when manual probes were applied for evaluation of soils, the results were descriptive, i.e. soils were divided into dense, medium, loose, etc. In general, depths of occurrence of "strong" or "weak" soil layers as well as borders of "strong" or "weak" lentils were determined; density of bulks was estimated, etc. The same approach was applied when the first mechanized CPT rigs appeared. However, numerous attempts were taken to quantitative estimation of soil properties including both theoretical and experimental research. The latter were of empirical character, i.e. presented comparison of CPT results with the results of standard soil tests with the following statistical treatment of comparisons. The obtained empirical dependencies were presented in formulae or tables applied in practice then. They were partly included in the Normative Documents, when reviewing the Documents they were defined.

As it was discussed in section 2.4, theoretical research did not reveal any formulae capable to exclude empirical dependencies; however, they allowed an order in a diversity of empirical dependencies and could reveal the causes of this diversity. Below are presented the widely applied empirical dependencies given in the Normative Documents and Recommendations, the analysis of their principal features involved.

Peculiar features of non-Russian and Russian approaches to studies of soil properties by CPT. In the former USSR the greatest attention was paid to application of CPT in resolving the problems connected with mass building especially applying pile foundations. In practical evaluation of soil properties by CPT data, attention was paid to evaluation of limited nomenclature of characteristics, namely: parameters of shear φ and c, modulus of deformation E. Some of the physical properties were also evaluated (liquidity index of clayey soils, density of sand strata), specific types of soils (collapsible, permafrost, etc.) were paid less attention to. In 1970s…1990s modification of CPT data application technique was made so much due to extension of the range of soil conditions where one could evaluate the characteristics, as application of new measuring operations and extension of the mentioned nomenclature of characteristics. Even measurement of pore pressure being so popular in western European countries and America as well, was not widely applied in practice in Russia.

In this period of time, in non-Russian countries the researchers paid greater attention to extension of nomenclature of characteristics evaluated by CPT as well as solution of special problems; thus, they indirectly admitted the fact that CPT settled itself in evaluation of strengthening and deformation characteristics of soil (φ, c, E). The greatest attention was paid to such problems as calculations of both duration and rates of consolidation of bases, calculations of filtration, determination of overburden pressure, etc. At the same time, in Russia application of CPT in various geological conditions as well as in specific soils was paid greater attention to than worldwide. Even application of CPT in collapsible soils does not raise interest in non-Russian researchers. On the contrary, application of CPT in shelves has been studied on a larger scale in non-Russian countries than in Russia. Moreover, special penetrometers combining CPT with other mechanical soil tests, application of geophysical methods and solution of ecological problems, etc. are considered to be a special feature of non-Russian approach. As it was mentioned earlier, great attention has been paid to measurement of pore pressure in CPT.

Table 3.4 shows typical characteristics of soils evaluated by CPT both in Russia and worldwide.

As one can see from table 3.4, nomenclature of the evaluated characteristics is wider in non-Russian practice than in Russian Normative Documents; however, one needs to consider some additional factors in such a comparison.

Firstly, table 3.4 does not reflect the characteristics of soils evaluated by one or the other Recommendations, but not regulated by National Normative Documents. For this reason, in Russia factual number of evaluated characteristics is larger than given in table 3.4.

Secondly, issues on evaluation of various characteristics have not equally been studied; reliability of applied dependencies and the rate of their detailing have not been equal either. The characteristics are partly evaluated with regard to genetic type of soils, partly – regardless to this. Here, different Normative Documents can differentiate soils in different ways. Nevertheless, for principal mechanical characteristics (φ, c, E) the National Normative Documents embrace more detailed information of this kind than non-Russian ones.

It was mentioned in section 2.4.1 that Lunne et al. (2004) admitting unequal reliability of characteristics obtained by CPT, developed the applicability rating for each characteristic to be used in practical calculations (6 ranks). The highest rating (the first rank – the highest reliability) was given to shear strength of clays s_u

evaluated as a single value without dividing into two parameters φ, c ("one-parameter model"). For sands the highest ratings were given to angle of internal friction φ (the second rank), density of strata (the second rank) and modulus of deformation (the second-the fourth ranks).

Table 3.4. Typical characteristics of soils evaluated by CPT

Names of characteristics	Non-Russian practice [1]	Russia	
		SP 11-105-97, sec. I	MGSN 2.07-01
Angle of internal friction, φ	+	+	+
Cohesion, c		+	+
Undrained shear strength of clayey soils, s_u	+		
Density of sand strata	+	+	+
Liquidity index of clayey soils, I_L		+	+
"Compressive" modulus of deformation, E_c	+		
"Stamping" modulus of deformation, E		+	+
Shear modulus (initial), G	+		
Coefficients of consolidation (in vertical and horizontal directions), c_v, c_h	+		
Horizontal stresses	+		
Overconsolidation ratio (ratio of former maximum overburden stress to the present one)	+		+
Coefficient of Earth pressure at rest	+		
Coefficient of filtration	+		

[1] Lunne et al. (2004)

Peculiar features of empirical dependencies yielding to theoretical explanation. When analyzing theoretical formula (2.13) and figure 2.20 as well, one can notice that to a certain extent, the ratings after Lunne et al. (2004) agree with both the formula and the figure. Here, as it was mentioned, shear strength s_u attains more concrete expression as τ_g, i.e. shear strength in natural stress σ_g. It appears to be desirable to mention three factors following from the given theoretical solution.

Firstly, in accordance with the formula (2.13) and figure 2.20 in typical for most clayey soils values of "rigidity index" $E/\tau_g = 50...200$, angles of internal friction $\varphi = 10...25°$ and $\beta = 1$, curves showing changes of value ψ in this zone of the graph are the most gently sloping, i.e. value ψ shows the highest stability. In other words, in the formula (2.13)

$$q_{\tilde{n}} = \beta\psi\tau_g$$

in slight changes of rigidity index there is to be $\beta\psi \approx$ const, due to this q_c may approximately be considered as the function from one variable τ_g. This means that *in clays and loams rather close correlation between q_c & τ_g. must be observed.*

However, in higher values of "rigidity index" (e.g., $E/\tau_g > 400$) the connection shall be significantly weaker or absent at all. In other words, *with the increase of "rigidity" of clayey soils* (*i.e. with the decrease of liquidity index I_L*) *coefficient of correlation in $q_c \sim \tau_g$ is to decrease.*

Secondly, in high values of the "rigidity index" ($E/\tau_g = 200...800$) typical for sands, the angle of internal friction φ is to effect on q_c, thus, the depth of penetration is of greater importance than in clays. This means that *in sands the correlation between q_c & φ dependent upon the depth of penetration must be observed.* In fact, shear strength of sand in natural stress σ_g, i.e.

$$\tau_g = \sigma_g \, tg\varphi,$$

is to be "more sensitive" to depth (to natural stress σ_g) than analogous resistances of clays where the independent upon the depth parameter – cohesion c occurs. "Rigidity index" E/τ_g is to be dependent upon the depth as well. Given the ideal homogeneous soil (in constant φ, c & E) shear stress of sand τ_g is to increase with the increase of the depth, and "rigidity index" E/τ_g – to decrease. When analyzing the graph in fig. 2.20, one can see that effect of τ_g on q_c is to prevail over the decrease of E/τ_g. It is obvious that cone resistance q_c increases proportionally to the increase of τ_g, while the decrease of "rigidity index" E/τ_g has a lower effect (value ψ changes less than E/τ_g). This means that *in sands with the increase of depth smaller angles of internal friction φ correspond to one and the same cone resistance q_c.*

Thirdly, as it was mentioned in section 2.4, connection between cone resistance q_c and modulus of deformation E is conditioned so much by correlation between strengthening and deformation characteristics of soil, as the mechanical processes under the cone, diversity being a distinctive feature of this correlation. For this reason, *tables and formulae linking q_c with E are to be diversified as well, i.e. in different genetic types of soils unequal dependencies are to be found out.*

3.2.3 Determination of principal soil strength properties

Evaluation of shear stress τ_g of clayey soils deserves special attention in spite of imaginary distance of such a problem from standard design models. The analysis of soil strength evaluation reveals some factors showing the peculiar features of empirical dependencies discussed further and defines the spheres of their application as well.

Correlation between q_c & τ_g in clayey soils was studied in detail in 1960s in BashNIIstroy. The CPT procedure was performed by S-832 CPT rig, shear stress in overburden stress was evaluated by φ & c, determined in laboratory conditions applying standard "shear" device. The distance from the holes where monoliths were selected to CPT locations was 1...2 m in length. The comparisons were carried out on 19 sites with various clayey soils, namely: in Bashkortostan, Tyumen and Kuibyshev (at present, Samara) regions, partly in Moscow, Leningrad (at present, St. Petersburg) and Archangelsk region as well.

Figure 3.11 (a, b, c) shows the examples of the mentioned link between q_c & τ_g on three sites with alluvial, dealluvial and lacustrine-alluvial clayey soils, i.e. soils typical for Russia. The correlation happens to be rather close; this agrees with theoretical formula (2.13) and the research of Lunne et al. (2004). When analyzing the analogous data obtained in "more consolidated" soils (with lower liquidity index I_L) with higher value of rigidity index E/τ_g, less close correlation is usually observed. Figure 3.11d illustrates this tendency as a diagram of dissipation. It shows the general tendency of coefficient of correlation r to decreasing in increasing of mean value of cone resistance q_c (total number of pairs "$q_c \sim \tau_g$" – 3096; in them the coefficients of correlation were calculated).

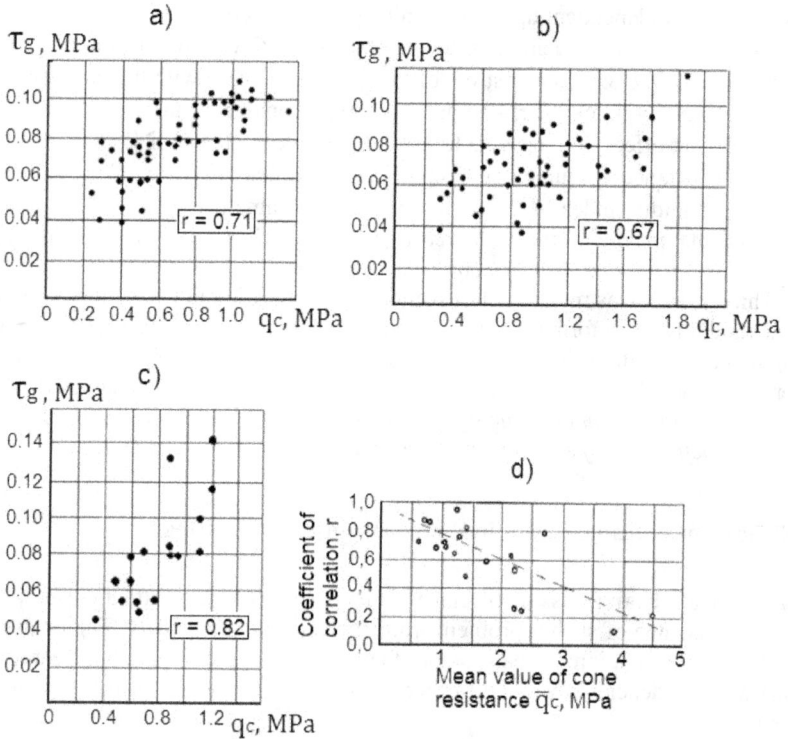

Fig 3.11 Schematics of correlations between cone resistances q_c in clayey soil and its undrained shear stress (in overburden stress) τ_g:

a – very soft – overconsolidated alluvial clays (test site of BashNIIstroy); b – dealluvial loams from very soft to semi-solid (test site in Lesoparkovy residential community in Ufa); c –very soft – overconsolidated lacustrine-alluvial clays (UOS plant, Ufa); d– comparison of coefficients of correlation r, showing the link between q_c & τ_g in clayey soils on 19 sites with mean values of q_c in these soils at each site

Thus, approximate solution of the problem on penetration of the cone into the medium being destructed in accordance with the Coulomb's law and deformed linearly (up to destruction) quite agrees with empirical regularities connecting cone resistance q_c with its shear stress τ_g in overburden stress. This allows τ_g in q_c to be reliably evaluated in most clayey soils (at least, overconsolidated, underconsolidated or very soft). Figure 3.12 shows regression lines $\tau_g = f(q_c)$, obtained at different sites of Bashkortostan with alluvial and dealluvial clayey soils as well as "envelope" curve $\tau_g = f(q_c)$, touching these lines at the bottom, i.e. ensuring the lowest values of τ_g to be obtained in given values of q_c.

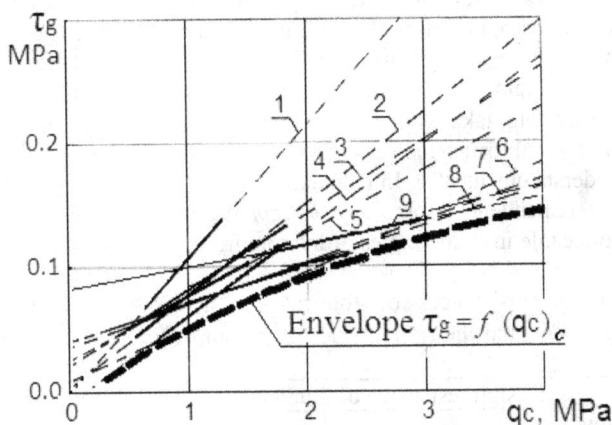

Fig. 3.12 Regression lines $\tau_g = f(q_c)$ at different sites of Bashkortostan and envelope curve showing the lowest values of τ_g in given values of q_c:

1 – alluvial clays, mostly very soft ones (an industrial enterprise in Ufa); 2 – alluvial clays and loams, mostly very soft ones (a residential community in Ufa); 3 – dealluvial loams, mostly overconsolidated ones (residential communities in Sterlitamak); 4 – overconsolidated and semi-solid dealluvial loams (residential buildings in Ufa); 5 – overconsolidated and semi-solid alluvial-dealluvial clays (test site of BashNIIstroy in Ufa); 6 – alluvial clays and loams, mostly overconsolidated ones (a residential community in Ufa); 7 – dealluvial loams, mostly semi-solid ones (a residential community in Ufa); 8 – overconsolidated and very soft alluvial clays and loams (a residential community in Neftekamsk); 9 – semi-solid and solid dealluvial loams (an industrial enterprise in Salavat)

However, the given evaluation of soil strength does not actually agree with the requirements of National Norms of designing, i.e. calculations of foundations, stability of slopes, supporting members, etc. where two parameters of shear φ & c must always be taken into account. CIS International Normative Documents in

128

geotechnical issues are also connected with model of soil possessing two parameters of shear. Nevertheless, the information on shear stress τ_g is, of course, useful owing to the following reasons.

Evaluation of shear stress τ_g may be considered as an intermediate stage in evaluation of soil strength presupposing the following determination of parameters φ & c. It appears to be possible in developing generalized tables and formulae for wide range of conditions as well as in particular cases applicable to concrete sites. In the first case interpretation of τ_g may be directed towards the most typical varieties of clayey soils in which the angle of internal friction φ may be given approximately. In the second case it is necessary to keep in mind that CPT is applied in combination with traditional ways of soil study (drilling, laboratory tests, and sometimes in-situ tests, etc.). In an extreme case, data from the adjacent sites are applied. Here, determination of the angle of internal friction for each engineering-geological element is not a problem.

It is necessary to take into account that in known value of τ_g the error in determination of φ will cause the error of opposite sign in value c, i.e. overstatement of φ means understatement of c. In calculations of foundations or stability of slopes these mutually dependent errors may *be compensated to a large extent*. It is particularly noticeable in small angles of internal friction (up to 15°).

Table 3.5 Design resistances of foundation soils R (see SNiP 2.02.01-83*) corresponding to one and the same values of τ_g, but different parameters of shear φ and c

Angle of internal friction φ	Design resistance of foundation R (kPa) in shear stress τ_g (kPa):					
	10	20	30	40	50	60
Basement depth 2 m						
0°	67.6	99.0	130.4	161.8	193.2	244.6
5°	75.5	111.5	147.5	183.5	219.5	255.5
10°	84.5	126.2	167.9	209.6	251.3	293.0
15°	96.0	144.2	192.2	240.2	288.2	336.2
20°	–	168.8	225.4	282.0	338.6	392.2
Basement depth 4 m						
0°	103.0	135.2	166.6	198.0	229.4	260.8
5°	111.1	147.8	183.8	219.8	255.8	291.8
10°	–	162.1	203.8	245.5	287.2	312.3
15°	–	179.4	229.4	277.3	325.4	373.4
20°	–	–	255.6	312.2	368.8	425.4

Table 3.5 illustrates this mutual compensation showing design resistances of foundation soils R calculated in accordance with §2.41 of National Building Codes SNiP 2.02.01-85* for different values of φ & c corresponding to one and the same

value of τ_g. Values of φ & c were evaluated with regard to the depths of 2 and 4 m, specific weight of soil equaled to 18.1 kN/m^3. For different values of shear stress (τ_g = 10...60 kPa) the following angles of internal friction were determined (φ = 0, 5, 10, 15 and 20°); cohesion c was calculated by the formula (2.16). Design resistance of foundation soil R of a shallow foundation was calculated taking into account the condition that the foundation was 2 m in width (smaller side of foundation), 2 m or 4 m in basement depth. For simplification all the conditions of behavior coefficients were taken to be equal to one. When cohesion (c) was obtained to be negative, this corresponds to dashes «–» in the table.

From table 3.5 one can see that in one and the same τ_g effect of the determined angle of internal friction φ on the value of design resistance of foundation R is significantly lower than one could imagine ignoring the calculations to be done. In alteration of the angle of internal friction in the limits of ± 5° design resistance R changes in the limits of ± 7...17 %, in most cases the interval of changes in R being ±10...13 %, i.e. is in the limits of accuracy of the most "standard" formula of evaluation of design resistance of foundation R given in the SNiP 2.02.01-83*. Here, alteration of φ even in the limits of ± 10° causes alteration of R just in the following limits – ±20...35 %.

Thus, in relevant estimation of τ_g the calculation under analysis does not make high demands to the accuracy of φ determination.

Another example is the calculation of stability of a slope in case the same characteristics of soil are known (as in the previous example), i.e. its specific weight and shear stress τ_g are known, but φ & c are unknown. Table 3.6 gives the coefficients of marginal stability of a slope 5 m in height, with basement depth – 3 m, calculated by the mode of circular cylindrical surfaces for mentioned soil conditions.

For simplicity ground massif of the slope was divided into two layers: higher layer – in the interval of depths – 0...2.5 m and lower one – 2.5...5 m. In both layers shear stress τ_g was considered to be equal. Three variants of calculation were used: "method of Cray-Bishop", "method of Terzagi" and "equilibrium pressure method" (after Chugaev). In all three cases the coefficient of marginal stability was admitted (as minimal) based on comparison of variants of calculation in different locations of slide surfaces.

As one can see from table 3.6, selection of the angle of internal friction φ in calculation of stability of slope has the same effect as in calculation of design resistance of foundation R. In most cases this effect is even weaker than in calculation of R. In alteration of the angle of internal friction in the limits of ±5° coefficient of marginal stability changes in the limits of ±1...15 %, in most cases the interval of alteration being ±3...7 %. In other words, in relevant estimation of τ_g the calculation of stability of the slope does not make high demands to accuracy of φ determination as well.

Analogical results can be obtained in many other calculations with application of shear parameters φ & c (calculations of pile bearing capacity, calculations of thrusts on supporting members, etc.).

Table 3.6 Coefficients of marginal stability of a slope ($h = 5$ m, $b = 3$ m) in different "decoding" of τ_g (i.e. in different combinations of φ & c)

Angle of internal friction φ	Minimal coefficient of marginal stability of a slope in shear stress τ_g (kPa):				
	10	30	40	50	70
Method of Cray-Bishop					
0°	0.292	0.876	1.168	1.461	2.045
5°	0.306	0.783	1.071	1.360	1.940
10°	0.232	0.718	1.000	1.282	2.045
15°	–	0.598	0.920	1.184	1.767
20°	–	0.527	0.811	1.136	1.714
Method of Terzagi					
0°	0.292	0.876	1.168	1.461	2.045
5°	0.314	0.816	1.107	1.400	1.984
10°	0.233	0.757	1.049	1.341	2.113
15°	–	0.655	0.974	1.250	1.852
20°	–	0.600	0.886	1.205	1.804
Equilibrium pressure method					
0°	0.292	0.876	1.168	1.461	2.045
5°	0.349	0.848	1.140	1.432	2.017
10°	0.316	0.822	1.115	1.407	2.179
15°	–	0.766	1.076	1.351	1.953
20°	–	0.768	1.036	1.355	1.940

Thus, despite the theoretical solution given in section 2.4 and corresponding considerations on mutual correlation of mechanical properties of soil that give no possibility to accurate determination of shear parameters φ & c in the value of q_c, approximate estimation of these parameters (with accuracy to be ±20...30 %) appears to be rather satisfactory in structural design due to mutual compensation of emerging errors. This fact allows the following to be concluded: CPT reflects parameters φ&c, if not of the examined soil as it is, then, at least, of its "equivalent" that will behave in the same way as the examined soil under the foundation.

In non-Russian practice determination of clayey soil resistance to shear is slightly different from the approach given.

Firstly, non-Russian professional engineers do not always consider shear strength (usually marked as s_u) as an auxiliary parameter for determination of φ & c, but use it as one of soil characteristics. As it is mentioned in the manual (Shutenko et al. 1989), "mode $\varphi = 0$" is widely applied in many countries; here, calculation of foundations is done in one parameter – s_u (in application of Coulomb's model s_u corresponds to cohesion c in assumption $\varphi = 0$). In the National Normative Documents, as it was mentioned, the calculations of foundations in strengthening properties of soil are anchored to two parameters of shear – φ & c.

Secondly, the majority of non-Russian authors do not anchor shear strength s_u to overburden stress. Although some authors identify s_u with soil resistance to shear in overburden stress (shear stress τ_g), (e.g. in Vesic 1965) rigidity index is given as G/τ_g), this statement is far from being generally accepted. For this reason, influence of overburden stress is taken into account by introducing an additional member in the formula connecting cone resistance q_c with shear strength s_u (Lunne et al. 2004):

$$q_c = Ns_u + \sigma_{vo}, \qquad (3.2)$$

where:

s_u = shear strength of soil;

σ_{vo} = vertical (overburden) stress in soil before penetration corresponding with σ_g in formulae (2.6) & (2.10);

N = cone factor which can be "theoretical" (N_c), depending upon a mathematical model or "empirical" (N_k) evaluated by parallel measurements of q_c & s_u.

If we compare the formula (3.2) with that (2.13), it becomes evident that they differ in occurrence or absence of the member reflecting overburden stress: in the formula (3.2) σ_{vo} occurs as a separate summand, in the formula (2.13) there is a lack of this summand. This is due to the fact that the parameter τ_g in the formula (2.13) corresponds to overburden stress ($\tau_g = \sigma_g \cdot tg\varphi + c$), i.e. reflects its effect on cone resistance q_c automatically; hence, the additional member σ_{vo} (or σ_g) is not necessary.

In connection to the mentioned difference in understanding of shear stress, non-Russian specialists usually apply "cone factor" N in resolving geotechnical problems but not cone resistance q_c.

$$N = \frac{q_c - \sigma_{vo}}{s_u}, \qquad (3.2a)$$

where the notations are the same as in the formula (3.2).

In the table 2.2, where s_u was marked as τ, and overburden stress as σ, values of "theoretical cone factor" N_c were shown in square brackets (in the formula after M.M. Baligh – in round ones). The product of $\beta\psi$ was the analog of N_c in the formula (2.13).

Non-Russian specialists apply "empirical cone factor" N_k in their practical work; this, being of the same essence is determined experimentally by comparison CPT data and those of laboratory tests of concrete soil types. In determination shear strength s_u the same formula (3.2) is used as a converted one

$$s_u = \frac{q_c - \sigma_{vo}}{N_k}, \qquad (3.2\text{б})$$

The book (Lunne et al. 2004) analyzes the results of such tests carried out by different authors in various engineering-geological conditions. The most typical values of N_k for different types of soil are given; correlation of N_k with various characteristics of soil is determined (by plasticity index, overconsolidation ratio, etc.).

For example, Kjekstad et al. (Lunne et al. 2004) carrying out the research in overconsolidated marine clays (not cracked) obtained mean value of empirical cone factor $N_k = 17$; here, s_u was evaluated by stabilometer (triaxial compression). Lunne and A. Kleven when studying the issue in normally consolidated marine clays obtained mean value $N_k = 15$, with the range of variability $N_k = 11...19$; here, s_u was evaluated by impeller (rotary shear) (Lunne et al. 2004).

In case piezocones are applied (piezometers), the corrected value of cone resistance q_t, is used instead of q_c; thus, effect on pore pressure resistances obtained and some other distinctive features of penetrometers is taken into account. It is obvious that other values of the cone factor N_{kt} may be considered as well

$$q_t = q_c + (1-a)u_2, \qquad (3.3)$$

where:

$u_2 =$ pore pressure measured behind the cone (between the cone and the friction sleeve);

$a = A_n/A_c$;

$A_n =$ cross sectional area of the inner surface (connected with the cone) with sensors mounted in it;

$A_c =$ cross sectional area of the cone base.

As it is underlined in the book (Lunne et al. 2004), in most non-Russian piezocones $a = 0.55...0.9$, but in weak soils it is possible to use piezocones with lower values of a. Special calibration hydraulic chambers allowing a to be determined in laboratory conditions in effect of pore pressure are applied.

For piezocones, empirical cone factor N_{kt} is determined practically in the same way as for "standard" penetrometers, but cone resistance is used as a corrected value:

$$N_{kt} = \frac{q_t - \sigma_{vo}}{s_u}, \qquad (3.4)$$

where:

$q_t =$ corrected value of cone resistance in accordance with (3.3);

$\sigma_{vo}, s_u =$ the same as in the formula (3.2).

Aas et al. (1986) carefully examined the connection between the empirical cone factor N_{kt} and various characteristics of soil (plasticity index, overconsolidation ratio, etc.). Undrained shear stress of soil s_u was determined as mean value after the results of three types of laboratory tests: triaxial compression (stabilometer), flat and triaxial tension. The parameter N_{kt} was in the limits of $8...16$.

Figure 3.13 illustrates the dependencies obtained by Aas et al. (1986). The authors analyzed the data related to Norwegian coastal clays and clays of North Sea; in them plasticity indexes were in the limits of $I_p = 3...50$. When analyzing the results presented, it is important to consider differences in terminology admitted in Russian classification of soils, namely: soils are considered to be clays with $I_p > 17$, and in $I_p = 7...17$ soils are believed to be loams, in $I_p < 7$ – sandy loams.

133

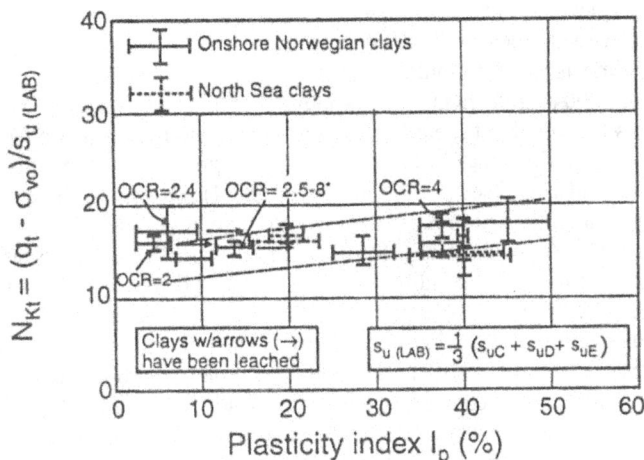

Fig. 3.13 Results of research after Aas et al. (1986):
OCR – overconsolidation ratio, q_t, σ_{vo}, s_u – the same as in the formula (3.3)

As one can see from fig. 3.13, with increase of plasticity index of clayey soil I_p empirical cone factor N_{kt} slightly increases. When expressing shear strength as

$$s_u = \frac{q_t - \sigma_{vo}}{N_{kt}} \qquad (3.4a)$$

one can draw the conclusion that in one and the same values of cone resistance q_t and overburden stress σ_{vo} (i.e. in one and the same depth) shear strength s_u in loams will be a little bit higher than that in clays, and in sandy loams it will be higher than in loams.

It follows from fig. 3.13 that *OCR* corresponding to one and the same values of N_{kt} also increase with the increase of plasticity index, i.e. in other things being equal, overconsolidation ratio in clays will be higher than that in loams, and in loams it will be higher than in sandy loams.

To determine shear strength of clayey soils s_u after CPT with piezocones, other techniques are applied in non-Russian practice as well. In them *effective* stresses under the cone, but not general ones are considered; due to this, other values of cone factor ("piezocone factor") are used. Campanella et al. (1982) apply the following dependency:

$$s_u = \frac{q_t - u_2}{N_{ke}}, \qquad (3.5)$$

134

where:

$N_{ke}=$ cone (piezocone) factor;

$u_2=$ pore pressure measured above the cone;

$q_t=$ the same as in the formula (3.4).

Lunne et al. (2004) give ranges of variability of piezocone factor N_{ke} after a number of authors (6…12, 1…13, etc.), occurrence of correlation between N_{ke} and B_q in accordance with the formula (3.1) is pointed out. Figure 3.14 shows dissipation diagram after Karlsrud et al. (see in T. Lunne et al. 2004) illustrating the correlation between piezocone factor and pore pressure ratio. Nevertheless, Lunne et al. point out that such an approach (consideration of soil resistivity in effective stresses) gives positive results not in each soil condition. They recommend its application in weak water-saturated clayey soils.

When comparing the techniques of shear strength s_u evaluation applied worldwide by the formulae (3.2b), (3.4a) and (3.5), one can note the following features. In all the techniques cone resistance is used not in "the initial variant" (as q_c), but in the corrected one, i.e. involving one or the other concrete factor (overburden stress, pore pressure, features of the penetrometer's geometry). In compliance with this, each technique applies its own values of cone factor (piezocone) – N_k, N_{kt} or N_{ke}.

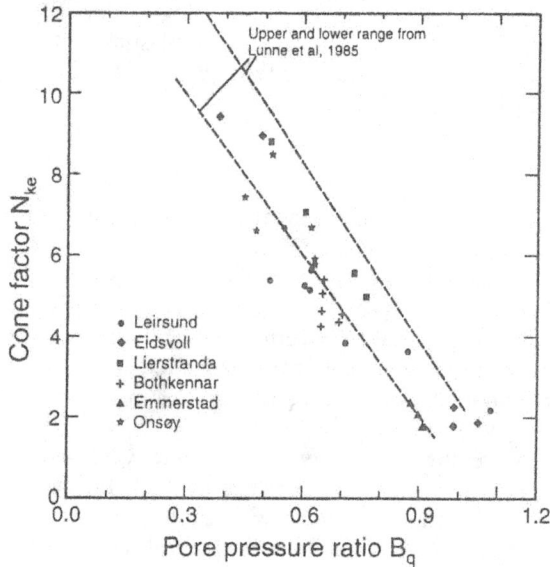

Fig. 3.14 Schematic illustration of correlation between cone factor N_{ke} and pore pressure ratio B_q after Karlsrud et al. (Lunne et al. 2004) (in comparison to the original this figure is slightly simplified)

135

When analyzing publications on evaluation of s_u of clayey soils in CPT data, Lunne et al. (2004) recommend the following:

- In order to evaluate s_u in deposits with lack of CPT results, it is recommended to admit cone factor N_{kt} to be equal to 15…20 for preliminary calculations, the upper limit is to be corresponded with the conservative estimate. It is important to take into account that in normally consolidated and underconsolidated clays the lower limit of N_{kt} may decrease up to 10 and, vice versa, in overconsolidated cracked clays the upper limit of N_{kt} may increase up to 30. In "very soft" clays it is important to consider pore pressure as well, i.e. to apply piezocones.

- In availability of results on estimation of soil strength in CPT data in concrete regional conditions the values of N_{kt} given above are to be corrected to the conditions of this region.

- In investigations for construction of large responsible structures one should pay the greatest attention to correctness in choosing "reference" values of shear strength s_u; the correlated dependencies are anchored to them.

The recommendations presented show the care of non-Russian specialists to "one-parametric" model of soil destruction at every stage of investigations and for structures of any level of responsibility.

In whole, when analyzing these techniques on estimation of soil resistance to shear by one parameter (s_u or τ_g) there appear to be sufficient reasons for the following to be underlined.

In Russian as well as in non-Russian practice unequal parameters of soil strength estimation are applied, the influencing factors are considered unequally as well. While in Russian practice (BashNIIstroy et al.) τ_g – shear stress in overburden stress is applied directly in cone resistance q_c (see fig. 3.12), in non-Russian practice shear strength s_u ignoring anchorage to normal stresses is used. Readings of piezometers (measurements of pore pressure) are generally applied for its estimation. When applying penetrometers without piezometers there still exists a difference between non-Russian and Russian approaches, since the modes of overburden stress recordings are different as well.

Table 3.7 Resistances of soil to shear determined by CPT data obtained in Russia (τ_g) and world wide (s_u)

Cone resistance, q_c (MPa)	Shear stress in overburden stress τ_g (MPa)	Shear strength s_u (MPa) with given depth in m			
		5	10	15	20
1	0.045	0.045	0.041	0.036	0.032
2	0.095	0.095	0.091	0.086	0.082
3	0.123	0.145	0.141	0.136	0.132
4	0.148	0.195	0.191	0.186	0.182

Table 3.7 presents the results of estimation of soil strength to shear (τ_g or s_u) by techniques applied in Russia and world wide, i.e. after fig. 3.12 (theoretical formula

$\tau_g = \dfrac{q_c}{\beta \psi}$) and by the formula $s_u = \dfrac{q_c - \sigma_{vo}}{N_k}$. Values of shear stress in overburden stress (τ_g) were determined by "an envelope" in fig. 3.12, cone factor N_k was admitted to be 20, specific weight of soil – 18 kN/m³.

As one can see from the table 3.7, in the most typical cases for clayey soils ($q_c = 1...2$ MPa in depths up to 10...12 m) the results prove to be almost equal, i.e. $\tau_g \approx s_u$. However, in larger depths and higher cone resistances q_c values of τ_g happen to be slightly lower than s_u.

Evaluation of shear parameters (φ & c) of clayey soils. The discussed estimation techniques of strength of clayey soils by one parameter (shear stress) are of a limited importance in Russian practice, since in Russia calculations of foundations and soil pressures are anchored to the model of soil with two shear parameters – φ and c. Nevertheless, possibility of τ_g evaluation by CPT data as well as approximate determination of φ and c in given τ_g (see tables 3.5 & 3.6) can serve a sufficient reason of the conclusion on availability of CPT application for separate evaluation of φ & c with accuracy admissible for many calculations in construction. For this reason, both the approach discussed when in given τ_g shear parameters φ & c are taken approximately and the research when the immediate connection between cone resistance q_c and each parameter are studied, appear to be of great importance.

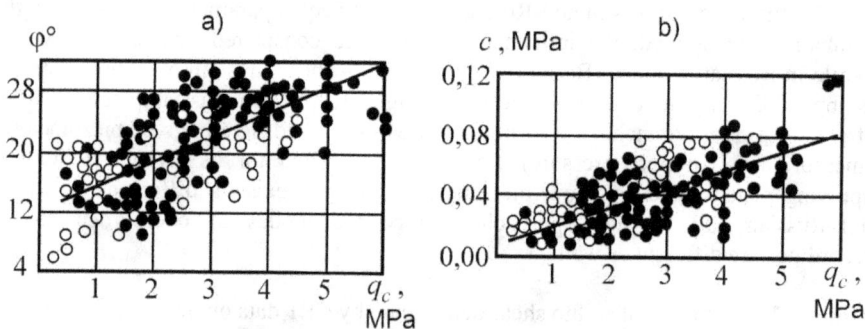

Fig. 3.15 Schematics of correlations between values of cone resistance q_c and shear parameters (φ & c) in clays and loams after Trofimenkov et al. (1977): a – connection between q_c and the angle of internal friction φ, b – connection between q_c and the cohesion c; white (not shaded) bubbles – clays, black ones (shaded) - loams

As opposed to the correlation "$q_c \sim \tau_g$", the correlation between q_c and separate parameters of shear of clayey soils, i.e. "$q_c \sim \varphi$" and "$q_c \sim c$" is rather weak. In small amount of initial data (less than 15...20 locations in diagram of dissipation) and small interval of alteration in q_c (e.g. less than 2 MPa), one can fail to notice it, as a rule. Perhaps, due to this, Lunne et al. (2004) ignore these determinations; they do not give any ratings of reliability and applicability to determination of φ & c of clays to be applied in practice. Nevertheless, in large amount of initial data as well as in wide range of these data the correlation can show itself quite clearly. Figure 3.15 gives example of such a research carried out by Trofimenkov et al. (1977) in 1970s in Fundamentproject.

The authors processed 202 laboratory tests of shear stress parameters (parameters φ & c) determination of quaternary clayey soils and CPT carried out in immediate proximity to the holes (up to 5 m) where the laboratory samplers were taken. CPT was carried out by S-979 CPT rig or SP-36, (i.e. type I mechanical cone penetrometer was applied), laboratory tests of φ & c were carried out by standard one-plane shear in normal pressures of 0.1...0.3 MPa. The tested soils were clays and loams of four different lithological-genetic types (alluvial, dealluvial, non-collapsible, fluvioglacial and glacial deposits). The experiments were carried out in different cities and regions of the former USSR: in Moscow, Moscow region, Rybinsk, Stary Oscol, Zheleznogorsk, Novokuznetsk, Krasnoyarsk, etc.

As one can notice from fig. 3.15, the occurrence of correlation between the compared parameters "$q_c \sim \varphi$" and "$q_c \sim c$" does not give rise to doubt; this happened to be possible due to large amount of initial data and wide range of alteration of cone resistance ($q_c = 0.2...6$ MPa). Perhaps, some rejection of data was done (at least, in $q_c = 6$ MPa the number of locations related to φ, is higher than those related to c), though the results obtained show that the non-Russian specialists (Lunne et al. 2004) do underestimate the possibility of shear parameters φ & c to be evaluated by CPT data. Coefficients of correlations of "$q_c \sim \varphi$" and "$q_c \sim c$" were 0.71 and 0.70 correspondingly; in heterogeneity of soils (see section 3.2.1) this is to be considered as a conclusive proof of correlation occurrence. It is necessary to point out that not any significant difference in correlation link of clays and loams was found.

The authors (Trofimenkov et al. 1977) also compare cone resistances q_c with shear stress τ, corresponding to normal pressures of 0.1 MPa and 0.3 MPa but not to overburden stress as in research of BashNIIstroy (fig. 3.11). In a small range of considered depths (up to 10...15 m) correlation of parameters "$q_c \sim \tau$" must be preserved, parameters of empirical dependency will naturally be different in $\sigma_1 = 0.1$ MPa and $\sigma_3 = 0.3$ MPa. Thus, the authors (Trofimenkov et al. 1977) applied the original mode allowing parameters of "$q_c \sim \varphi$" and "$q_c \sim c$" dependencies to be determined. They analyzed two equations of empirical link obtained after statistical treatment of diagrams of dissipation – "$q_c \sim \tau_1$" and "$q_c \sim \tau_3$", where τ_1 corresponded to the following stress: $\sigma_1 = 0.1$ MPa, and τ_3 – to the stress – $\sigma_3 = 0.3$ MPa:

$$\begin{cases} \tau_1 = 0{,}016q_c + 0.39 \\ \tau_3 = 0{,}025q_c + 0.92 \end{cases}$$

Taking into account that $\tau_1 = \sigma_1\, tg\varphi + c$ and $\tau_3 = \sigma_3\, tg\varphi + c$, the authors (Trofimenkov et al. 1977) obtained the formulae to evaluate φ & c:

$$tg\varphi = 0.045q_c + 0.26, \qquad (3.6)$$

138

$$c = 0.016q_c + 0.125. \tag{3.7}$$

On the basis of evaluations by the formulae the authors give the table of the angle of internal friction values as well as the values of cohesion of clays and loams (ignoring the distinction between them) corresponding to the following values of cone resistance: $q_c = 0.5...6$ MPa (table 3.8).

Table 3.8 Angles of internal friction and cohesion of clays and loams corresponding to different values of cone resistance after the research of Fundamentproject (Trofimenkov et al. 1977)

Values of cone resistance q_c (MPa)	Angle of internal friction φ (°)	Cohesion c (kPa)
0.5	16	18
1.0	17	24
1.5	18	30
2.0	19	36
2.5	20	41
3.0	22	47
3.5	23	53
4.0	24	58
4.5	25	64
5.0	26	70
5.5	27	76
6.0	28	82

The table generalizes the vast experimental material, although it aims at the most probable averaged conditions ignoring diversified particular cases deviating from these conditions. As it was discussed in section 2.4.4 (see fig.2.22c) the correlation between φ & c is observed, i.e. definite (the most probable) value of c corresponds with each value of φ. However, the correlation is rather weak, thus each value of c always corresponds with a number of values of φ. The table under consideration gives stable correlation between φ & c, therefore the accuracy of estimations obtained can vary in a wide range then.

This drawback can be reduced in separate consideration of different types of clayey soils, at least, in separate consideration of clays and loams. The authors (Trofimenkov et al. 1977) ignored this, believing that regression equations obtained did not depend upon lithological-genetic type of soil. However, one can argue against this, since white bubbles (clays) in fig. 3.15 tend to be located on the diagram of dissipation "$q_c \sim \varphi$" below the black ones (loams), and on the diagram of dissipation "$q_c \sim c$" – vice versa.

Nevertheless, despite the mentioned drawbacks, tables 3.5 and 3.6 reveal that for significant number of calculations done in construction (foundations, stability of slopes, etc.) the accuracy ensured by application of table 3.7 is to be quite acceptable. The possible errors may always be compensated by introduction of corresponding "resources". Increase of reliability of such determinations is necessary for reduction of unjustified "resources".

The data given in table 3.8 were repeatedly specified, then in 1980s the specified tables for evaluation of φ & c of clays and loams were included in the Normative Documents (first, they were given in SNiP 1.02.07-87, then after abolition of this Code they were taken to set of rules – SP 11-105-97. Table 3.9 prersents values of φ & c corresponding to different values of cone resistance q_c in accordance with SP 11-105-97.

Table 3.9 Evaluation of shear parameters φ & c by cone resistance q_c after SP 11-105-97 (SP 47.13330.2012)

| q_c, (MPa) | Normative values of the angle of internal friction φ and cohesion c of loams and clays (apart from glacial soils) | | | |
| | Loams | | Clays | |
	φ (°)	c (kPa)	φ (°)	c (kPa)
0.5	16	14	14	25
1	19	17	17	30
2	21	23	18	35
3	23	29	20	40
4	25	35	22	45
5	26	41	24	50
6	27	47	25	55

The data in table 3.9 refer to quaternary clayey soils with organic substances to be less than 10 %. Nevertheless, the sphere of application of the table given is rather wide, as it embraces the whole territory of Russia, these soils prevailing here. Due to this, the data in table 3.9 possess the certain "resources" of reliability ensuring safety of designing for Russia, but one should not expect high accuracy to be obtained.

Table 3.10 Evaluation of shear parameters φ & c by cone resistance q_c after MGSN 2.07-01

| Values of cone resistance q_c (MPa) | Normative values of the angle of internal friction φ and cohesion c of loams and clays (apart from glacial soils) | |
	φ (°)	c (kPa)
1.0	20	25
2.0	21	28
3.0	22	32
4.0	23	35
5.0	24	40

The higher accuracy may be attained in generalization of narrower ranges of soil conditions, i.e. conditions characterized by genetic soil types of concrete region.

Consideration of specific character of deposits in the region becomes possible. Table 3.10 gives example of connection between shear parameters φ & c and cone resistance q_c appearing in Moscow Urban Building Codes MGSN 2.07-01 relevant just for conditions of Moscow only.

In table 3.10 as well as in the article (Trofimenkov et al. 1977) clays and loams are considered jointly. As opposed to the table SP 11-105-97, the tables MGSN 2.07-01 present glacial soils as well (table 3.11).

Table 3.11 Evaluation of the angle of internal friction φ and cohesion c of glacial loams and clays after MGSN 2.07-01

q_c (MPa)	Normative values of the angle of internal friction φ and cohesion c for the following soils							
	Moraine, lacustrine-glacial and integumentary				Fluvioglacial			
	Loams		Clays		Loams		Clays	
	φ (°)	c (kPa)	φ (°)	c (kPa)	φ (°)	c (kPa)	φ (°)	c (kPa)
1	15	22	13	35	14	20	12	29
2	17	43	16	57	16	35	15	46
3	20	63	19	79	19	50	18	63
4	23	83	22	101	22	65	21	80

Specialists of Belarus made a thorough division of dependencies "$q_c \sim \varphi$" and "$q_c \sim c$" by genetic criteria. Table 3.12 shows connection between shear parameters φ & c and cone resistance q_c, given in the Manual P2-2000 of the Building Codes of Belarus (SNB 5.01-99) (for simplicity the table was shaped by analogy to those of tables 3.9...3.11). Table 3.12 embraces the soil types including loams given in tables 3.9 and 3.10 (the column "Pulverescent-clayey quaternary deposits"). However, in table 3.12 these loams possess lower strengthening characteristics than those in tables 3.9 and 3.10 in $q_c = 1...5$ MPa, i.e. regional specific character of soils is not reduced by genesis only. The calculations prove the relative stability of shear stress τ_g corresponding to one and the same cone resistance q_c of similar soils in different regions. In spite of rather great differences between shear parameters φ & c of these soils in tables 3.9, 3.10 and 3.12, the evaluated values of τ_g in one and the same q_c vary insignificantly – $\pm 10...20$ %.

The research carried out in BashNIIstroy in 1980s revealed that the accuracy of shear parameters φ & c evaluation could be raised in case both cone resistances q_c and unit sleeve friction resistances f_s are taken into account. Table 3.13 gives such dependency for quaternary clays and loams of Bashkir Urals (Workbook 1991). Values of q_c & f_s corresponding to "relaxation-creep" CPT are given in brackets (see section 2.3); here, the resistances of soil are measured in conditional limit equilibrium of the penetrometer. In table 3.13 clays and loams are not divided, since in Bashkir

Urals clayey soils possessing the plasticity index I_p = 12...25 prevail, i.e. the difference between clays and loams is not high.

Table 3.12 Evaluation of the angle of intenal friction φ and cohesion c of clayey soils after the Manual P2-2000 of the Building Codes of Belarus (SNB 5.01-99)

q_c, MPa	Values of the angle of internal friction φ and cohesion c for the following soils													
	Glacial (moraine)			Lacustrine -glacial	Loess-like (non-collapsible)			Pulverescent-clayey quaternary deposits (besides mentioned earlier) with organic substances up to 10%						
	Sandy loams		Loams		Loams and clays		Sandy loams		Loams		Sandy loams		Loams	
	φ, deg.	c, kPa	φ, deg.	c, kPa	φ, deg.	c, kPa	φ, deg.	c, kPa	φ, deg.	c, kPa	φ, deg.	c, kPa	φ, deg.	c, kPa
1	26	23	25	30	14	36	22	18	21	25	18	11	16	16
2	27	27	26	36	14	43	26	22	23	30	21	13	18	23
3	27	31	26	40	13	56	27	26	24	34	24	15	20	25
5	28	36	27	45	12	66	27	30	25	36	27	17	22	28
8	29	40	28	49	11	87	28	32	26	41	29	19	24	35
10	30	42	29	52	10	102	28	34	27	46	29	21	25	39
>12	31	48	29	56	9	130	29	36	28	52	30	24	26	47

Table 3.13 Evaluation of the angle of internal friction φ and cohesion c of quaternary clays and loams in Bashkir Urals after the Recommendations of Ufa NIIpromstroy (Workbook 1991)

q_c (q_{cs}), MPa	φ, degrees	Cohesion c, kPa, in f_s (f_{ss}), in kPa				
		20 (15)	40 (30)	70 (45)	120 (80)	190 (130)
0.6 (0.4)	11	12	18	22		
1.0 (0.7)	14	15	21	25	29	
1.5 (1.0)	16	19	23	30	34	40
2.0 (1.4)	18	23	26	34	39	45
2.5 (1.7)	20			39	44	50
3.0 (2.0)	21			43	49	55
3.5 (2.3)	22				55	60
4.0 (2.7)	23				60	65

If to compare the data from table 3.13 with those from the previous tables showing the analogous dependences, it seems easy to notice that in table 3.13 definite angle of internal friction φ corresponds with each value of cone resistance q_c, and cohesion c is to be taken according to unit sleeve friction resistance f_s. In here, the cohesion c may be higher or lower than the values given in tables 3.9, 3.10 & 3.12.

Values of shear stress τ_g evaluated by table 3.13 (in mean values of c) possess the values close to those evaluated by tables 3.9, 3.10 & 3.12.

In general, the analysis of empirical dependences given in tables 3.9…3.13 does not reveal any opposition between these dependences and theoretical solution of the problem on penetration of the cone in normally consolidated medium (section 2.4.3). There are some doubts just in the role of the overburden stress. According to theoretical formulae 2.13…2.16, in constant mechanical characteristics of elastic-plastic medium (φ, c, E & v) cone resistance of this medium q_c shall increase with the depth, though rather weakly. Lower values of shear parameters φ & c are to correspond with one and the same value of q_c in larger depth. Nevertheless, the authors of the tables (tables 3.9…3.13) while working separately in different regional conditions did not manage to find out this dependency. Probably, this was caused by a small range of cone penetration depths, as in the 20th century in the former USSR CPT was carried out in depth up to 14…15 m. The depth reached 20 m only in weak clayey soils (30 m – in special cases).

Evaluation of the angle of internal friction φ of sands is more agreed issue than that of evaluation of clayey soil strength. Specialists in CPT agree on possibility of such evaluation, the disagreements are connected with some special issues. Correlation between cone resistance q_c and the angle of internal friction φ is observed in all sands, though the forms of empirical dependencies may vary (sometimes significantly) due to the type of sand and range of depths.

The majority of researchers anchor the empirical dependencies to theoretical solutions of the problem on penetration of the cone in loose medium, non-deformation to destruction ("axially symmetric problem" of limit equilibrium theory). Such an approach does not cause particular opposition, but it fails to explain the variety of the dependencies "$q_c \sim \varphi$" obtained. The solution of the "mixed problem" taking the medium deforming linearly to destruction and destructed in accordance with the Coulomb's law as a soil model proves to be more effective means of empirical dependencies to be explained. As it was mentioned earlier, according to the solutions the value of cone resistance q_c depends upon both strength and compressibility of this soil, i.e. this refer to $c = 0$ as well. Thus, in different sands one and the same cone resistance q_c may correspond to different combinations of the angle of internal friction φ and modulus of deformation E. Effect of compressibility of sands on the character of the dependency "$q_c \sim \varphi$" is admitted as an empirical fact by many specialists directing toward solutions of limit equilibrium theory. It was informed in section 3.2.2 that the link "$q_c \sim \varphi$" shall depend upon overburden stress, i.e. upon the depth of penetration, hence, "lower values of φ are to correspond with q_c in larger depths".

Figure 3.16 illustrates the widely applied empirical dependency popular in a number of non-Russian countries; the dependency between cone resistances q_c and the angles of internal friction φ of sands was obtained by P.K. Robertson and R.G.Campanella (see in Trofimenkov 1995). The dependency was obtained in normally consolidated sands of average compressibility. Overburden stress σ_{vo} is the factor reflecting the depth. Rather high values of the angles of internal friction φ (up to 48°) deserve attention. Values of $\varphi > 42°$ are hardly ever observed in Russian practice, they being typical for large fragmental soils in which penetration is usually a

143

failure, but not for sands. Perhaps, the authors applied the values of φ determined by some indirect methods.

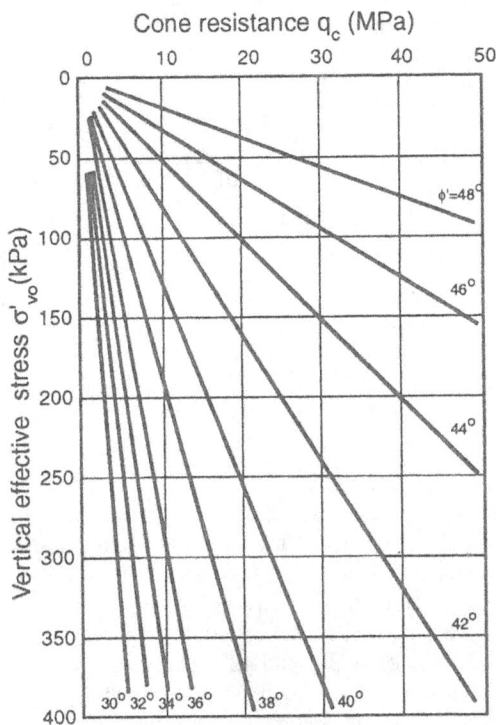

Fig. 3.16 Schematic illustration of empirical dependency of cone resistance q_c of sands upon their angle of internal friction φ and overburden stress σ_{vo} after P.K. Robertson and R.G. Campanella (see Trofimenkov 1995); normally consolidated sands of average compressibility, mainly quartz ones

Table 3.14 presents the dependency "$q_c \sim \varphi$" for sands admitted in the National Normative Document of Russia SP 11-105-97 (SP 47.13330.2012). Values of the angle of internal friction φ in the depths from 2 to 5 m are determined by interpolation. In depth exceeding 5 m the link "$q_c \sim \varphi$" is thought to be independent upon the depth of the layer under study; this is likely to explain the authors' orientation to the known solutions of the problems of limit equilibrium theory (Berezantsev V.G. & Yaroshenko V.A. 1962, Yaroshenko V.A. 1964, Berezantsev V.G. 1966). In the mentioned solutions it is admitted that from the depths (5 m or less) penetration of the cone is possible due to interaction between the zones of

144

destruction and consolidation without projecting of the penetrometer on the surface. At the same time, the empirical dependency in fig. 3.16 indicates the effect of overburden stress in the whole depths of penetration. Theoretical formula (2.13) is also supposed to indicate the effect of overburden stress in the whole depths (τ_g is to indefinitely increase with depth). Final solution of this issue is likely to be obtained in accumulation of greater amount of CPT data and laboratory tests in wide range of depths (up to 35...40 m).

Table 3.14 Evaluation of the angle of internal friction φ of sands by cone resistance q_c after SP 11-105-97

q_c (MPa)	Normative angle of internal friction of sandy soils φ (°) in CPT depth (m)	
	2	5 and more
1,5	28	26
3	30	28
5	32	30
8	34	32
12	36	34
18	38	36
26	40	38

If to compare the angles of internal friction obtained in fig.3.16 and table 3.14, one can reveal significant divergences. In one and the same values of q_c angles φ in table 3.14 are considerably lower than those in figure 3.16. For example, in $q_c = 5$ MPa, 5 m in depth (overburden stress 90...100 kPa), $\varphi = 30°$ in accordance with table 3.14, $\varphi = 38...39°$ in accordance with figure 3.16. In $q_c = 12$ MPa (in the same depth) the angles of internal friction φ are 34° and 42° correspondingly, in $q_c = 26$ MPa $\varphi = 38°$ and $\varphi = 45°$ correspondingly. The reasons of such divergences seem to be difficult to be explained, as both dependencies are empirical reflecting the results of statistic treatment of great amount of data.

Table 3.15 gives the dependency "$q_c \sim \varphi$" for sands admitted in Moscow Urban Building Codes MGSN 2.07-01.

Table 3.15 Evaluation of the angle of internal friction φ of sands by cone resistance q_c after MGSN 2.07-01

Depth of penetration (m)	Values of φ (°) in q_c (MPa) equal to						
	1	2	3	4	5	6	10 and more
2	30	32	34	36	38	40	42
5 and more	28	30	32	34	36	38	40

As one can see from tables 3.14 & 3.15, the angles of internal friction φ after MGSN 2.07-01 are considerably higher than those in SP 11-105-97; in here, Moscow norms are closer to the recommendations of P.K. Robertson and R.G. Campanella.

145

Table 3.16 gives the dependency "$q_c \sim \varphi$" for sands admitted in the Normative Document of Belarus – the Manual P2-2000 to the Building Codes SNB 5.01-99 (for simplicity the shape of table is changed).

Table 3.16 Evaluation of the angle of internal friction φ and cohesion c of sandy soils after the Manual P2-2000 to the Building Codes of Belarus SNB 5.01.01-99

q_c (MPa)	Values of the angle of internal friction φ and cohesion c for sandy soils							
	Coarse		Medium fineness		Fine		Pulverscent	
	φ (°)	c (kPa)	φ (°)	c (kPa)	φ (°)	c (kPa)	φ (°)	c (kPa)
1	30	-	28	-	26	-	24	-
2	32	-	30	-	29	-	26	2
3	34	-	32	-	30	1	28	3
5	36	-	35	1	32	2	30	4
8	38	-	36	1	34	2	32	4
10	39	1	37	2	35	3	33	5
15	40	1	38	2	36	4	34	6
20	41	1	38	2	36	4	34	6
30	42	2	39	3	37	5	35	7
>30	43	2	40	3	38	6	36	8

As one can see from table 3.16, according to the norms of Belarus, the angles of internal friction φ corresponding to one and the same values of q_c may significantly differ due to coarseness of sand, but as opposed to other norms given they do not depend on the depth. For sands of medium fineness φ is slightly higher than in SP 11-105-97 of Russia (for depth of 5 m); they are quite close to Moscow norms MGSN 2.07-01, although in $q_c > 5$ MPa the norms of Belarus give lower values of φ than MGSN 2.07-01.

Application of table 3.16 requires additional information on coarseness of sandy soils to be available, i.e. it assumes application of CPT in combination with drilling and laboratory treatment of sand samplers (with determination of particle-size distribution of sands).

In general, when considering dependencies in tables 3.14, 3.15 & 3.16, it is necessary to take into account that table 3.14 is intended for the widest range of conditions (the whole of Russia); thus, it contains rather larger "resources".

In the European Pre-standard ENV-1997-3 the table for evaluation of density of gravels, sands and silty soils by CPT data is presented. The density is estimated by one parameter – angle of shearing resistance marked as N' ($N' = arctg(\tau/\sigma)$); this seems unusual for Russian specialists. It is obvious that for loose and not enough cohesive soils N' nearly agrees with the angle of internal friction ($N' \approx \varphi$). Table 3.17 gives values of N' corresponding to different cone resistances q_c.

The given angles of shearing resistance are related to sands (quartz or feldspar) for gravels and silty soils the corrections are introduced:

146

- For gravels the angle of shearing resistance is increased in 2°.
- For silty soils the angle of shearing resistance is decreased in 3°.

Table 3.17 Evaluation of density of gravel, sandy and silty soils by CPT data after ENV-1997-3 (Eurocod 7)

Relative density	Cone resistance q_c (MPa)	Angle of shearing resistance N' (°)
Very low	0...2.5	29...32
Low	2.5...5	32...35
Medium	5...10	35...37
High	10...20	37...40
Very high	> 20	40...42

When comparing this table with those of the National Normative Documents given earlier, one can easily notice that the difference between them is not great. To a large extent, it is connected with the fact that in the National Normative Documents the depth of penetration is taken into account, as opposed to ENV-1997-3 where it is ignored. Besides, in the National Normative Documents the angle of shearing resistance $arctg(\tau/\sigma)$ is not meant, but the angle of internal friction φ presupposing the occurrence of the cohesion (c) in silty soils is meant, i.e. in silty soils the angles φ must be lower than N', given in table 3.17 (even with regard to the corrections).

3.2.4 Determination of soil stiffness moduli

In Russian and world practice the following characteristics corresponding to the parameters of linearly deformable medium are applied as the principal deformation properties of soil, namely: modulus of deformation E and Poisson ratio v (lateral deformation). T. Lunne et al. (2004) pay attention to the modulus of deformation of real soils (especially clayey ones) to be dependent on a great number of factors including the history of consolidating loads (underconsolidated, normally consolidated and overconsolidated soils), stages of foundation soil behavior, drainage conditions and directions of stresses. In accordance with this, in world practice the following parameters are usually distinguished:

- Modulus of deformation in drained loading E.
- Modulus of deformation in undrained (usually rapid) loading E_u.
- Shear modulus in low deformations G_o.

Along with these characteristics the coefficient of compressibility m_o functionally connected with the modulus of deformation in drained loading may be used. If it necessary to determine the speed and duration of settings, the coefficient of consolidation c_v is applied. In computation of elastic foundation beds "the coefficient of soil reaction" C is often applied: it reflects either deformability of bed or peculiar features of overground and underground structures or features of design model, etc.;

thus, it is determined by a design engineer during designing period, but not by a surveyor then.

Both in non-Russian and Russian practice the coefficient of lateral deformation v is hardly ever determined by means of instruments (stabilometer): in most cases it is determined indirectly by tables, ignoring special tests.

Evaluation of the modulus of deformation E in drained loading is the most important task, as it is the modulus that is sufficient in designing of most structures. In Russian surveys in estimation of deformability of soils CPT is applied in evaluation of "drained" modulus of deformation, as the necessity in determination of both the modulus of elasticity and shear modulus hardly ever occurs. Nevertheless, in non-Russian practice CPT is sometimes used for evaluation of both the modulus of elasticity and shear modulus, and in availability of piezocones the coefficients of consolidation are determined as well. The issue will be considered further.

Empirical dependencies combining CPT results with the modulus of deformation are diversified. As it was mentioned in section 2.4.1, the number of the dependencies is calculated by dozens. Each empirical dependency appears to be rather reliable only in definite range of soil conditions, though generally accepted criteria of delimitation of such ranges have not been made yet. In Russia and CIS the evident preference is given to genetic show (e.g. quaternary, alluvial, fluvioglacial, eluvial, Jurassic, etc.). Non-Russian specialists do not usually deny the approach, but they tend to consider more obvious indications, e.g. territorial location of soils (clays of North Sea, coastal Norwegian clays, etc.), composition of soil concentrating on its specific character (pulverescent sandy loams, sensitive clays, silty sands, etc.). The greatest attention is paid to the history of consolidation of soils, i.e. overburden stresses in the process of geological history (normally consolidated soils, overconsolidated soils, etc.). Both granulometric and mineralogical makeup (quartz, feldspar, etc.) is often used as the criterion for sands.

The reason of the variety of the empirical formulae "$E \sim q_c$" is in insufficient definition of the modulus of deformation itself which is different in different tests. For example, moduli of deformation obtained in testing in odometer (compressive device) prove to be several times less than those obtained in in-situ tests by a stamp. Here, the parameters of the testing equipment applied can also effect on the results obtained (e.g. stamp dimensions). In Russian practice reference evaluation of the modulus of deformation of soil is obtained by the stamp of the area of 5000 cm^2. Thus, the results of other tests are corrected; in the first place - laboratory tests in the odometer. For structures of reduced level of reliability (the third level of reliability) the National Norms of Russia allow correction of ""compressive" moduli by table coefficients ignoring "stamp" tests to be performed SNiP 11-105-97. It is obvious that the empirical formulae based on comparison of cone resistances q_c with moduli of deformation obtained by various methods, may significantly differ.

Compressive tests in odometer are believed to be the most popular method of evaluation of the modulus of deformation in non-Russian and Russian practice. These methods excel the other ones in their simplicity. Adjustment of "compressive" moduli being compulsory in Russia in accordance with SP 11-105-97 is not compulsory in non-Russian practice and is not always performed. For this reason, "compressive" moduli are paid great attention worldwide; a lot of work has been done in revealing the empirical dependencies between these moduli and cone resistances q_c. The dependencies obtained are usually given as (Lunne et al. 2004)

$$M = \alpha_m q_c,$$ (3.8)

where M = "compressive" (odometer) modulus of deformation, α_m = conversion factor.

In the National Normative Documents SNiP 11-105-97 the moduli of deformation evaluated by the stamp and the compressive device (odometer) are usually equally marked as E (in publications they are sometimes marked as E_{st} & E_{comp}). Due to the fact that all the discussed investigations on evaluation of the compressive modulus refer to non-Russian publications only, the symbols of primary sources are preserved in the chapter, i.e. M – compressive modulus, E – stamp one.

Figure 3.17 illustrates the results of research after G. Richeri et al. given in the book of Yu.G. Trofimenkov (1995) as a diagram of dissipation. G. Richeri et al. compared the compressive moduli M of clayey soils with cone resistances q_c.

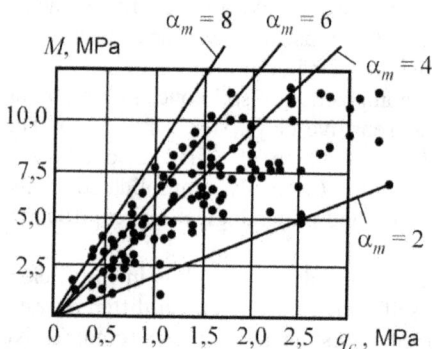

Fig. 3.17 Results of comparison of compressive moduli of deformation M of clayey soils with cone resistances q_c (after G. Richeri et al.): $\alpha_m = 2, 4, 6, 8$ – lines corresponding to the coefficients α_m in the formula (3.8)

Table 3.18 gives the values of α_m recommended in the books of G. Sanglerat, J.K. Mitchell and W.S. Gardner (see Lunne et al. 2004). The distinction between the classifications of soils given by the authors and the Russian classification slightly complicates the estimation of the given values, thus, types of soil (in the third column) are duplicated by the original names.

As one can see from table 3.18, in most cases "compressive" modulus of deformation of clayey soil equals to twofold-sixfold value of cone resistance q_c. Occurrence of organic admixtures in the soils is of great importance: in one and the same values of q_c modulus of deformation of peat soils may be two-three times lower than that of non-peat ones.

149

Table 3.18 Values of the coefficient α_m in the formula (3.8) after G. Sanglera, J.K. Mitchell and W.S. Gardner

Cone resistance q_c (MPa)	α_m	Type of soil
< 0.7	3...8	Clays of low plasticity (clay of low
0.7...2.0	2...5	plasticity)
> 2.0	1...2.5	
> 2	3...6	Pulverscent sandy loams and loams of
< 2	1...3	low plasticity (silts of low plasticity)
< 2	2...6	Pulverescent clayey soils of high
		plasticity (highly plastic silts and clays)
< 1.2	2...8	Pulverescent sandy loams and peat
		loams (organic silts)
< 0.7		Peat and peat clay (peat and organic clay)
water content: 50...100 %	1.5...4	
– » – 100...200 %	1...1.5	
– » – > 200 %	0.4...1	

Yu. G. Trofimenkov (1995) gives data on α_m of some other authors as well. In order of magnitudes they are not considerably different from α_m in table 3.18. Some authors reveal the connection between α_m and plasticity index I_p (slightly lower values of α_m correspond to higher indexes of plasticity), there are the results denying connections between α_m & q_c. Nevertheless, in authors mentioned in (Trofimenkov 1995) the values of α_m are not significantly different being generally in the limits of 3...7.

When applying the piezocones, instead of q_c the corrected value q_t is used (Lunne T. et al. 2004):

$$M = \alpha_i (q_t - \sigma_{vo}), \tag{3.9}$$

where:

$M =$ "compressive" modulus of deformation, i.e. the same as in the formula (3.8);

q_t & $\sigma_{vo} =$ the same as in the formula (3.3), i.e. the corrected value of cone resistance and overburden stress;

$\alpha_i =$ conversion factor taken for overconsolidated soils in the limits $\alpha_i = 5...15$, for normally consolidated ones - $\alpha_i = 4...8$.

The discussed books of non-Russian authors contain very few information on evaluation of the modulus of deformation of sands. Perhaps, this is due to the lack of experimental data, as selection of monoliths from sandy layers is no easy task requiring application of costly apparatuses.

In general, when analyzing the books mentioned earlier, the authors (Lunne et al. 2004) point out that evaluation of the "drained" moduli of deformation after CPT results which correspond to "undrained" loading of soil is quite a reliable procedure.

It is acceptable only in availability of local empirical dependencies as "$M \sim q_c$" and collaboration of highly skilled personnel as well.

During last decades in Russian practice the attention has been paid to study of connection between CPT results and the modulus of deformation corresponding to "stamp" tests rather than "compressive" ones. Figure 3.18 illustrates the results of comparison of "stamp" moduli of deformation E and cone resistances q_c, obtained in Fundamentproject in 1960s...1070s (Trofimenkov & Vorobkov 1981).

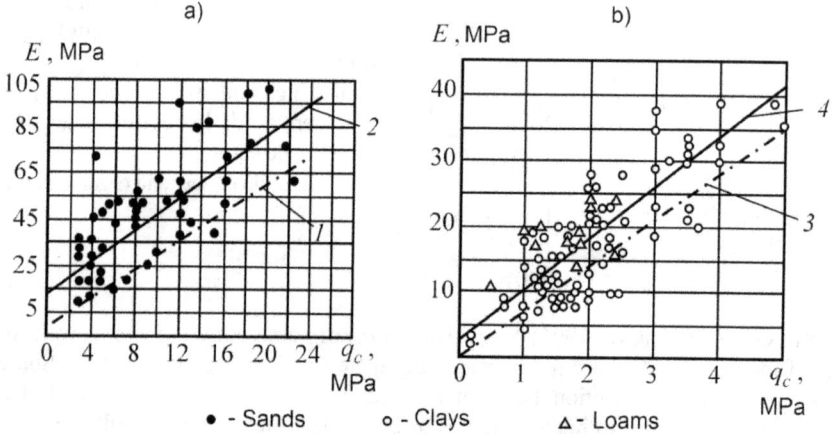

● - Sands   ○ - Clays   ∆ - Loams

Fig. 3.18 Schematics of comparisons between "stamp" moduli of deformation E and cone resistances q_c after "Fundamentproject" data (Trofimenkov & Vorobkov 1981):

a – research in sands, b – research in clayey soils;

approximating lines: *for sands* 1 – line $E = 3q_c$, 2 – line $E = 3.4q_c + 13$; *for clayey soils* 3 – line $E = 7q_c$, 4 – line $E = 7.8q_c + 2$ (E & q_c in MPa in given empirical formulae)

The data given in fig. 3.18a refer to fine, moist, medium dense and dense quaternary alluvial sands. The modulus of deformation was evaluated by stamp testing of the area of 5000 cm² in the depths from 2 to 15 m (in pits and shafts). The data in fig. 3.18b refer to quaternary clayey soils; the modulus of deformation was evaluated by the same stamp. *Type I cone penetrometers* were applied in both cases.

Figure 3.19 illustrates the diagrams of dissipation obtained for moraine loams and Jurassic clays on the territory of Moscow as an example of results of later research on connection "$E \sim q_c$" with regard to *type II cone penetrometers* (Ziangirov & Kashirsky 2006). The modulus of deformation was evaluated by stamp soil test of the area of 600 cm² in holes.

As one can see from fig. 3.19, dependencies "$E \sim q_c$" obtained in soils of different genesis (quaternary moraine and Jurassic) are not equal, though the difference is not quite great. In order to compare them the line $E = 7q_c$, (position 2) is drawn on the diagram of dissipation; the line is often used in estimation of modulus of deformation of clayey soils. It is worse in characterizing the link "$E \sim q_c$" then the approximated lines for moraine loams $E = 6.4q_c + 7$ (position 1) and for Jurassic clays $E = 8q_c$ (position 3) as well.

On both diagrams of dissipation there are "springing out" points deviating both to understating and overstating of E. Perhaps, they appear due to methodological deficiency in experiments (effect of heterogeneity of soil, failures in equipment operation, etc. – see section 3.2.1), although one should be careful with such assumptions and allow rejection of results only after the testing conditions have been examined. One should keep in mind that in such experiments the random factors are capable to increase divergences between the parameters compared and decrease them as well.

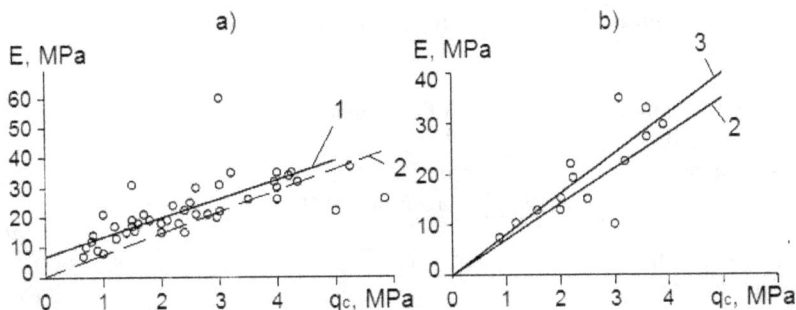

Fig.3.19 Schematics of comparisons between "stamp" moduli of deformation E and cone resistances q_c after R.S. Ziangirov & V.I. Kashirsky (2006):

a – moraine loams, b – Jurassic clays;
1 – line $E = 6.4q_c + 7$; 2 – line $E = 7q_c$; 3 – line $E = 8q_c + 2$ (E & q_c in MPa in given empirical formulae)

The article (Ziangirov & Kashirsky 2006) generalizes the research connected with the study of connection between the "stamp" moduli of deformation and cone resistances deposited on the territory of Moscow. More than 20 empirical formulae contained in Moscow Normative Documents published in different papers and monographs and obtained by the authors themselves are given. For example, the following formulae (E & q_c in MPa) are related to the latter:

- Moraine loams: $E = 6.4q_c + 7$.
- Jurassic sandy loams in native bedding: $E = 3.4q_c + 2$.
- Water-saturated sandy loam, pulverescent sand in the depth up to 5 m ($q_c = 0.5...5$ MPa): $E = 1.1q_c + 3$.
- Sandy loams and light solid loams of aeration zone ($q_c = 1.0...5$ MPa): $E = 3.1 q_c + 8$.
- Coarse sands, dense sands of medium fineness ($q_c = 4.0...20$ MPa): $E = 16q_c + 1.4$.

152

- Loose sands of medium fineness ($q_c = 2.0...10$ MPa): $E = 1.7q_c + 5$.
- Fine sands, pulverescent sands- dense and medium dense ($q_c = 1.5...2.5$ MPa): $E = 1.4q_c + 13$.
- Jurassic clays: $E = 8q_c$.
  In all cases that are not defined, the soils are of quaternary age.

At present, the whole of dependencies "$E \sim q_c$" given in the Normative Documents of Russia and CIS are anchored to the "stamp" moduli of deformation.

Tables 3.19 & 3.20 present the dependencies admitted in the Normative Document of Russia – SP11-105-97 (SP 47.13330.2012). As in estimation of soil strength, the dependencies $E \sim q_c$ given in this Document are intended for quaternary soils with less than 10 % content of organic substances. SP 11-105-97 does not give any values of modulus of deformation for clays and loams of glacial origin.

Table 3.19 Evaluation of the modulus of deformation E by cone resistance q_c after SP 11-105-97 (SP 47.13330.2012)

Names of sands	Normative modulus of deformation of sandy soils E in q_c (MPa)									
	2	4	6	8	10	12	14	16	18	20
The whole of genetic types apart from alluvial and fluvioglacial ones	6	12	18	24	30	36	42	48	54	60
Alluvial and fluvioglacial	17	20	22	25	28	30	33	36	38	41

Table 3.20 Evaluation of the modulus of deformation E by cone resistance q_c of clays and loams after SP 11-105-97 (SP 47.13330.2012)

Names of clays and loams	Normative modulus of deformation of clays and loams E in q_c (MPa)						
	0.5	1	2	3	4	5	6
The whole of genetic types apart from soils of glacial complex	3.5	7	14	21	28	35	42

It could be seen from tables 3.19 & 3.20 that the admitted dependencies "$E \sim q_c$" correspond to the empirical formulae: for clays and loams $E = 7q_c$, for sands (apart from alluvial and fluvioglacial ones) $E = 3q_c$, for alluvial and fluvioglacial sands another link is admitted.

Table 3.21 shows the analogical dependencies admitted in Moscow Urban Building Codes MGSN 2.07-01 embracing wider range of conditions.

Table 3.21 Formulae to evaluate the modulus of deformation by cone resistance q_c after MGSN 2.07-01

Origin and age of soils	Dependence of E (MPa) on q_c (MPa)
Sands:	
Modern alluvial (a-Q_4) and lacustrine - marsh (l_1h - Q_4)	$E = 3 \, q_c$
Ancient alluvial (a-Q_3), fluvioglacial (f-Q_2) and inner moraine	$E = 2.5 \, q_c + 10$
Loams and clays:	
Modern alluvial (a - Q_4) and lacustrine - marsh (l_1h - Q_4)	$E = 7 \, q_c$
Integumentary (Pr-Q_{2-3}), lacustrine – marsh (l_1h - Q_3) and lacustrine-glacial (lg-Q_2)	$E = 7.8 \, q_c + 2$
Moraine (g-Q_2)	$E = 8 \, q_c + 7.5$
Fluvioglacial (f-Q_2)	$E = 5.4 \, q_c + 7.4$

Table 3.22 shows the dependencies "$E \sim q_c$" for sands admitted in the Normative Document of Belarus – the Manual P2-2000 to the Building Codes of Belarus – SNB 5.01-99.

Table 3.22 Evaluation of the modulus of deformation E by cone resistance q_c of sands after the Manual P2-2000 to the Building Codes of Belarus – SNB 5.01-99

Sandy soils of various genesis	Values of the modulus of deformation E (MPa) in q_c (MPa)								
	1	2	4	6	8	10	12	15	20
Coarse and medium gravel sands regardless of water content	10	15	21	25	32	38	45	50	60
Fine sands regardless of water content	8	12	18	22	26	30	36	42	50
Water-unsaturated pulverescent sands	7	10	14	18	21	25	30	35	40
Water-saturated pulverescent sands	6	8	10	14	18	21	25	30	35

In values of q_c exceeding those given in table 3.22, the Manual P2-2000 recommends to take the values of E in maximum tabular values of q_c, (i.e. ignoring extrapolation). According to the Manual in clayey soils the modulus of deformation E is recommended to be evaluated by the formula

$$E = \frac{3.14\alpha(1+v)(3-4v)}{16(1-v)} q_c,$$

(3. 10)

where:

α = empirical coefficient equal to the following:
- for moraine sandy loams $\alpha = 8$,
- for moraine loams $\alpha = 8.5$,

- for other genetic types of soils: sandy loams $\alpha = 8.8$, loams $\alpha = 9.5$, clays $\alpha = 11$;

v = coefficient of lateral deformation (Poisson ratio) determined on triaxial compression devices in laboratory conditions or approximately equal to the following:
  - for sands $v = 0.30$,
  - for sandy loams $v = 0.35$,
  - for loams $v = 0.40$,
  - for clays $v = 0.45$,
  - for clays possessing up to 25 % of organic substances $v = 0.50$.

It is allowed to apply the formula (3.10) for sands as well in case $q_c = 3...10$ MPa, here, $\alpha = 7$.

The formula (3.10) is an improved variant of V.I. Ferronsky's (1969) solution (see section 2.4.2). The authors of the Manual introduced the elevating empirical coefficient α in V.I. Ferronsky's formula; its value was in the limits of 7...11. If to do a simple calculation by the formula (3.10) using the values of α and v, recommended by the Manual (i.e. given earlier, in decoding), then the formula (3.10) will be the following for "non-moraine" soils (not of glacial origin):
  - for clays $E = 6.82q_c$, i.e. $E \approx 7q_c$;
  - for sandy loams $E = 5.73q_c$;
  - for sands $E = 4.59q_c$.

Thus, in applying the corrected (empirical) coefficients α, admitted in the Belorussian Normative Document, the results obtained by the formula of V.I.Ferronsky are the same as those of other analogous empirical formulae.

In the European Pre-standard ENV-1997-3 the table for evaluation of drained Young's modulus of gravels, sands and silty soils by CPT data is given (table 3.23).

Table 3.23 Evaluation of drained Young's modulus of gravels, sands and silty soils by CPT data after ENV-1997-3

Relative density	Cone resistance q_c (MPa)	Modulus of deformation E (MPa)
Very low	0...2.5	< 10
Low	2.5...5	10...20
Medium	5...10	20...30
High	10...20	30...60
Very high	> 20	60...90

The given values of the modulus of deformation E are related to the same sands as in table 3.17. They correspond to deformations of foundation which have been under the load for 10 years, i.e. anchored to observations after the settings of real structures rather than soil tests. It is necessary to point out that the calculation of the moduli of deformation has been done with some volitional assumptions on intensity of dissipation of vertical stresses in the foundation (2:1). It is supposed that in sandy loams the moduli of deformation may be 50 % lower than those given in table 3.23,

155

and in gravels – 50 % higher than the tabular ones. Here, if the load acting on the foundation exceeds 2/3 from the limit strength, moduli of deformation are recommended to be 50 % reduced.

Besides the given table, ENV-1997-3 gives another table for a wider range of soils; the table shows the connection between the modulus of deformation E and cone resistance q_c as $E = \alpha q_c$. The table presents the values of coefficient α (see table 3.24). Table 3.24 is the same as table 3.18 supplemented with two lines.

Table 3.24 Evaluation of the modulus of deformation of clayey soils by CPT data after ENV-1997-3 (Eurocod 7)

Type of soil	Cone resistance q_c (MPa)	Value of α
Low-plasticity clay	< 0.7	3.0...8.0
	0.7...2	2.0...5.0
	> 2	1...2.5
Low-plasticity silt	< 2.0	3...6
	≥ 2.0	1...2
Very plastic clay	⎱ < 2.0	2...6
Very plastic silt	⎰ ≥ 2.0	1...2
Very organic silt	< 1.2	2...8
Peat and very organic clay	$q_c < 0.7 MPa$ ⎰ w= 50...100%	1.5...4
	w= 100...200%	1...1.5
	w > 300%	< 0.4
Chalks	2.0...3.0	2...4
	> 3.0	1.5...3.0
Sands	< 5.0	2
	> 10	1.5

When comparing the tables of ENV-1997-3 with the data suggested by the Russian authors, one need to underline that there are not any significant oppositions between them. Nevertheless, the comparison proves that one fails to rely upon the reliable determination of the modulus of deformation just only by the value of q_c regardless of lithological property of soil.

In general, when analyzing the given formulae and the tables as well, it is important to underline the following:

• The empirical dependencies linking the modulus of deformation of soil and CPT results greatly vary and are dependent upon lithological type of soil, its genesis and other factors as well, reflecting local specific character of engineering-geological conditions. For these reasons, the most reliable information obtained is ensured when applying CPT in combination with the traditional methods of soil study (drilling, laboratory and in-situ tests, etc.). The amount of work by these traditional techniques may be considerably reduced.

• In one and the same type of soil (clays, sands, etc.) the greatest differences are observed between the glacial deposits and other types of quaternary soils.

Although the obtained empirical data are not sufficient, one can notice that in moraine soils higher values of strengthening properties (φ, c) correspond to one and the same value of q_c, and slightly lower ones – to deformation (E). For example, it could be seen from comparison between table 3.12 and calculations by the formula (3.10). Thus, according to table 3.12 in $q_c = 1$ MPa, in moraine sandy loams $\varphi = 26°$, $c = 23$ MPa and according to the formula (3.10) $E = 5.2$ MPa; in pulverescent clayey "non-glacial", "non-loessial" sandy loams (table 3.12) $\varphi = 18°$, $c = 11$ MPa, $E = 5.7$ MPa. By analogy, for moraine loams – $\varphi = 25°$, $c = 30$ MPa, $E = 5.4$ MPa; for "non-glacial", "non-loessial" loams $\varphi = 16°$, $c = 16$ MPa, $E = 6.1$ MPa, correspondingly, etc. All these testify to benefit of mixed "elastoplastic" model of soil where both strengthening and deformation properties are determining factors.

- When applying CPT in virgin regions it is important to include comparison of CPT results with traditional soil tests in investigations; this will allow estimating the applicability of known dependences "$E \sim q_c$" for local conditions.

Correction techniques to evaluate the parameters obtained by CPT data are discussed in chapter 6.

Evaluation of the modulus of deformation E_u in undrained loading is hardly ever performed in Russian practice; CPT is not usually applied here. It is more popular among non-Russian specialists. T. Lunne et al. (2004) generalize the most popular publications on the issue. The solutions under analysis are based on correlation of strengthening and deformation properties of soil. Undrained Young's modulus E_u is evaluated by the following formula:

$$E_u = n s_u , \qquad (3.11)$$

where:

$s_u =$ the same as in the formula (3.2), i.e. maximum shear strength of soil

$n =$ parameter depending upon the stage of soil deformation, overconsolidation ratio, sensitivity (thixotropy) and other factors as well.

The technique of evaluation of E_u involves the following stages:

- Evaluation of undrained shear strength of soil s_u by the value of its cone resistance q_c, see section 3.2.3 and formula (3.2).
- Selection of parameter n by means of corresponding tables or graphs.
- Evaluation of E_u by the formula (3.11).

Owing to the fact that in any deformations soil behaves as nonlinearly deformed medium, the modulus of elasticity is anchored to concrete stage of soil deformations, characterized by the parameter τ/s_u (τ – acting shearing stresses in the soil). According to the research of C.C. Ladd et al. (1977) the values of n in the formula (3.11) are in the wide limits ($n = 25...1500$ after the authors). In alteration of τ/s_u from 0.2 to 0.8 the values of n in normally consolidated soils are as follows:

Portsmouth, loams $I_p{=}15$, $w_L{=}35$ %, $\tau / s_u = 0.2 \rightarrow n = 1500$,

	The same	$\tau / s_u = 0.8 \rightarrow n = 550,$
Boston, clays	I_p=22, w_L=41 %,	$\tau / s_u = 0.2 \rightarrow n = 1100,$
	The same	$\tau / s_u = 0.8 \rightarrow n = 270,$
Bangkok, clays	I_p=41, w_L=65 %,	$\tau / s_u = 0.2 \rightarrow n = 1000,$
	The same	$\tau / s_u = 0.8 \rightarrow n = 180,$
Maine, clays	I_p=38, w_L=65 %,	$\tau / s_u = 0.2 \rightarrow n = 700,$
	The same	$\tau / s_u = 0.8 \rightarrow n = 140,$
Tailor River, peat	w =500 %,	$\tau / s_u = 0.2 \rightarrow n = 110,$
	The same	$\tau / s_u = 0.8 \rightarrow n = 25.$

For overconsolidated clays the value of n 2…4 times reduced with the increase of overconsolidation ratio from 2 to 8 (in *OCR* less than 2, the value of n was insignificantly reduced, and for loams in *OCR* to be 1.5…10 % increase of n was observed).

Evaluation of shear modulus G_o in small deformations is a task which may be resolved by CPT. Most authors identify this modulus with *a dynamic shear modulus*. Non-Russian specialists pay attention to nonlinearity of shear deformations: in the process of loading shear modulus decreases; it is considered to be stable only in deformations less than 10^{-3} % (Lunne et al. 2004). It is the *initial* shear modulus that is evaluated; it is marked as G_o. P.W. Mayne and J.G. Rix (1993) suggested the following empirical formulae for evaluation of G_o

$$G_o = 99.5 \cdot (p_a)^{0,305} \frac{(q_t)^{0,695}}{(e_o)^{1,130}}, \qquad (3.12)$$

where:

p_a = atmospheric pressure in the same units as G_o (0.0981 *MPa*);

e_o = initial porosity factor of soil;

q_t = the same as in the formula (3.3), i.e. corrected value of cone resistance.

In order to evaluate G_o by the formula (3.12) it is necessary to apply piezocones, since CPT results as q_t, depending upon pore pressure are required, see formula (3.3).

The technique of evaluation of shear modulus G_o by estimation results of shear waves velocity distribution in the soil is developed; for this seismic penetrometers are applied (see section 1.3.4 and fig. 1.12) (Lunne et al. 2004). The calculation is done by the formula:

$$G_o = \rho \cdot V_s^2, \qquad (3.13)$$

where:

$\rho = \gamma/g$ = soil density

γ = specific weight of soil

g = gravity acceleration (9.81 m/s²)

V_s = shear wave velocity.

On these considerations P.K. Robertson et al. (1995) developed an identification chart of soils allowing *initial* shear modulus to be evaluated (fig. 3.20).

"Normalized cone resistance" Q_t as well as the type of soil determined by data of drilling or the charts presented in figures 3.2 or 3.3 are the initial data in application of the chart in figure 3.20.

"Normalized cone resistance" Q_t is a relative value (non-dimensional) evaluated by the formula

$$Q_t = \frac{q_t - \sigma_{vo}}{\sigma'_{vo}} , \qquad (3.14)$$

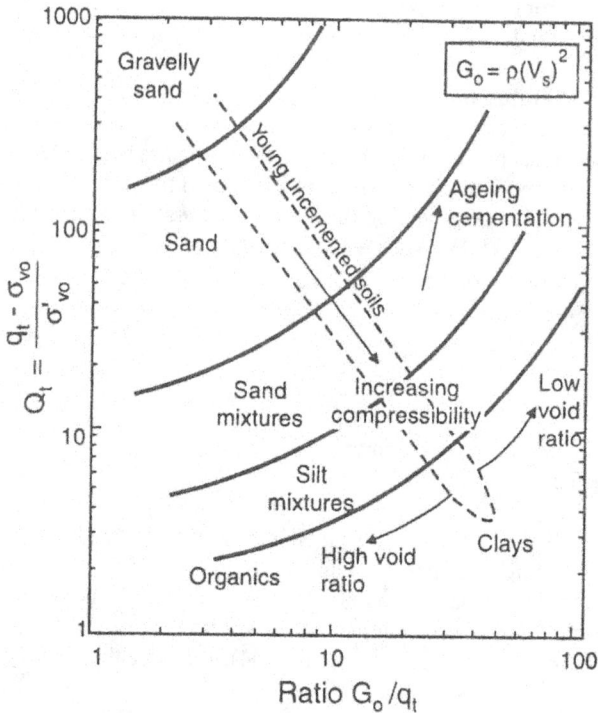

Fig.3.20 Chart after P.K. Robertson et al. (1995) to evaluate the initial shear modulus

where:

$q_t =$ the same as in the formula (3.3), i.e. corrected value of cone resistance;

$\sigma_{vo} =$ overburden stress in the soil (stress due to net weight) expressed through total vertical stress;

$\sigma'_{vo} =$ the same expressed through effective vertical stress.

159

With regard to sands, simpler dependences are also applied in non-Russian practice. G. Baldi et al. (1988) revealed rather close link between G_o & q_c and stated that the ratio G_o/q_c decreases with the increase of density index I_D

$$I_D = \frac{e_{max} - e}{e_{max} - e_{min}} \ [\times 100\%],$$ (3.15)

where e, e_{max} and e_{min} = factual, maximal and minimal values of porosity factor.
They suggest the following dependency for pulverescent quartz sands:

I_D, %	55	65	76
G_o/q_c	8.7	7.1	5.1

It is important to know the initial shear modulus G_o (as well as the modulus of elasticity mentioned earlier E_u) in estimation of dynamic effects on the soil (vibrations, impact loads, etc.). The initial shear modulus G_o should not be confused with shear modulus G in significant deformations in drained loading

$$G = \frac{E}{2(1 + v)},$$ (3.16)

where E = modulus of deformation (drained), v = coefficient of lateral deformation (Poisson ratio).

Parameter G may sometimes be lower than G_0 by an order of magnitude; this can be seen from the chart in fig. 3.20: the ratio G_o/q_t may attain 100, while G/q_c does not usually exceed 10.

3.2.5 Determination of other soil parameters

CPT can be successfully applied for estimation of physical properties of soil – density of sand composition, consistency of clayey soils (flow index) alongside with evaluation of mechanical properties φ, c & E. Successful attempts are being taken to evaluate fineness (granulometric composition) of sands approximately by CPT data. Some special properties are also estimated by CPT data: overconsolidation ratio, horizontal earth pressure at rest, coefficient of consolidation, coefficient of filtration, thixotropy (sensitivity), etc. In non-Russian and Russian practice CPT is used for studying specific soils (permafrost, collapsible, thixotropic, etc.). Application of CPT in specific conditions (in specific soils, etc.) is discussed in chapter 5; here, the data on "ordinary" soils are given, i.e. applicable in most engineering-geological conditions.

As opposed to evaluation of the principal mechanical properties φ, c & E, estimation of physical properties of soil by CPT has always been based on purely empirical dependencies; attempts have not been taken to interpret them theoretically. It was thought that higher density of soil and lower water content (in clayey soils) corresponded to higher strength, since mechanical properties of soil depended upon the amount and interaction force of contacting particles.

Density of sands is characterized by two parameters in accordance with the International Standard GOST 25100-95:

- Density of composition: sands are divided into dense, medium dense and loose (dependent upon porosity factor and fineness of particles).
- Density index I_D (see formula (3.15)): sands are divided into overconsolidated ($I_D = 0.66...1.0$), normally consolidated ($I_D = 0.33...0.66$) and underconsolidated ($I_D = 0...0.33$).

In the National Norms SP 11-105-97 (SP 47.13330.2012) only determination of density of sand composition by CPT data is considered. The dependency given in SP (see table 3.25) presupposes fineness of sand to be known, the depth of the layer being ignored.

Table 3.25 Evaluation of density of sand composition by cone resistance q_c after SP 11-105-97 (SP 47.13330.2012)

Sands	Density of composition in q_c (MPa)		
	Dense	Medium dense	Loose
Coarse and medium sized regardless of water content	> 15	5 ... 15	< 5
Fine regardless of water content	> 12	4 ... 12	< 4
Pulverescent water-unsaturated	> 10	3 ... 10	< 3
Pulverescent water-saturated	> 7	2 ... 7	< 2

On the contrary, in Moscow Urban Building Codes MGSN 2.07-01, density of composition is evaluated with regard to the depth of the layer, fineness of sand being ignored (table 3.26).

Table 3.26 Evaluation of density of sand composition (coarse, medium-sized and fine regardless of water content) by cone resistance q_c after MGSN 2.07-01

Depth of penetration, m	Values of q_c (MPa) for sands		
	Dense	Medium dense	Loose
3 and less	> 7	2.5... 7	< 2.5
5	>10	3 ...10	< 3
10 and more	> 15	5 ... 15	< 5

When comparing tables 3.25 and 3.26 it is easy to notice that the accuracy of the tables is not high; there are not many differences in the given accuracy due to uncertainty in each table. For example, $q_c = 5$ MPa may correspond to the sand of any fineness mentioned in table 3.25, but according to both tables it is the sand of medium density. In the depth of more than 10 m both tables give the same results for coarse and medium sized sands. Table 3.26 appears to be more convenient, as it allows applying just the initial data obtained by CPT rig. Nevertheless, taking any responsible decision based on CPT, ignoring control drilling or former in-situ investigations is an inadmissible situation for a skilled professional.

When estimating density of sands *density index* I_D corresponding to the formula (3.15) is applied in non-Russian practice. Non-Russian specialists pay special attention to evaluation of this parameter by CPT data.

J.K. Mitchell & R.K. Katti (1981) suggest the table for approximate estimation of I_D (table 3.27). Their classification of sands by density does not agree with the national one, but here, it is not of principal importance.

Different authors suggest rather complicated formulae and graphs for more accurate estimation of I_D; here, stresses in the soil before penetration are required to be known. For normally consolidated sands it is effective vertical stress before penetration σ_{vo}' (overburden stress expressed through effective stresses). For overconsolidated sands mean (octahedral) effective stress σ_{mean}' is used but not vertical one.

Table 3.27 Approximate evaluation of density index I_D of sands by their cone resistance q_c (Mitchell & Katti 1981)

Sand	q_c (MPa)	I_D (%)
Very loose	≤ 5	≤ 15
Loose	5...10	15...35
Medium dense	10...15	35...65
Dense	15...20	65...85
Very dense	≥ 20	≥ 85

In general, these are the empirical formulae obtained by treatment of unique laboratory results in calibration cells of large size. The cone was penetrated in the sand placed in such cells; appearance, density and stresses in the sand were "controlled".

G. Baldi et al. (1981) obtained the formula for evaluation of density index I_D

$$I_D = \frac{1}{C_2} \cdot \ln \left| \frac{q_c}{C_o (\sigma')^{C_1}} \right|, \qquad (3.17)$$

where:

C_o, C_1, C_2 = constants of foundation (experimental coefficients);

σ' = effective stress in the soil before penetration (vertical or mean – see above), kPa;

q_c = cone resistance of sand, kPa.

M. Jamiolkowski et al. (1985) give another formula for normally consolidated sands obtained by R. Lancelotta by data acquisition after 144 experiments of different authors with five types of soil:

$$I_D = 66 \cdot ln\left(\frac{q_c}{\sqrt{\sigma'_{vo}}}\right) - 98, \qquad (3.18)$$

where σ_{vo}' = effective vertical stress in the soil before penetration expressed (as q_c) in t/m^2.

The authors point out that the accuracy of the formula is very high, although in practical application they recommend to decrease q_c introducing lowering coefficients.

Fineness of solid soil particles (generally, sands) is also possible to study by CPT. In accordance with the International Standard of CIS GOST 25100-95 sands are divided into gravel, coarse, medium sized, fine and pulverescent due to its granulometric composition. Besides, in Russian and particularly non-Russian practice soil is sometimes characterized by a quantitative parameter – mean diameter of its particles D_{50}.

In Russian practice CPT is applied mainly for qualitative delimitation of sands into coarse, fine, etc. according to the GOST 25100-95. R.S. Ziangirov & V.I. Kashirsky (2006) developed the table (see table 3.28) to estimate fineness of sands, types of gravel-pebble soils and types of sandy loams deposited on the territory of Moscow by CPT data. The authors apply both cone resistance q_c and friction ratio f_s/q_c as the initial data (see section 3.1.1). The tables are partially reflected in the chart in figure 3.4a. In all the lines except 8, soils are related to quaternary alluvial and fluvioglacial deposits; in line 8 the sands are chalky and marine. Higher values of q_c are typical for denser soils, lower ones – for looser soils. Higher values of f/q_c are typical for heterogeneous soils, lower ones – for more homogeneous soils with regard to granulometric composition.

Perhaps, the depth of penetration did not influence much, thus, the authors combined the results referred to the wide range of depths – 2…18 m.

In Russian and particularly non-Russian practice successful attempts are being taken in estimating the coarseness of sands (and partially clayey soils) quantitatively by *mean diameter of particles* D_{50} applying CPT. To solve the tasks non-Russian specialists usually apply CPT in combination with "Standard Penetration Test" by soil sampler   (SPT) being typical for dynamic CPT (especially in the USA).

It was the ratio between static and dynamic CPT results that was taken as the initial information, but not cone resistances q_c, i.e. the ratio q_c/N, where N is a number of impacts required for soil sampler to be penetrated in 30 cm. In here, the number of impacts N is corrected so that it corresponds to the case when the efficiency of energy consumption in driving in the soil sampler (marked as ER_r) is 60 % (the rest is related to deformations of a push rod during driving). If the applied equipment gives the efficiency of driving ER_r different from the standard one (i.e. $ER_r \neq 60\%$), the number of impacts N obtained during driving in the soil sampler is multiplied $ER_r/60$.

Table 3.28 Identification of non-cohesive soils and sandy loams by CPT data on the territory of Moscow (Ziangirov & Kashirsky 2006)

№	Cone resistance q_c (MPa)	Ratio f/q_c (%)	Types of soil in the depth of 2…18 m
Gravel-pebble soils			
1	25…35	0.5…0.6	Homogeneous
	18…25	0.6…1.2	With sandy filler
Sands			
2	16…25	0.5…1.0	Gravel
3	16…22	0.5…1.2	Coarse
4	10…20	0.6…2.0	Medium coarseness
	2.5	0.8	Medium coarseness, loose
5	3.5…8	1.0…1.18	Fine
	1.1	–	Fine loose (singular values)
6	2…3	2…3	Pulverescent
	2	3.5	Pulverescent loose
7	9	0.5	Medium coarseness saturated with fuels and lubricants
8	26…28	0.9…1.2	Pulverescent and fine, very dense in the depth of 25 m ($e = 0.4…0.45$)
Sandy loams			
9	1…2	2…3	Pulverescent, plastic
10	2…3	1.2	Sandy

Dimensionless value q_c/p_a, where p_a = atmospheric pressure in the same units as q_c is often used instead of q_c; this ensures independence of measurements upon admitted dimensions (kPa, MPa, etc.).

Figure 3.21 illustrates the results of research after F.H. Kulhawy & P.H. Mayne (1990). They generalized experimental data obtained by different authors who had studied the dependence of $(q_c/p_a)/N$ upon mean diameter of soil particles D_{50}. As one can see from fig. 3.21 $(q_c/p_a)/N$ increases with the increase of mean size of particles; this can be characterized by the empirical dependency

$$\frac{q_c}{p_a \cdot N} = 5{,}44 \cdot D_{50}^{0,26}, \tag{3.19}$$

where p_a = atmospheric pressure in the same units as q_c, N = corrected (multiplied $ER_r/60$) number of impacts necessary to penetrate the soil sampler.

In availability of the results of both CPT and SPT non-Russian specialists often apply the parameter $(q_c/p_a)/N$ in addition to the charts (e.g., those given in fig. 3.2 and 3.3) to specify the type of soil. P.K. Robertson et al. (1986) suggest the approximate values of $(q_c/p_a)/N$ to be used as the additional information to the chart in fig. 3.3 (table 3.29).

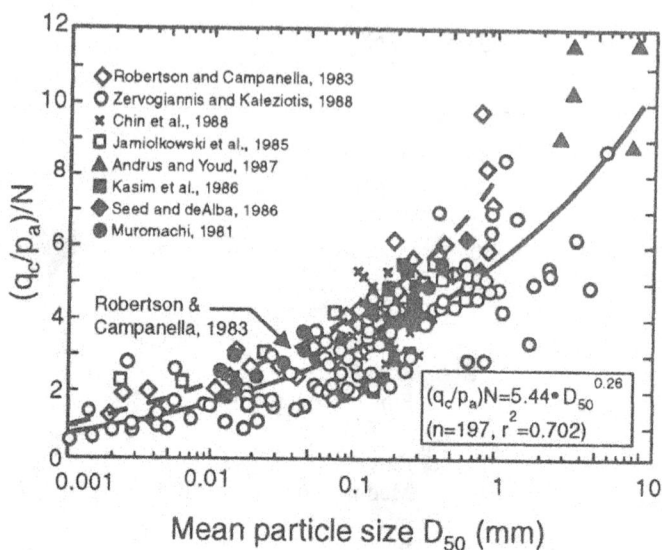

Fig. 3.21 Connection between mean diameter of soil particles and $(q_c/p_a)/N$ (Kulhawy & Mayne 1990)
where p_a – atmospheric pressure in the same units as q_c, N – corrected number of impacts in dynamic penetration test by soil sampler (SPT); different forms of "points" show the results of different authors

Table 3.29 Specification of soil types given in the chart (fig. 3.3) by the parameters $(q_c/p_a)/N$ (Robertson et al. 1986)

N of zone on the chart	Type of soil	Parameter $(q_c/p_a)/N$
1	Sensitive fine grained	2
2	Peaty clayey soils	1
3	Clays	1
4	Silty clay to clay	1.5
5	Clayey silt to silty clay	2
6	Sandy silt to clayey silt	2.5
7	Silty sand to sandy silt	3
8	Sand to silty sand	4
9	Sands	5
10	Gravel and coarse sands	6
11	Very dense finely dispersed	1
12	Cemented or overconsolidated sands	2

The Russian professional engineers R.S. Ziangirov and V.I. Kashirsky (2006) have a simpler solution to the issue on evaluation of mean diameter of soil particles restricting themselves by application of CPT only. This approach is of interest for a Russian specialist since in Russia SPT has not been widely applied.

Figure 3.22 illustrates the results of research (Ziangirov & Kashirsky 2006) of correlation link between mean diameter of soil particles D_{50} and friction ratio (friction index) $R_f = f_s/q_c$ for soil conditions of Moscow.

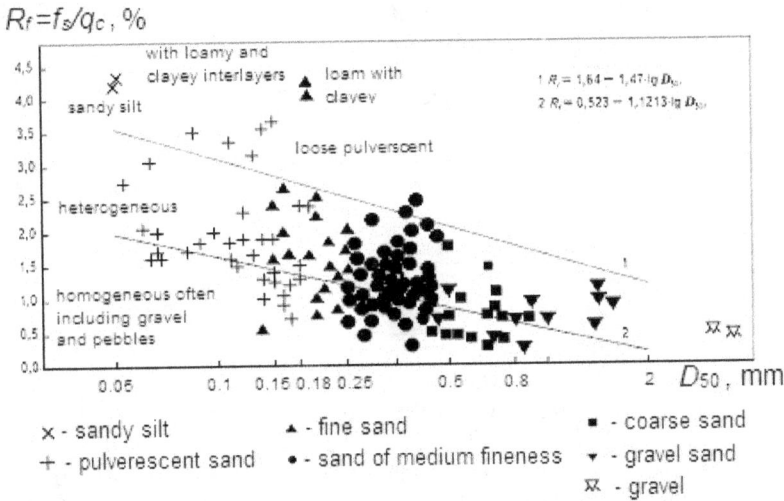

Fig. 3.22 Dependence of friction index $R_f = f_s/q_c$ on mean size of particles D_{50}

As one can notice from figure 3.22, the connection between D_{50} & f_s/q_c is not doubtful, although the accuracy of D_{50} evaluation by the parameter f_s/q_c seems to be rather low. For example, pulverescent sand may be observed in in the whole of ranges of f_s/q_c, i.e. from 0.7 to 4.5 %, but in typical diameters of particles (D_{50} = 0.06...0.2 mm) one can fail to find out any appropriate connection between D_{50} & f_s/q_c. Nevertheless, in availability of drilling data the obtained dependencies may be useful for identification of sands, i.e. their relation to one or the other classification group (coarse, fine, pulverescent, etc.).

Consistency of clayey soils. Determination of flow index I_L of clayey soils by CPT data is based on purely empirical dependencies as well as determination of density of sands analyzed earlier. At present, both indexes measured by the penetrometer – cone resistance q_c and unit sleeve friction resistance f_s are applied as the initial data. According to SP 11-105-97 when applying I type cone penetrometers, side friction resistance Q_s is translated for every Engineering-Geological Element (EGE) into unit side friction resistance f_s. Thus,

166

$$f_s = \frac{Q_i}{A_i} \, , \tag{3.20}$$

where Q_i = total side friction resistance related to i-engineering-geological element, A_i = lateral area of the cone related to this element.

If EGE is totally passed by the cone, Q_i & A_i are taken from the footing to the roof of this EGE.

Such an approach should be considered to be rather approximate, as cone resistance evaluated by the formula (3.20) is one and a half - two times lower than that measured by type II cone penetrometer on the friction sleeve. The issue is discussed in section 4.2 in detail.

As in all empirical dependencies linking CPT data and properties of soil, the connection "$I_L \sim q_c$" in different regions as well as in soils of different origins may vary. The National Normative Document of Russia gives the table to evaluate flow index I_L for quaternary clayey soils with the content of organic substances less than 10 % (table 3.30).

Table 3.30 Evaluation of flow index I_L of clayey soils by cone resistance q_c after SP 11-105-97 (SP 47.13330.2012)

q_c (MPa)	Flow index I_L of clayey soils in f_s (MPa)										
	0.02	0.04	0.06	0.08	0.10	0.12	0.15	0.20	0.30	0.40	$\geq$0.50
1	0.50	0.39	0.33	0.29	0.26	0.23	0.20	0.16	-	-	-
2	0.37	0.27	0.20	0.16	0.12	0.10	0.06	0.02	-0.05	-	-
3	0.22	0.16	0.12	0.09	0.07	0.05	0.03	0.01	-0.03	-0.06	-
5	0.09	0.04	0.01	0.00	-0.02	-0.03	-0.05	-0.07	-0.09	-0.11	-0.13
8	0.01	-0.02	-0.04	-0.06	-0.07	-0.08	-0.09	-0.11	-0.13	-0.14	-0.15
10	-	-0.05	-0.07	-0.08	-0.09	-0.10	-0.11	-0.13	-0.14	-0.16	-0.17
12	-	-	-0.09	-0.11	-0.11	-0.12	-0.13	-0.14	-0.16	-0.17	-0.18
15	-	-	-	-0.13	-0.14	-0.15	-0.16	-0.17	-0.18	-0.19	-0.20
20	-	-	-	-	-0.17	-0.18	-0.18	-0.19	-0.20	-0.20	-0.21

The same table is given in Belorussian Normative Document P2-2000 to the Building Codes of Belarus – SNB 5.01-99.

Despite rather high "status" of the table, the reliability of the data given, related to solid soils ($I_L < 0$) cause some doubts, as it was mentioned earlier. In quaternary clays and loams cone resistances q_c do not usually exceed 5...6 MPa. Higher resistances may be anticipated only in very dense solid sandy loams possessing low plasticity index I_p, but in such soils flow index reflects the mechanical properties poorly; thus, its evaluation is not of practical importance.

As an example of "regional" connection between CPT results and flow index it is desirable to present the table given in the Recommendations of NIIpromstroy (BashNIIstroy) for alluvial and dealluvial clayey soils in Bashkir Urals (table 3.31). As in table 3.13 cone resistances are given in brackets; they correspond to conditional equilibrium of the penetrometer ("relaxation-creep" CPT). Empty boxes correspond to combinations of q_c & f_s being non-typical for Bashkir Urals.

Table 3.31 Evaluation of flow index I_L of clayey soils by cone resistance q_c after the Recommendations of BashNIIstroy (Workbook 1991)

q_c (MPa)	Flow index I_L of clayey soils in f_s (MPa)					
	0.02 (0.015)	0.04 (0.030)	0.06 (0.040)	0.08 (0.050)	0.12 (0.08)	0.19 (0.13)
0.3 (0.2)	0.78	0.69	-	-	-	-
0.6 (0.4)	0.68	0.61	0.54	0.49	-	-
1.0 (0.7)	0.58	0.52	0.45	0.42	0.38	-
1.5 (1.0)	0.49	0.43	0.37	0.34	0.29	0.27
2.0 (1.4)	0.42	0.38	0.30	0.27	0.22	0.20
2.5 (1.7)	-	-	0.25	0.20	0.15	0.13
3.0 (2.0)	-	-	-	0.15	0.10	0.07
3.5 (2.3)	-	-	-	-	0.07	0.04
4.0 (2.7)	-	-	-	-	0.05	0.02

If to compare tables 3.30 and 3.31, it is easy to notice that in one and the same values of q_c & f_s flow index I_L in soils of Bashkir Urals is higher than on average in Russia, i.e. Bashkir clays and loams possess higher cone resistances, all other things being equal.

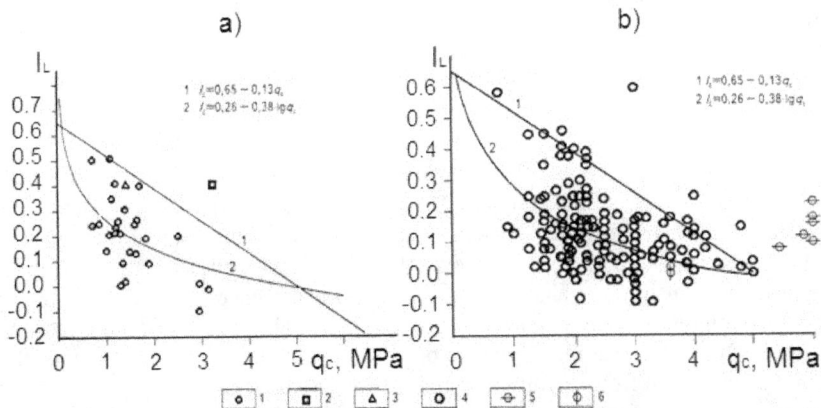

Fig. 3.23 Schematics of dependency between flow index I_L and cone resistance q_c in some soils of Moscow (Ziangirov & Kashirsky 2006):

a – moraine clayey soils: 1 – loams, 2 – sandy loam, 3 – clay; b – Jurassic deposits: 4 – Jurassic clays, 5 – phosphorite inclusions of fragmental rocks with sand and clay, 6 – solid clay in the depth of 20…30 m

In a number of cases the specialists working out the Regional Recommendations or Normative Documents apply simplified dependencies connecting flow index I_L with one parameter only – q_c ignoring f_s (until 1980s the approach was generally accepted). Figure 3.23 illustrates the dependencies "$I_L \sim q_c$" for some clayey soils typical for Moscow (quaternary moraine and Jurassic deposits) (Ziangirov & Kashirsky 2006).

When comparing the diagrams of dissipation (fig. 3.23) for moraine (*a*) and Jurassic deposits (*b*), one can fail to reveal any significant divergences. The identical empirical lines are drawn on both diagrams (1 & 2). The lines are nearly in the same position concerning the experimental "points" on both diagram - *a* and *b*. In both cases the correlation "$I_L \sim q_c$" is rather weak.

Overconsolidation ratio *OCR* is becoming generally accepted in Russia; it is the soil property that is being paid great attention to both world wide (Lunne 2004, Burns & Mayne 1998) and in Russia as well. Many non-Russian professional engineers consider *OCR* as one of the most important properties reflecting geological history of deposits, and in many cases they apply it as a systematic index dividing the deposits into overconsolidated and normally consolidated. *OCR* is understood to be a relationship of maximum overburden stress at one or the other stage of geological formation of soil (lithification) to the present overburden stress, i.e

$$OCR = \frac{\sigma_p'}{\sigma_{zg}} , \qquad (3.21)$$

where:

$\sigma_p' =$ maximum overburden stress during geological formation of soil (overconsolidation stress),

$\sigma_{zg} =$ present overburden stress.

Maximum stress σ_p' is usually determined by compressive tests with regard to the location of the transition point of the initial (more "gently sloping") sector of compressive curve towards the following "steeper" sector. The results of in-situ tests by rotary shear are used as well. However, the reliability of these evaluations gives rise to doubts in many specialists; these are due to occurrence of the gently sloping sector on compressive curve that can be caused by absolutely different reasons: repeated drying and/or wetting, chemical processes, etc. Some specialists believe that the point of transition from "gently sloping" sector of compressive curve to "steeper" one does not characterize the former pressures but characterizes "structural strength" of the soil under study. Over half a century ago the American specialist G.P. Chebotarev highlighted the conditionality of *OCR* evaluation warning the professional engineers against "excessive practical importance" of load rates of preliminary consolidation obtained in laboratory conditions (Chebotarev 1968). T. Lunne et al. (2004) pay attention to careful application of overconsolidation ratio with regard to soils possessing case-hardening bonds or foundations of prolonged operation. Nevertheless, the majority of professional engineers do follow the opinion on acceptability of OCR estimation methods as well as the usefulness of OCR application. Thus, great amount of work is being done on estimation of *OCR* by CPT.

In Moscow Urban Building Codes MGSN 2.07-01 the following formula to evaluate maximum overconsolidation stress of clayey soils is given:

$$\sigma_p' = \lambda q_c,$$ (3.22)

where:

σ_p' = maximum overconsolidation stress,
q_c = cone resistance,
λ = coefficient admitted to be equal to
  in $I_p = 10$ $\lambda = 0.45$
  in $I_p = 20$ $\lambda = 0.35$
  in $I_p = 30$ $\lambda = 0.30$.

If to substitute σ_p' in the formula (3.21), one can obtain the required coefficient, i.e. OCR.

As it was said earlier, evaluation of *OCR* is paid greater attention worldwide than in Russia; thus, a considerable part of regularities was obtained by non-Russian professional engineers, then. Piezocones are the engineering tools that are used for evaluation of *OCR* rather than "standard" cone penetrometers. When applying standard cones the mode of evaluation of *OCR* is analogous to that described earlier, although the quantitative parameters may vary. When applying piezocones, a lot of solutions of the problem have been developed. The empirical formulae linking *OCR* with different indexes obtained by piezocones have been developed as well.

In their general report at the 12th Congress on Soil Mechanics and Foundation Engineering T. Lunne et al. (1989) give empirical dependencies linking *OCR* with various parameters obtained by piezocones. The diagrams of dissipation after T. Lunne et al. are given in figure 3.24. These diagrams do not reduce the whole of diversity of approaches to evaluation of *OCR*; e.g. the authors (Lunne et al. 2004) give 12 parameters applied in solution of the problem by piezocones.

As one can see from fig. 3.24, the correlation between the compared parameters is rather close, particularly in comparison of *OCR* with normalized cone resistance and normalized variation value of pore pressure. This fact is undoubtedly of practical importance regardless of the situation whether the obtained "coefficient of overconsolidation" is connected with the former overburden stress or it characterizes some other structural features of soil.

In non-Russian practice the most popular mode of evaluation of *OCR* is based on application of the following dependency (see in Lunne et al. 2004):

$$OCR = k \cdot [(q_t - \sigma_{vo}) / \sigma_{vo}'],$$ (3.23)

where:

$(q_t - \sigma_{vo})/\sigma_{vo}'$ = "normalized" value of cone resistance:
  q_t = corrected value of cone resistance (see formula (3.3)),

σ_{vo} and $\sigma_{vo}' =$ total and effective vertical overburden stresses,

$k =$ coefficient varying in the limits 0.2...0.5 (mean value – 0.3), due to plasticity index I_p and other factors.

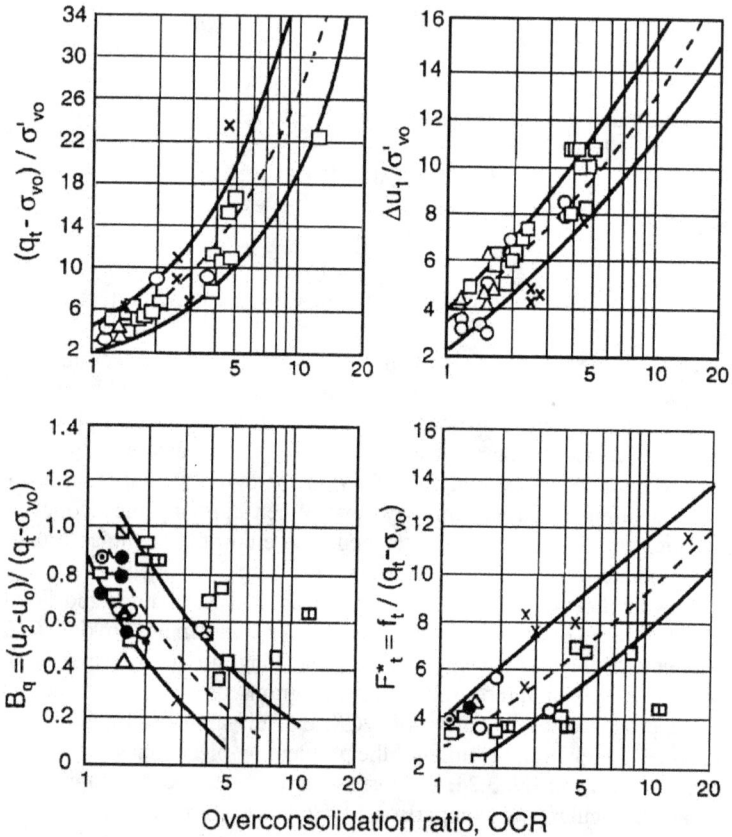

Overconsolidation ratio, OCR

Fig.3.24 Empirical dependencies linking OCR with various indexes evaluated by piezocones (Lunne et al. 1989)

Different marks show concrete terrain where the initial data were obtained. X-axes present OCR on the whole of diagrams of dissipation, Y-axes: $(q_t - \sigma_{vo})/\sigma_{vo}'$ – normalized cone resistance, where q_t – corrected cone resistance, see formula (3.2), $\sigma_{vo})$ & σ_{vo}' – total and effective vertical overburden stresses; $\Delta u_1/\sigma_{vo}'$ – normalized variation value of pore pressure, where Δu_1 – variation of pore pressure measured on the cone; $B_q = (u_2 - u_0)/(q_t - \sigma_{vo}')$ – pore pressure parameter, see formula (3.1); $F_t^* = f_t/(q_t - \sigma_{vo})$ – normalized friction ratio, where q_t and f_t – corrected (with regard to concrete piezocone) cone resistance and unit sleeve friction resistance

171

The highest values of k are recommended to be taken in heavy clays.

Despite the fact that in the analyzed mode piezocones are supposed to be applied, and the technique of MGSN 2.07-01aims at "standard" cone penetrometers, there are some similarities between the approaches. In both cases higher OCR correspond to higher values on cone resistance. The higher the plasticity index I_p is, the higher OCR will be. The principal difference is that in MGSN 2.07-01, as in all National Normative Documents of Russia directly measured cone resistances q_c are used, while in the non-Russian mode relative ("normalized") values corrected with regard to concrete overburden stress and measured pore pressure are applied. Perhaps, application of relative values is somehow beneficial, but one can fail to obtain any high accuracy in both cases. The authors (Lunne et al. 2004) pay the greatest attention to the necessity of considering the local experience and application of the regional dependencies connecting OCR with other indexes.

Horizontal earth pressure at rest (in absence of outer loads) can also be evaluated by CPT. The problem emerges when designing underground structures when it is important to evaluate the stress on walling. This problem is closely connected with that of evaluation of OCR. The same as for OCR initial data are used, i.e. the results obtained by type II cone penetrometers or piezocones, plasticity index I_p is taken into account, etc. In some cases evaluation of OCR is the first stage in evaluation of horizontal earth pressure at rest.

Most of research has been carried out by non-Russian specialists (see Lunne et al. 2004, Burns & Mayne 1998), while in Russia interest to this tendency of CPT application is not great, though urgent.

Various techniques of evaluation of horizontal earth pressure at rest have been developed; they are discussed in detail in T. Lunne et al. (2004). The accuracy of the techniques is not high; knowledge and skills of the professional engineers resolving similar problems are of principal importance.

Since the estimation of vertical overburden stress is not a problem, in order to evaluate horizontal earth pressure at rest, the coefficient of side pressure at rest K_o is determined; it is the relation between effective stresses acting in the soil – horizontal to vertical:

$$K_o = \frac{\sigma'_{ho}}{\sigma'_{vo}} , \qquad (3.24)$$

where σ'_{ho} & σ'_{vo} = horizontal and vertical stresses in the soil correspondingly.

Coefficient of side pressure at rest K_o corresponds to the situation when the surface of the structure taking soil pressure (outer side of underground structure, breast wall at covering side, etc.) is absolutely rigid, i.e. shifts neither to the direction of soil pressure nor to the opposite direction.

The further stage of evaluation of horizontal stresses in given K_o and vertical (overburden) stress does not need any particular explanations.

The developed techniques require piezocones to be used, but nevertheless there are techniques for standard cone penetrometers to evaluate K_o, for example, the technique after T. Masood & J.K. Mitchell (1993). The authors suggest unit sleeve friction resistance f_s expressed in relative form to be applied, i.e. as f_s/σ'_{vo}, where σ'_{vo} – effective vertical stress in the soil before penetration (overburden stress). The second determining parameter is the overconsolidation ratio being evaluated beforehand. Figure 3.25 shows the graph to evaluate K_o by T. Masood & J.K. Mitchell's technique (1993).

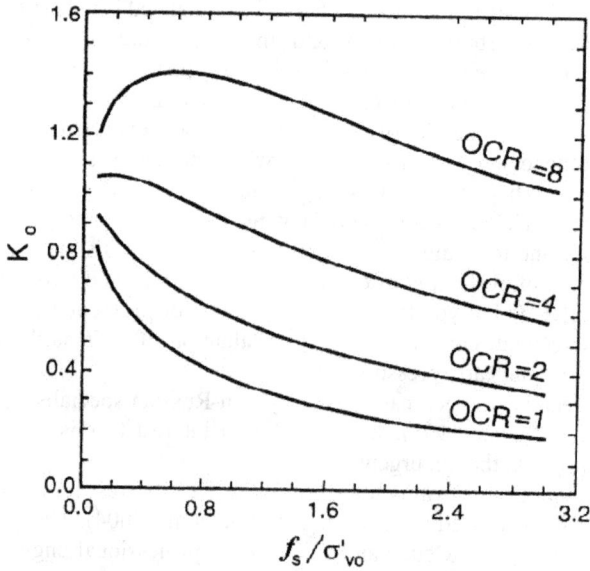

Fig. 3.25 Schematic illustration of relationship between the coefficient of side pressure at rest K_o, relative value of sleeve friction resistance f_s/σ_{vo} and OCR after T.Masood & J.K.Mitchell (1993)

As one can see from fig. 3.25, for normally consolidated soils, i.e. in $OCR = 1$, the most typical values of coefficient of side pressure at rest K_o are to be in the following limits: 0.3...0.7. This is due to the fact that in the depth up to 10...15 m sleeve friction resistances are as follows: $f_s = 0.03...0.1$ MPa. For overconsolidated soils K_o may be significantly higher. With the increase of soil strength values of K_o (in $OCR < 4$) decrease in most soils; in $OCR > 4$ dependence of K_o upon soil strength is of more complicated curvilinear character.

There are techniques where instead of f_s/σ_{vo} relative value of shear strength s_u/σ'_{vo}, is used; plasticity index I_p is taken into account as the additional factor. Here, s_u is evaluated by CPT data (see section 3.2.3). The nomograms are developed to evaluate K_o (see Lunne et al. 2004 for details).

In application of piezocones the empirical dependency after F.H. Kulhawy & P.W. Mayne (see in Lunne et al. 2004) is applied in non-Russian practice:

$$K_o = 0.1 \cdot \left(\frac{q_t - \sigma_{vo}}{\sigma'_{vo}} \right),$$ (3.25)

where the symbols are the same as in the formula (3.24).

The techniques based on assumption that distribution of pore pressure around the cone is the function of horizontal stress at rest σ_{ho} have been developed. The empirical dependencies between K_o and relative difference in pore pressure values measured on the cone and between the cone and the friction sleeve are applied as well. According to J.P. Sully and R.G. Campanella (see in Lunne et al. 2004) linear dependency with the following parameters happens to be acceptable for most soils (but not all!):

$$K_o = 0.11 \cdot \left(\frac{u_1 - u_2}{\sigma'_{vo}} \right) + 0.5,$$ (3.26)

where u_1 and u_2 = pore pressure values measured on the cone and between the cone and the friction sleeve; σ'_{vo} = effective vertical stress in the soil before penetration (overburden stress).

T. Lunne et al. (2004) consider that the techniques based on formulae (3.25) and (3.26) possess rather low accuracy and can be applied only for preliminary estimation.

For *sands* the applied techniques for evaluation of horizontal pressure at rest are usually simpler than for clayey soils. For normally consolidated sands the formula of J. Jaky (1944) has been applied for more than half a century:

$$K_o = 1 - sin \, \varphi,$$ (3.27)

where φ = angle of internal friction (in consolidated-drained shear).

Connection between the coefficient of side pressure at rest K_o and OCR being observed in many sands is widely applied worldwide. P.W. Mayne (1992) gives the formula for approximate estimation:

$$K_o = 0.35 \cdot OCR^{0.65},$$ (3.28)

More complicated dependencies are also used, though there is a lack in information on their reliability.

In the National Normative Documents of Russia there are not any techniques of evaluation of the coefficient of side pressure at rest K_o by CPT data; there is an indication in MGSN 2.07-01 that in $OCR > 6$, the value of K_o may exceed 2; this is to be taken into account in calculations of underground structures.

Coefficient of consolidation c_v is applied by CPT data mostly worldwide (Lunne et al. 2004, Burns & Mayne 1998). It is an integrated index showing the behavior of water-saturated soil under loading from the point of view of the theory of *filtration (initial) consolidation*.

$$c_v = k \frac{(1 + e_o)}{a \cdot \gamma_w},$$ (3.29)

where:

$c_v =$ coefficient of consolidation,
$k =$ coefficient of permeability (filtration coefficient),
$e_o =$ initial void ratio (porosity factor),
$a =$ coefficient of compressibility,
$\gamma_w =$ specific weight of water.

If to use the parameter $a_0 =$ coefficient of relative compressibility instead of coefficient of compressibility ($a_o = a/(1+e_o)$), the formula (3.29) is of simpler kind:

$$c_v = \frac{k}{a_o \cdot \gamma_w}.$$ (3.29a)

Coefficient of consolidation is used in Russian and non-Russian practice as one of the principal parameters in the formulae for calculation of deformation time and estimation of alterations in time. These calculations are needed in designing of large hydraulic structures where awareness of final settings is not sufficient and one needs reliable forecasting of their development in time. Nevertheless, accuracy of calculations based on the theory of filtration consolidation is not high, especially with regard to clayey soils of solid, semisolid and even overconsolidated consistency. Besides the processes of filtration (initial) consolidation, the processes *of secondary consolidation (creep of skeleton)* occur in soil under loading; their role has been underestimated until recently.

At present, in the National Codes of Design both factors determining soil behavior are taken into account in foundations of hydraulic structures SNiP 2.02.02-85*. Calculations of unstabilized setting to temporal value t are done with regard to both initial and secondary consolidation. Here, coefficient of consolidation c_v lacks its dominating role in the calculations becoming only one of the parameters determining the development of deformations of soil. Nevertheless, in weak water-saturated soils calculations based on interpretations of the theory of initial consolidation give acceptable results; thus, evaluation of coefficient of consolidation c_v for such conditions remains an urgent task.

The technique of evaluation of coefficient of consolidation c_v is based on "*dissipation tests*" (see section 2.3); here, the penetrometer equipped with piezometer (piezocone) is stopped in given depth, then pore pressure dissipation in the adjacent soil is recorded. Coefficient of consolidation is estimated by the speed and behavior of pore pressure. As opposed to most solutions of the problems analyzed in section 3.2.5, evaluation of c_v is based not only on empirical research but theoretical results are also considered.

175

The experiments of a number of non-Russian specialists (see in Lunne et al. 2004) reveal that the process of dissipation of pore pressure around the cone may continue for several hours in clays, but its intensity decreases rapidly; thus, in 10 minutes one can observe 50 % decrease of pore pressure in most soils. It is this degree of pore pressure dissipation (i.e. 50 %) that in opinion of non-Russian professional engineers must be referred to in evaluation of coefficient of consolidation.

Therefore, the process of pore pressure dissipation is affected by a lot of factors, including initial pore pressure as well. As a rule, dissipation occurs horizontally, therefore, non-Russian specialists consider CPT being capable to evaluate "horizontal" coefficient of consolidation (usually marked as c_h).

P.K. Robertson et al. (1992) suggested a chart-nomogram for evaluation of "horizontal" coefficient of consolidation c_h by the duration of the process of pore pressure dissipation (attainment of 50 % realization of the process, to be more precise). The chart after P.K. Robertson et al. is given in figure 3.26.

As one can see from fig. 3.26, coefficient of consolidation decreases with the increase of consolidation time, the connection between the logarithms of these parameters is linear.

The part of realization of pore pressure dissipation U is expressed by the formula

$$U = \frac{u_t - u_o}{u_i - u_o} \times 100\% = \frac{\Delta u_t}{\Delta u_i} \times 100\%,$$ (3.30)

where:

$u_t =$ pore pressure at present temporal value t,

$u_i =$ initial pore pressure, i.e. in $t = 0$ (immediately after application of force),

$u_o =$ pore pressure in the location of the ground massif before penetration of the cone.

In $t = 0$, i.e. in initial pore pressure dissipation $\Delta u_t/\Delta u_i = 1$, since $u_t = u_i$. In equilibrium condition ($t \to \infty$) $\Delta u_t/\Delta u_i = 0$, since $u_t = u_o$. As it was mentioned earlier, horizontal coefficient of consolidation c_h is evaluated by time t_{50}, corresponding to $\Delta u_t/\Delta u_i = 0.5$, i.e. 50 % of realization of the process of pore pressure dissipation. The parameter is evaluated in "relaxation-creep" CPT with piezocone, i.e. during repeated measurements of pore pressure (u_i, u_t) in given intervals. Here, evaluation of u_o is not a problem, as it is a "standard" hydrostatic pressure in non-forcing ground water depending upon the depth of soil layer (in relation to ground water level) and specific weight of water.

Dashed lines $c_h = f(t_{50})$ in fig. 3.26 correspond to location of filters (piezometers) immediately behind the cone, continuous ones – on the cone. The lines depend on the rigidity index (shear) $I_r = G/s_u$ (or $I_r \approx E/3s_u$).

The arrows show evaluation procedure of c_h by t_{50} (here $t_{50} = 12$ min $\to c_h = 0.45$ cm²/min in rigidity index – $I_r = 50$ applying piezocone with filters located behind the cone).

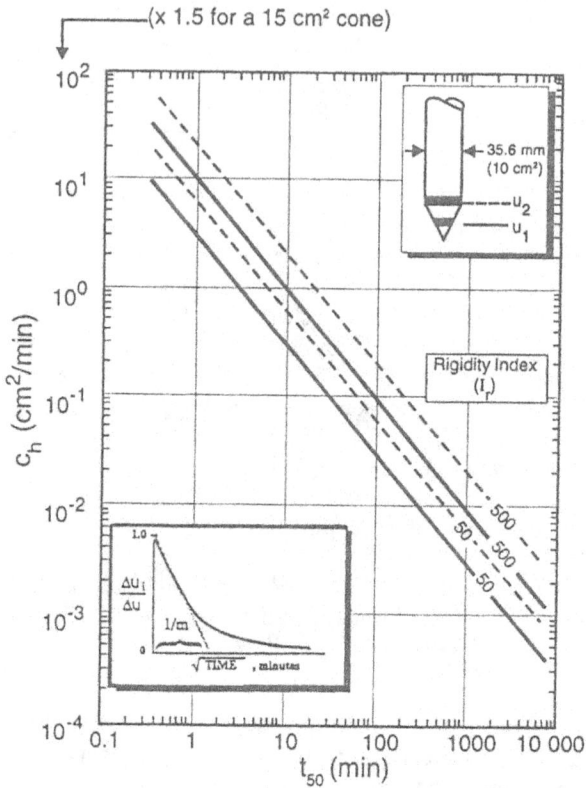

Fig. 3.26 Chart-nomogram after P.K. Robertson et al. (1992) for evaluation of "horizontal" coefficient of consolidation c_h by t_{50}, characterizing the duration of pore pressure dissipation in "relaxation-creep" CPT with piezocone.

The graph in smaller scale refers to evaluation of coefficient of consolidation by dependence of $\Delta u_t / \Delta u_i$ upon time t (imaged in scales of $\sqrt{}$ or lg) applying the formula (3.29)

Besides t_{50} other parameters are also applied; they characterize the process of pore pressure dissipation in relaxation-creep CPT by piezocone. For example, the article (Robertson et al. 1992) presents the mode of evaluation of coefficient of consolidation in which the process of pore pressure dissipation is characterized by special parameter m determined by graphical processing of dependency $\Delta u_t / \Delta u_i$ from square root of time ($\sqrt{t}$) but not by the parameter t_{50}. lgt is occasionally used instead of $\sqrt{t}$. Figure 3.26 illustrates (in rectangular window) the principle of evaluation of

177

parameter m. The following empirical formula for evaluation of "horizontal" coefficient of consolidation c_h is applied:

$$c_h = (m/M)^2 \cdot r^2 \cdot \sqrt{I_r} , \qquad (3.31)$$

where:

$m =$ gradient of the initial sector of the curve "$\Delta u_t / \Delta u_i \sim \sqrt{t}$" obtained after "relaxation-creep" soil tests; it is determined by the graph given in fig. 3.26, $\min^{-0,5}$;

$M =$ nondimensional parameter (gradient corresponding to the theoretical curve of pore pressure dissipation and corrected with regard to concrete geometry of the piezocone) evaluated by table 3.30;

Table 3.30 Parameters M in the formula (3.30)

Disposition of filters	On the cone	On the friction sleeve (on the rod) behind the cone	On the push rod, 5 diameters away from the cone
Values of M	1.63	1.15	0.62

$I_r =$ rigidity index, i.e. relation of shear modulus to shear strength ($I_r = G/s_u$);

$r =$ cone radius.

Instead of the parameter $(m/M)^2$ G.T. Houlsby and C.I. Teh (1988) use the ratio T^*/t, where t – time spent on attainment of examined degree of consolidation, T^* – "modified time factor".

$$T^* = \frac{c_h \cdot t}{r^2 \cdot \sqrt{I_r}} , \qquad (3.32)$$

where c_h, r & I_r = the same as in the formula (3.27).

T^* is evaluated by table 3.31, and the obtained (tabular) values of T^* are substituted in the formula (3.32) equated to

$$c_h = \frac{T^*}{t} \cdot r^2 \cdot \sqrt{I_r} , \qquad (3.32a)$$

where the symbols are the same as in the formula (3.32).

When analyzing the techniques of evaluation of coefficient of consolidation T.Lunne et al. (2004) point out that they are rather approximate, since there are a lot of factors complicating the evaluations. These are the as follows: complexity of the initial pore pressure u_i evaluation, destruction of soil structure in CPT, anisotropy of soil, etc. When considering possible ways of filter dispositions, the authors (Lunne et al. 2004) give preference to their disposition behind the cone. The vertical coefficient of consolidation c_v is suggested to be evaluated by the formula

$$c_v = c_h \cdot \frac{k_v}{k_h}, \qquad (3.33)$$

where:

c_v & c_h = vertical and horizontal coefficients of consolidation correspondingly,
k_v & k_h = vertical and horizontal coefficients of permeability.
The most typical values of k_v/k_h for clayey soils are given in table 3.32 (Jamiolkowski et al.1985).

Table 3.31 Values of "modified time factor" T^* in the formula (3.32a)

Degree of consolidation (%)	Disposition of filters			
	On the cone	On the friction sleeve (on the rod) behind the cone	On the push rod, 5 diameters away from the cone	On the push rod. 10 diameters away from the cone
20	0.014	0.038	0.294	0.378
30	0.032	0.078	0.503	0.662
40	0.063	0.142	0.756	0.995
50	0.118	0.245	1.11	1.458
60	0.226	0.439	1.65	2.139
70	0.463	0.804	2.43	3.238
80	1.040	1.60	4.10	5.24

Table 3.32 Typical values of k_h/k_v ratios in "soft" clays after M. Jamiolkowski

Type of clays	Ratio k_h/k_v
Macrofabrics are either absent or are extremely weak; in general, it is typical for homogeneous deposits	1...1.5
Macrofabrics are noticeable; e.g. clayey layers alternate with other soil layers being water permeable or they include lens of these soils	2...4
Clear stratified macrofabric, occurrence of great number of high water permeable spreading layers	3...15

Coefficient of permeability (filtration) k_h is occasionally evaluated by CPT, mostly in non-Russian practice (Lunne et al. 2004, Burns & Mayne 1998), and rarely – in Russian one GOST 25260-82. The simplest way of such evaluation is based on identification of soil types by the charts discussed in section 3.1.1 and further evaluation by special tables of coefficient of permeability for every soil type.

Thus, P.K. Robertson et al. (1986) suggest the chart given in fig. 3.3 and the additional table of coefficients of permeability for every soil type highlighted on the chart to be applied. For example, for clays (position 3 in fig. 3.3) they recommend to admit the coefficient of permeability in the following limits: $10^{-9}...10^{-10}$ m/c ($\approx 10^{-2}...10^{-3}$ cm/day), for sands (position 9) – $10^{-3}...10^{-4}$ m/c ($\approx 10^{4}...10^{3}$ cm/day) correspondingly, etc. In the National Standard GOST 25260-82 coefficient of permeability is ordered to be evaluated by the ratio q_c/f_s:

q_c/f_s	< 25	46...25	46...67	> 67
k_h, m/day	0.001	0.001...0.1	0.1...2	5...50

It is obvious that the accuracy of this evaluation is far from being high, especially if the tables and the ratio q_c/f_s are not anchored to concrete soil types (by lithological and genetic features) in specific region.

Theoretically, the techniques based on applications of "dissipation" soil tests by piezocones appear to be more appropriate (though not very reliable). Here, the empirical dependencies between the values of t_{50} and horizontal coefficients of permeability k_h are used. Figure 3.27 illustrates the dependency between t_{50} & k_h, obtained by P.K. Robertson et al. (1992).

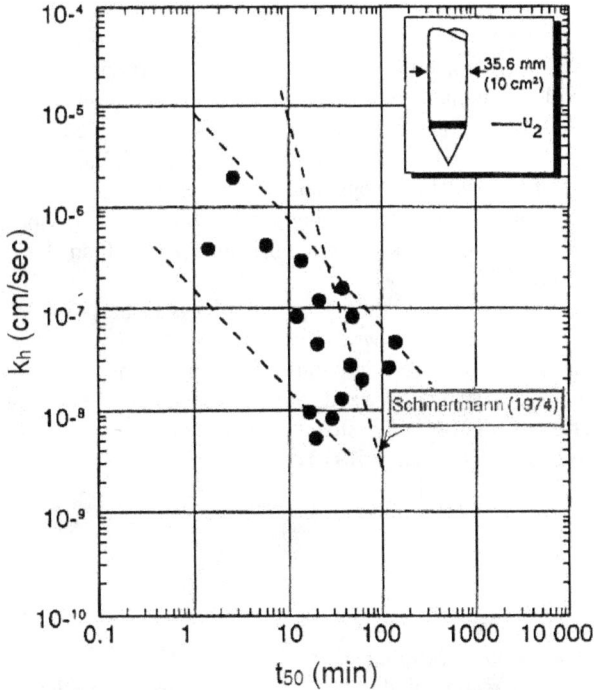

Fig.3.27 Schematic illustration of connection between "horizontal" coefficient of permeability k_h and parameter t_{50} characterizing the duration of pore pressure dissipation in dissipation soil tests by piezocone

As it is shown in fig. 3.27, connection between t_{50} & k_h is rather weak. Nevertheless, it is evident that lower coefficients of permeability correspond to longer pore pressure dissipation. In $t_{50} = 10$ min the coefficient of permeability is on the average $k_h = 10^{-7}$ cm/s ($0.8 \cdot 10^{-2}$ cm/day), in $t_{50} = 1$ min $k_h = 10^{-6}$ cm/s ($0.8 \cdot 10^{-1}$ cm/day), in $t_{50} = 0.5$ min $k_h = 5 \cdot 10^{-6}$ cm/s ($4 \cdot 10^{-1}$ cm/day).

Thixotropy (sensitivity) S_t of clayey soils is occasionally measured in practice, since one needs the parameter to solve special problems. The problems may emerge in exploration the territories with weak water-saturated clayey soils, in search of clays as the components of "clayey mortars" intended for strengthening of walls of deep excavations, in estimation of soil liquefaction hazards in underground workings, etc.

The traditional approach of evaluation of sensitivity of clayey soils is based on evaluation of the parameter

$$S_t = \frac{s_u}{s_u'},$$ (3.34)

where s_u & $s_u' =$ shear strengths in natural state of soil and in its broken structure.

In application of type II cone penetrometers there is the possibility to estimate sensitivity of clayey soils, since the cone is penetrated in the soil of natural structure, while the friction sleeve contacts with broken soil. For this reason, *"friction ratio" ("friction index")* $R_f = f_s/q_c$ in "extremely sensitive" soils is to be lower than in soils of small sensitivity. The difficulty appears when "friction ratio" f_s/q_c reflects not only the sensitivity of soil but absolutely opposite properties as well, e.g. increase of soil particles or decrease of its connection ("clayiness"); due to this in sands and gravels with lack of thixotropy it happens to be significantly lower than that in clays (see sections 2.4.4 & 3.1.1).

For this reason estimation of thixotropy of soils is to be possible in assurance that *the examined soil is really clayey, water-saturated and very soft or soft* ($I_L >$ 0.75). The assurance is to be based on parallel drilling data as well as prevalence of low values of q_c being not typical for sands (e.g. $q_c < 0.5$ MPa).

J.H. Schmertmann (1978) suggested the empirical formula for evaluation of "sensitivity" of soils S_t by CPT data in 1970s

$$S_t = \frac{N_S}{R_f},$$ (3.35)

where:

$R_f = f_s/q_c =$ friction ratio,

$N_S =$ the parameter evaluated empirically.

J.H. Schmertmann recommended $N_S = 15$, although in their later research the authors recommended N_S to be slightly lower: P.K. Robertson and R.G. Campanella suggested $N_S = 6$; N.S. Rad and T. Lunne recommended $N_S = 5...10$ (mean value $N_S =$ 7.5) (Lunne et al. 2004).

Figure 3.28 illustrates the results of research after N.S. Rad and T. Lunne (1986) who studied the connection between friction ratio R_f and sensitivity S_t.

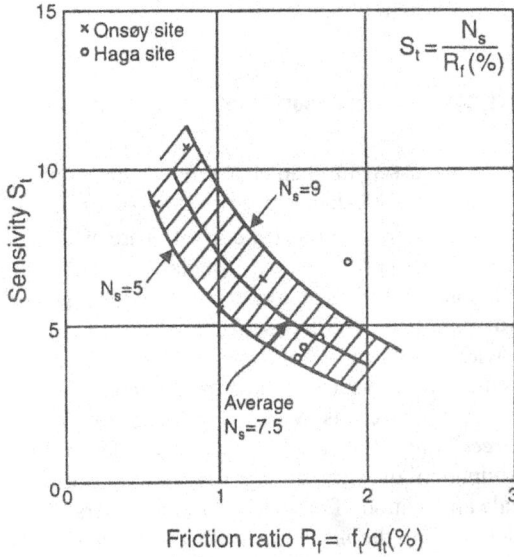

Fig.3.28 Schematic illustration of connection between sensitivity and friction ratio f_s/q_c after N.S. Rad and T. Lunne (1986)

As it is underlined in the book (Lunne et al. 2004), the parameter N_s depends upon a number of factors (mineralogical composition, overconsolidation ratio, etc.) and must be defined for every concrete region.

There are some problems in the necessity of measurements of low values of f_s and particularly q_c. The accuracy and sensitivity of measuring equipment of most tensometric cone penetrometers is such that in $q_c < 0.3$ MPa and $f_s < 0.02$ MPa the measurements are of very low accuracy or are not "detected" at all. This has an effect on reliability of friction ratio R_f application highlighted by T. Lunne et al. (2004).

182

4 "DIRECT" METHODS OF CPT DATA APPLICATION

4.1 SPECIFIC FEATURES OF "DIRECT" AND "INDIRECT" METHODS

The methods of CPT data application discussed in chapter 3 refer to the group of *'indirect'* methods (after the classification of M. Jamiolkowski 1995 – see section 1.2), i.e. the ones in which first the characteristics of soils (φ, c, E, etc.) are evaluated, and then they are used as the initial data in corresponding calculations. These are various computations of bases, foundations as well as underground structures, slopes' stability determination, estimation of quality of the bulks made, etc. However, in world practice another approach is widely applied; here, standard properties of soil are not evaluated, but CPT results as they are, serve as the initial data in computations (q_c, f_s, Q_s). These methods combining CPT results with the solution of definite geotechnical problem are called *"direct"* (after the same classification). During last decades, in Russian practice determination of pile bearing capacity has been of principal importance due to large-scale application of these structural members.

It is universality that is considered to be the *merit of "indirect" methods*: the results obtained may be applied in solution of various problems including computations of bases, foundations, underground structures, stability of slopes, control of subsurface soil compaction, etc. *The disadvantages of "indirect" methods* are concerned with the necessity of resolving of two different problems by turn: evaluation of soil properties as well as the computations (including foundations). The errors in evaluation of soil properties are added to significant inaccuracies in computations resulting in inaccuracy of the final result.

The simplicity and, hence, the possibility of obtaining higher accuracy of the final result are *the merits of "direct" methods*. *The disadvantage* is their limitedness by specific features of geotechnical problems: definite types of foundations, definite technological processes, etc. Thus, for every type of foundation as well as for definite range of values of their parameters one has to apply his empirical formulae connecting CPT data and resistivity of this foundation (bearing capacity). Adjustment of the computation results is to be carried out when comparing the test results of as-build foundations of certain types, behavior of definite slopes, etc. In other words, in comparison to "indirect" methods, the "direct" methods show their empirical nature to a large extent.

The possibility to obtain higher accuracy of the final result in application of "direct" methods is confirmed by T. Lunne et al. (2004); the data generalize non-Russian experience gained. Table 4.1 shows reliability ratings (efficiency) determined by T. Lunne et al. (after a survey made among a number of specialists). The reliability ratings of CPT application are applicable in resolving the problems of structural design or construction operations by "direct" methods. As in rating of the reliability of estimation techniques of soil properties (see section 3.2.2), the maximum reliability corresponds to minimal number, i.e. "1" – high reliability, "2" – reliability from high to medium, "3" – medium reliability, "4" – from medium to low one and "5" – low reliability.

Table 4.1 Perceived applicability of the CPT/CPTU for various direct design problems (Lunne et al. 2004)

Soils	Pile design	Bearing capacity	Settlement	Compaction control	Liquefaction
Sand	1-2	1-2	2-3	1-2	1-2
Clay	1-2	1-2	3-4	3-4	
Intermediate soils	1-2	2-3	3-4	3-4	

As one can see from table 4.1, in all soil conditions the highest reliability (1-2) is attained in determination of pile bearing capacity, and in lithological homogeneous soils – in determination of shallow foundations. In sands, density control in deep compaction and estimation of liquefaction (in seismic effects) appear to be the most effective. Reliability of computations of shallow foundations in the "mixed soils" (in alternation of sandy and clayey strata) is less effective. The medium rating (in all five problems) is the following: in sands – 1.7, in clays – 2.5, in mixed soils (including clayey and sandy deposits) – 2.7.

According to the same authors, in estimation of soil properties by CPT (see section 2.4.1), i.e. in the "indirect" approach, the ratings happen to be significantly lower: in sands, the medium rating is 3.0, in clays – 3.4. In here, the rating of deformation characteristics determination ("compressive" modulus of deformation, modulus of elasticity, shear modulus in large prolonged deformations and shear modulus in small "momentary" deformations) is, on average, 4.5 – in sands and 2.8 – in clays. As it was underlined in section 3.2.2, non-Russian professional engineers gave high rating "1-2" to evaluation of shear strength (parameter s_u) in "indirect" approach. When taking into account that in "indirect" methods of CPT data application errors are added followed from inaccuracies of design models of foundations, the advantages of "direct" methods become much more significant.

Thus, the applied "direct" methods of CPT data application are recognized as more reliable than the "indirect" ones, though they rebate the latter in universality.

4.2 EVALUATION OF SHALLOW FOUNDATIONS RESISTANCE

Designing of shallow foundations (foundations on ground bed) applying "direct" methods of CPT data application is hardly ever used in practice. Diversity of parameters of such foundations (forms, dimensions, depth, etc.) as well as differences in conditions of their application (availability or lack of basements, action of various loads, moments, etc.) complicates the link between cone resistances q_c and foundation resistances R_o and makes it inconvenient for practical application. For this reason "indirect" methods appear to be more preferable.

Nevertheless, for a limited range of conditions "direct" methods happen to be rather suitable; due to this, simple empirical dependencies have been applied worldwide for decades. It is typical for designing of shallow structures, low-rise buildings of reduced responsibility. In designing the structures of normal and enhanced responsibility on shallow foundations, direct methods are usually applied for *preliminary* computations.

When considering Russian and non-Russian computation techniques of shallow foundations by CPT data, it is important to take into account some conceptual differences in realization of the computations in different countries. In general case, designing of foundations includes the solution of two particular problems:

- Evaluation of pressure under the foundation which shall not exceed the resistive capacity of the foundation.
- Evaluation of deformations which shall not exceed the permissible values for the given structure.

Both in non-Russian and Russian practice evaluation of deformations is realized more or less equally, although the pressure is interpreted in different ways. In many countries the selection of the pressure is understood as the evaluation of foundation *bearing capacity* allowing the dimensions of the foundation to be accepted. As a rule, high "safety factor" (usually threefold) is introduced.

In the National Building Codes of Russia it is allowed to design foundations by deformations, while computations of foundations' bearing capacity are carried out just only in particular cases. Nevertheless, the first stage in computation by deformations is the selection of pressure acting on the foundation; here, breaking strains (plastic strains) are limited, thus, allowing the foundation to be considered as a linearly deformable medium. The pressure called "design resistance of foundation soils" R is usually several times lower than bearing capacity of the foundation resulting in its complete failure. The subsequent "standard" computation of settlements is possible only in pressures under the foundation being lower than the parameter R. Besides, the National Norms (SNiP 2.02.01-83*, SP 50-101-2004) allow completing computations at this stage, since the requirements on deformations are considered to be met. The "reserve" of foundation resistive capacity is evaluated here by the "reserves" taken in selection of design values of soil properties (φ, c and γ). As a rule, it happens to be significantly lower than the "reserve" by bearing capacity (it is in the limits of 20...30 %).

Therefore, the parameters of most foundations being designed in Russia are taken in agreement with the design resistance of foundation R which is *not considered to be the bearing capacity*. When selecting the empirical formulae the Russian specialists usually compare cone resistances q_c with the mentioned "design resistance R", while non-Russian specialists tend to apply corrected (reduced) values of bearing capacity of foundations. It is necessary to point out that in most cases these two parameters ("design resistance" and reduced "bearing capacity") prove to be rather close. As one can observe in practice, despite different approaches the pressures under the foundations of shallow foundations taken in analogical soil conditions in different countries are hardly ever differ from each other. Here, for

185

structures of low responsibility the pressures under the foundations of foundations are often taken based on the construction experience ignoring the computations; the corresponding tables allowing pressures under the foundations to be evaluated approximately are applied (e.g. proceeding from physical properties of soil). These approximate values of foundation soil resistances may serve as a reference in obtaining the empirical dependencies combining CPT data with the resistive capacity of foundation as well.

"Direct" method of CPT data application in designing of shallow foundations usually indicates evaluation of foundation resistance (realized in accordance with conceptions admitted in the specific country) by the value of cone resistance q_c. The simplest empirical dependencies (formulae or tables) are applied here.

In 1956 G.G. Meyerhof (1956) suggested the formula to evaluate bearing capacity of shallow foundations by cone resistance q_c with regard to sands

$$q_{ult} = \overline{q}_c \frac{b}{C}\left(1 + \frac{d}{b}\right), \tag{4.1}$$

where:

q_{ult} = bearing capacity of foundation (in accordance with the original symbols), kPa;

$\overline{q}_c$ = average cone resistance in the depths from d to $d+b$, kPa;

d = depth of foundation, m;

b = width of foundation, m;

C = empirical coefficient equal to $C = 12.2$ m.

Later, a number of other analogical formulae were applied in non-Russian practice (in general, for sandy soils). The formula after K.E. Tand et al. (1995) aimed to estimate the foundations composed of weakly cemented sands of mean density can serve as an example:

$$q_{ult} = R_k q_c + \sigma_{vo}, \tag{4.2}$$

where:

R_k = value changing in the limits 0.14...0.2 dependent on the foundation form and the depth of foundation, kPa;

σ_{vo} = overburden (vertical) stress of soil in the given depth, kPa.

In Russian practice clayey soils were paid greater attention to. The Norms dated from 1970s SN-448-72 (1972) contain the table (see table 4.2) allowing design resistance of shallow foundation to be approximately evaluated in the following conditions:

• Foundation soils – clays or loams.
• Width of foundation 0.6...1.5 m.
• Depth of foundation 1...2.5 m.

186

Table 4.2 Connection between cone resistance q_c and design resistance of foundation R_o after SN-448-72 (1972)

q_c (MPa)	1	2	3	4	5	6
R_o (MPa)	0.12	0.22	0.31	0.40	0.49	0.58

If to compare design resistances obtained by the "direct" method (by table 4.2) with the analogical values obtained by the "indirect" method, i.e. by evaluation of φ & c by table 3.9 (in accordance with SP 11-105-97 (1998) and subsequent calculation of foundation resistance R (by formula (7) in sec. 2.41 of the National Building Codes of Russia SNiP 2.02.01-83*), we obtain the results given in table 4.3.

Table 4.3 Comparison of design resistances of foundations R_0 obtained by the "direct" method (after table 4.2) with the analogous resistances R obtained by the "indirect" method, i.e. by evaluation of φ & c

Method of evaluation	Soils	Foundation parameters: width b (m) depth d (m)	Design resistances of foundations R (R_0) (MPa) in cone resistances q_c (MPa)		
			1	3	5
"direct" (after table 4.2), R_o	Loams and clays	b=0.6...1.5 m, d=1.0...2.5 m	0.12	0.31	0.49
"indirect" (by evaluation of φ & c), R	Loams	b=0.6 m, d=1.0 m	0.150	0.254	0.370
		b=1.5 m, d=2.5 m	0.235	0.363	0.502
	Clays	b=0.6 m, d=1.0 m	0.204	0.286	0.400
		b=1.5 m, d=2.5 m	0.280	0.380	0.516

As one can see from table 4.3, in most cases divergences between R_0 and R are taken to "reserve" (if the results of the "indirect" method are considered as a "reference"). The simplified ("direct") method gives lower resistances of foundations than the "indirect" method embracing larger numbers of factors. However, for small foundations (b = 0.6 m, d = 1 m) the following situations when the simplified method enhances the resistive capacity of the foundation are possible, i.e. $R_0 < R$; e.g. this is typical for loams.

The given comparison reveals the considerable effect of shallow foundation parameters (dimensions of foundation, depth of foundation) on foundation resistance R even in rather limited range of alterations of the parameters. Later, table 4.2 was excluded from the National Normative Documents, thus, designing of shallow foundations was directed towards the "indirect" method which allowed specific features of shallow foundations to be taken into account. Nevertheless, the research continued.

In 1990s N.B. Gareeva and B.V. Goncharov (Ryzhkov & Goncharov 2006) carried out a vast experiment embracing CPT and soil testing by the stamp; here cone resistance q_c was compared with "initial critical pressure" p_{ic} (i.e. the maximal possible pressure before the formation of plasticity zones). "Relaxation-creep" CPT

187

was performed, i.e. the penetrometer was in equilibrium ignoring measurements of pore pressure ("relaxation-creep" CPT – see section 2.3). Figure 4.1 illustrates the results of comparison obtained at 26 sites possessing clayey soils in Bashkir Urals (cone resistance in equilibrium is marked as "q_{ce}"). Analytically the empirical dependency of the "initial critical pressure" (p_{ic}) upon cone resistance (q_{ce}) is presented by the authors in the following formula (p_{ic} & q_{ce} – in MPa) (Ryzhkov & Goncharov 2006, Gareeva & Gorbatova 1985):

$$p_{ic} = 0.14 q_{ce}^{0,5}, \qquad (4.3)$$

where p_{ic}, q_{ce} = "initial critical pressure" and cone resistance in conditional equilibrium, MPa.

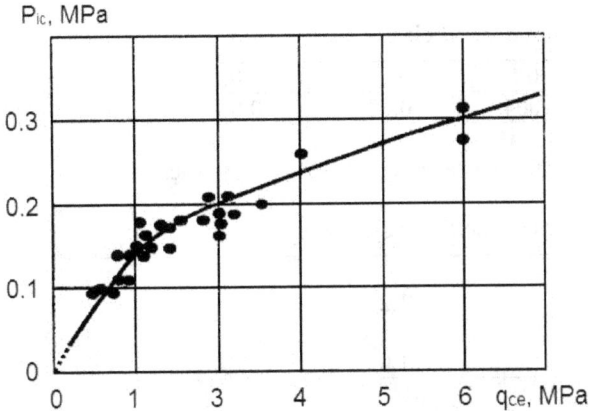

Fig.4.1 Results of comparison between cone resistance and "initial critical pressures" obtained by N.B. Gareeva and B.V. Goncharov (Ryzhkov & Goncharov 2006):
P_{ic} – "initial critical pressure" under the standard stamp (5000 cm^2), q_{ce} – cone resistance in conditional equilibrium

In the dependency given earlier (4.3) the value of the "initial critical pressure" p_{ic} differs from the bearing capacity q_{ult} in the formulae of non-Russian specialists and design resistance R_0 in SN-448-72 as well. The notion "bearing capacity of the foundation" means the transition of foundation to plasticity (limit equilibrium condition), and "design resistance of foundation soils" (both R_0 in accordance with SN-448-72 and R in accordance with SNiP 2.02.01-83* means the development of plasticity zones at a depth of ¼ b (b – width of foundation), "initial critical pressure" p_{ic} corresponds to the situation when plasticity zones do not occur. The value of "initial critical pressure $p_{i.c}$ is usually only 10...20 % lower than that of "design resistance of foundation".

To evaluate the design resistance of soils R the authors (Ryzhkov & Goncharov 2006, Gareeva & Gorbatova 1985) suggest the formula

$$R = m_1 m_2 (p_{ic} + kb) = m_1 m_2 \left(0.14 q_{ce}^{0,5} + kb \right), \qquad (4.4)$$

where:

$p_{i.c.}$ = "initial critical pressure" under the standard stamp (5,000 cm^2), MPa;
q_{ce} = cone resistance in equilibrium condition, MPa;
b = width of foundation, m;
k = coefficient considering the width of foundation effect MPa/m, taken with regard to q_{ce} after table 4.4:

Table 4.4 Values of coefficient k

q_{ce} (MPa)	< 0.5	0.5...1.0	1.0...3.0	3.0...5.0	> 5.0
k (MPa/m)	0.004	0.008	0.013	0.015	0.020

m_1, m_2 = coefficients taken with regard to $q_{c(e)}$ after table 4.5:

Table 4.5 Values of coefficients m_1 & m_2

$q_{c(e)}$ (MPa)	m_1	m_2
> 1.5	1.2	1.05
1.0...1.5	1.1	1.00
< 1.0	1.0	1.00

Table 4.6 shows the results of design resistance of foundations R evaluated by the formula (4.4) for the most typical cases. Taking into account that in "relaxation-creep" CPT cone resistances q_{ce} are nearly 30 % lower in most clays and loams than those in standard CPT (q_c), the most probable values of q_c corresponding with those of q_{ce} are given in table 4.6.

Table 4.6 Results of design resistance of foundations R evaluated by the formula (4.4)

q_{ce} (MPa)		0.7	1.4	2.1	2.8	3.5	4.2
Value of q_c, approximately corresponding with q_{ce} (MPa)		1.0	2.0	3.0	4.0	5.0	6.0
R (MPa)	in b=1 m	0.12	0.19	0.26	0.31	0.35	0.38
	in b=4 m	0.15	0.25	0.32	0.38	0.43	0.46

One can easily notice that design resistances of foundation evaluated by the formula (4.4) for the foundations of a small width are quite close to the values obtained by table 4.3, but given significant strength of clayey soil they happen to be reasonably lower than the latter ones. For example, for the foundation of 1 m in width in q_{ce} = 0.7 MPa (this approximately corresponds to q_c = 1 MPa) the values of design resistance coincide ($R = R_0$ = 0.12 MPa), but in q_{ce} = 2.8 MPa ($q_c \approx$ 4 MPa) the difference is rather significant (R = 0.31 MPa and R_0 = 0.4 MPa). Besides, as opposed to table 4.3, the formula (4.4) allows the width of foundation to be taken into account.

It is important to pay attention to the mode of presentation of foundation resistance estimation results given by N.B. Gareeva and B.V. Goncharov. As it was mentioned earlier, inaccuracy in estimations by CPT is greatly compensated by their large number; this allows examining changeability of one or the other parameter by depth and spread of the ground massif as well. N.B. Gareeva and B.V. Goncharov insist on making the "numerical models" of R, i.e. sections or charts showing the changeability of R in horizontal and vertical directions, but not limit oneself to "point" foundation resistance R estimations. The principles of their making are the same as in making the section by CPT results shown in fig. 3.9.

The charts (horizontal sections) showing various resistances R as isolines in the contour of the designed structure or the whole of the developed territory seem to be the most convenient in practical applications. These charts allow rational location of structures in plan; simplify the estimation of settlement irregularities and solving of a number of other problems as well.

Computations of settlements of shallow foundations are usually done by the "indirect" method, although in a number of cases the "direct" method appears to be appropriate. It is convenient in computations of settlements in sandy beds equal in depths or in small dimensions of foundations. The formulae used here are to be rather simple, but their efficiency is limited by quite narrow range of conditions. It is obvious that in more complicated cases one has to apply more complicated models (design models) of foundations and to consider the distribution of stresses in them as well as the differences of compressibility of soils in different zones, etc. However, the distinction between the "direct" method and the "indirect" one becomes formal in these cases: in one and the same design model instead of the modulus of deformation (E) its expressions though the values of cone resistance (q_c) appear in the formulae. It is easy to notice when considering two computation methods known worldwide – G.G. Meyerhof and J.H. Schmertmann (see in Trofimenkov 1995, Lunne et al. 2004). The following formulae are given to evaluate the settlements of shallow foundations in sands:

G.G. Meyerhof –

$$s = \Delta p \frac{B}{2q_c} , \qquad (4.5)$$

J.H. Schmertmann –

$$s = c_1 c_2 \, \Delta p \sum_{i=1}^{i=n} \frac{I_z}{kq_c} \Delta z_i , \qquad (4.6)$$

where:

s = settlement of foundation;

Δp = net foundation pressure, i.e. total pressure under the foundation bottom of the foundation regardless of the overburden stress at the depth of the foundation;

B = width of foundation bottom;

$$c_1 = 1 - 0.5\left(\frac{\sigma'_{vo}}{\Delta p}\right) =$$ coefficient with regard to the depth of the foundation:

σ'_{vo} = effective vertical stress (overburden stress);

$$c_2 = 1 + 0.2 \cdot lg\left(\frac{t}{0,1}\right) =$$ coefficient with regard to the time of duration of load acting t (years);

I_z = "strain influence factor" showing the changeability of deformations in depth (see fig. 4.2);

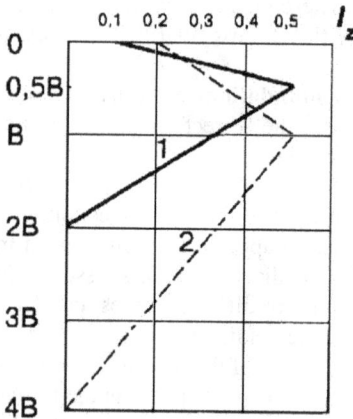

Fig. 4.2 Schematic illustration of changeability of "strain influence factor" I_z with regard to relative depth (B – width of foundation bottom):
1 – for separate foundation with square bottom ($L/B = 1$), 2 – for strip bottom ($L/B = 10$)

Δz_i = thickness of bed layer;

q_c = cone resistance in the given depth;

k = coefficient taken to be 2.5 in square foundation and 3.5 – in strip foundation.

When analyzing the formulae (4.4) & (4.5), one can notice that the formula (4.4) assumes a homogeneous bed, but the formula (4.5) – a multilayer bed incorporating layers of different deformability.

One and the same models (design models) as in computation of settlements by the "indirect" methods are used. If to take that for most sands $E \approx (2...3)q_c$ (see section 3.2.4), the formula (4.5) may be as follows:

$$s = \frac{(1,0...1,5)\cdot \Delta p \cdot B}{E}. \tag{4.7}$$

It becomes evident that this formula is analogous to the known formula of F. Shleiher:

$$s = \frac{\omega \cdot (1 - \nu^2)\cdot \Delta p \cdot B}{E}, \tag{4.8}$$

where:

$\omega =$ coefficient showing the form of the foundation bottom (for square one – $\omega=0.88$, for rectangular one with the ratio of dimensions $L/B=2$ $\omega=1.22$) (Tsytovich 1973);

$\nu =$ Poisson's ratioко (for sands $\nu =0.27...0.3$);

$s, \Delta p, B =$ the same as in the formula (4.4).

The same approach may be taken when analyzing the formula (4.6), in which the expression kq_c presents the modulus of deformation. The widely applied "layerwise summation method" is posited in computation; here the bed is divided into horizontal layers, the settlements of each layer are evaluated, the results are summed up and the total settlement is obtained. J.H. Schmertmann simplifies the pattern of depth-change of stresses by means of "strain influence factor" I_z. The development of settlements in time (by coefficient c_2) is also considered in a simplified way.

For clayey soils the computations of settlements by the "direct" methods are hardly ever done.

4.3 PILE BEARING CAPACITY EVALUATION

4.3.1 General information

Pile bearing capacity evaluation, especially of driven piles is the most important area of CPT practical application. The research has been carried out for last 3...4 decades in Russia, thus, the evaluation techniques have been obtained. They leave behind only CPT tests of full-sized piles in reliability of the results obtained. If to take into account that CPT is much cheaper and easier to produce, its principal importance in designing of pile foundations becomes obvious. The National Normative Document of Russia SP 50-102-2003 (2004) relates CPT to both compulsory and "the most preferable" ways of soil study in case pile are applied.

The efficiency of the "direct" method in CPT application for pile bearing capacity evaluation is significantly conditioned by similarity of the processes occurring in the soil under the cone and the pile as well, since the driven pile can be considered as a large cone. It was realized when CPT came into practical application. The first primitive rigs were successfully applied for specification of solid layer depth capable to withstand the column piles. Later, friction pile bearing capacity evaluation techniques appeared; originally, they were not of high accuracy, however, they became popular because of their technological advantages. A vast research on specification of the techniques was put on a large scale (see section 1.4).

192

In spite of the diversity of approaches to the solution of the problem, the idea of calculation of pile resistance to vertical load was unified – total pile resistance equaled to the sum of pile end bearing capacity and side bearing capacity.

$$F_u = Q_p + Q_f, \qquad (4.9)$$

where:

$F_u =$ pile bearing capacity to vertical load;
$Q_p =$ pile end bearing capacity;
$Q_f =$ pile side bearing capacity.

The whole of the "direct" methods have been based on this concept regardless of the type of cone, the types of piles or the type of soil conditions. They are the modes of evaluation of resistances Q_p & Q_f that are different. With regard to special piles (piles with friction reducers, bore injection ones, etc.) one needs to take into account a number of additional factors. For this reason, their bearing capacity is often evaluated by the "indirect" method, i.e. by preliminary evaluation of standard soil properties. As a rule, they assume the following correction of results by static testing of the piles. Although the "direct" method appears to be applicable for these piles, but here one needs the additional correction coefficients to be introduced in order to take into account their specific parameters.

When considering the publications on pile bearing capacity evaluation, one should keep in mind that specialists from different countries tend to use different terminological vocabulary. In non-Russian publications the term *"pile bearing capacity"* is interpreted in a wider sense than in Russian ones. It is usually supplied with the additional definitions (Somerville & Paul 1986):

- *"Ultimate bearing capacity"* means the lack of "reserves"; it corresponds with *"ultimate pile resistance"* in the National Normative Documents.

- *"Safe bearing capacity"* means presence of some "reserves"; it corresponds with the term *"bearing capacity"* in the National Normative Documents.

- *"Ultimate (or safe) net-capacity"* corresponds with the notions mentioned earlier but assuming the subtraction of the overburden stress portion, i.e. the weight of overlying soil massif (the term is ignored in the National Normative Documents on pile applications).

The term *"allowable load"* is also applied; it means safe load acting on piles. When evaluating it one must take into account *all* influencing factors including anticipated settings, "bush effect", etc.

"Safe bearing capacity" and "allowable load" assume generalization of results obtained in different locations of a site, although the transition from particular values to generalized ones is not regulated in most non-Russian Standards. "Safety factor" is usually taken as a stable parameter (usually in the interval 2...3) ignoring statistic calculations.

The term "pile bearing capacity" is sometimes used as the term reflecting the resistivity of a pile both in the definite location of a site and generally in the whole of the site (or its section).

In the Federal Standards of Russia (SNiP 2.02.03-85*, SP 50-102-2003) the transition from the particular values to generalized ones is interpreted more definitely. The following terminology is taken here:

- *"Particular value of ultimate pile resistance"* F_u – pile resistance in the definite location of a site, i.e. the load that a pile is capable to bear in the location (in i-location $F_{u,i}$).

- *"Normative value of ultimate pile resistance"* $F_{u,n}$ – generalized (mean) value of pile resistances $F_{u,i}$ in the site or its section ignoring the introduction of "reserves" compensating the "spread" of values $F_{u,i}$ in the site evaluated by the formula

$$F_{u,n} = \frac{\sum_{i=1}^{n} F_{u,i}}{n},$$ (4.10)

where:

$F_{u,i}$ = particular values of ultimate pile resistances in various locations of the site (i=1, 2, 3, ...n),

n = number of site locations (or its marked section) where CPT was carried out.

- *"Pile bearing capacity"* F_d = generalized value of pile resistances with "factor of safety" including the virtual "spread" of values $F_{u,i}$; F_d is evaluated as the result of diving the normative value $F_{u,n}$ into the safety factor in soil γ_g

$$F_d = \frac{F_{u,n}}{\gamma_g},$$ (4.11)

where γ_g = safety factor in soil evaluated by statistic treatment of the values $F_{u,i}$ obtained.

When designing a pile foundation (in selection of lengths and a number of piles) the additional lowering safety factor γ_k is introduced, i.e. the reduced ratio F_d/γ_k is applied rather than the bearing capacity F_d; in the former Standards it was called " *specified load admitted on a pile*" marked as P. This load nearly corresponds to the "allowable load" admitted in non-Russian Standards. It is important to take into account that the bearing capacity of the bed and its deformability are considered separately in the National Standards of Russia (calculations by groups I and II limit conditions).

In the Norms (SNiP 2.02.03-85*, SP 50-102-2003) the term "allowable load" is not used but there appears the compulsory condition to be observed; it is the same as "specified load admitted on a pile"

$$N \leq \frac{F_d}{\gamma_k},$$ (4.12)

194

where N = specified load transferred on the pile; γ_k = safety factor taken in calculations after CPT to be 1.25.

Thus, according to the National Norms of Russia (SNiP 2.02.03-85*, SP 50-102-2003) in designing a pile foundation a design engineer is to consider the ratio F_d/γ_k being the lowest value of the assumed *mean* pile resistance in the surveyed sector with regard to the effects of the whole of random factors. The approach implies pile resistances F_u in various locations of the site to be known; it is possible to be realized only in application of CPT. In absence of CPT, i.e. in small number of locations where pile resistances are to be evaluated, the normative resistance has to be determined in a simplified way (e.g. by minimal test result). The issue on transition from the particular values of pile resistances to the allowable load has not been studied thoroughly, especially in parallel application of pile resistance evaluation methods of different reliability. One can anticipate that significant reserves exist in this sphere, i.e. it is possible to consider the specific features of definite sites more effectively as well as the reliability of methods of F_u evaluation. The issues on realization of the reserves by specification of the safety factor γ_k concerning concrete conditions are discussed in chapter 6 in detail. During last decades the specialists in CPT have paid great attention to evaluation of "particular values" of ultimate pile resistances F_u. Sections 4.3.2 & 4.3.3 show the principal results obtained during this period by non-Russian and Russian professional engineers.

4.3.2 CPT research and its practical application for pile engineering in former USSR and in CIS states

Driven piles were of great importance in pile engineering of the former USSR; this has influenced the development of pile engineering in the former Soviet Republics. Out of 10 mln m^3 of reinforced concrete piles applied annually nearly 9.5…9.7 mln m^3 went to driven piles in 1980s in the former USSR. Here, 90 % of the piles were prismatic with the section 0.3×0.3 m. Metal piles were hardly ever applied, just in some unique structures in Moscow and Leningrad (now St. Petersburg). Rather wide application of metal was typical only for hydraulic engineering construction incorporating steel sheet piles. Wooden piles were not used in mass construction in the former USSR in the second half of the 20th century.

It is obvious that in these conditions the attention was paid to the evaluation of driven pile bearing capacity especially by CPT results. The issues were discussed at numerous scientific and technical conferences, seminars as well as in various publications. Special consultative body coordinating the research in this sphere – "Permanent Commission on CPT" – functioned in RSFSR for many years. The commission involved skilled researchers from scientific, research and designing institutions.

The research was carried out mainly in 1960s … the beginning of 1970s; in the following years specification of the developed methods was done with regard to different soil conditions. The results obtained were recorded in the National Building Codes of Russia, Departmental Norms and various Recommendations.

The research and further practice showed that the traditional approach to pile foundations as special members for unfavorable soil conditions did not reveal the capabilities of these foundations. It was pile foundation that raised the efficiency of a zero cycle in the diversified soil conditions in conversion to industrialized construction techniques and application of precast reinforced concrete. It is natural that this demanded the traditional views concerning designing of pile foundations, their geometries, the technology of their construction and particularly, the techniques of their bearing capacity evaluation to be reconsidered. It was application of CPT that helped solve the last problem. A number of scientific, designing and research organizations dealt with the problem, although the greatest attention to the evaluation of pile bearing capacity by CPT results has been paid by BashNIIstroy (Ufa) and Fundamentproject (Moscow) institutions. Later, NIIOSP after N.M. Gersevanov (Moscow) have carried out a great amount of research on generalization and specification of CPT applications and elaboration of the Norms as well.

BashNIIstroy carried out the research applying type II cone penetrometers (with friction sleeve), while Fundamentproject – type I cone penetrometers. The most significant results are given further.

Both BashNIIstroy and Fundamentproject studied pile and cone behavior experimentally, in-situ tests and data acquisition obtained by the organizations were paid the greatest attention. The performed theoretical research did not give the solutions applicable for real piles; however, they allowed the experiments to be rationally performed with consideration of the most significant factors.

As it was said, ultimate pile resistance F_u had been regarded as the sum of pile end bearing capacity and side bearing capacity; this showed the general approach mentioned in section 4.3.1. As to application of type II cone penetrometer (with friction sleeve) the general formula (4.9) was as follows (fig. 4.3):

$$F_u = R_s A + u \sum f_i h_i, \qquad (4.13)$$

where:

F_u = particular value of ultimate pile resistance;

A, u = area and perimeter of pile cross section correspondingly;

h_i = thickness of i-layer of soil;

R_s, f_i = pile end bearing capacity and side bearing capacity correspondingly (in i-layer):

$$R_s = \beta_1 q_c, \quad f_i = \beta_i f_{si}, \qquad (4.13a), (4.13b)$$

where:

q_c, f_{si} = cone resistance and unit side friction resistance of type II cone penetrometer (in i-layer);

β_1, β_i = coefficients of "transition from the cone to the pile" for the end of the pile and its side correspondingly (in i-soil layer).

196

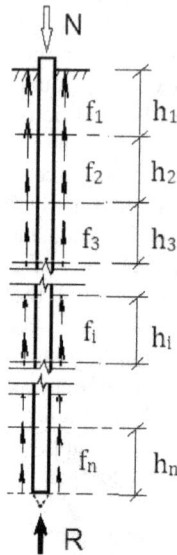

Fig. 4.3 Design model of pile resistance evaluation

With regard to type I cone penetrometers the general formula (4.9) was as follows (Trofimenkov & Vorobkov 1981)

$$F_u = R_s A + ufh, \qquad (4.14)$$

where:

$f =$ pile side bearing capacity evaluated as the function from f_s, i.e. from the resistance of the soil on the side of the embedded part of the push rod ($f_s = Q_s/A_s$, where $A_s =$ side surface area of the push rod immersed in depth h);

$h =$ the depth of pile immersion;

$F_u, R_s, A, u =$ the same as in formula (4.13).

Later, the unified representation of the dependency was admitted in the Norms (SNiP 2.02.03-85*, SP 50-102-2003): only the formula (4.14) was applied but the value of f in type II cone penetrometers was understood as

$$f = \frac{\sum \beta_i f_{si} h_i}{h}, \qquad (4.15)$$

where the symbols are the same as in formulae (4.13a) & (4.14).

In general, the research was connected with hunting of the ways of transition "from the cone to the pile"; this came to specification of coefficients β_l and β_i. Nevertheless, other design models were suggested at the first stage of the study; they differed from that one given in the formula (4.13). For example, the hypothesis ("ternary model" (Firestein 1968)) was suggested; in accordance with it, large radial (staving) pressures increasing pile side bearing capacity appear at the end of the pile in vertical load action on the pile. The hypothesis proved to have been contradictory to the views of pile behavior, although the developed calculation technique gave the results agreeable to a great number (!) of CPT of piles (over 100). This didactic story convincingly showed that the formulae insolvent theoretically were capable to be successfully applied in practice as the empirical dependencies. Apparently, the assumed and true physical processes being different may be presented in such a way that the consequences will agree with the reality. Further, these dependencies can reasonably be interpreted apart from the primary interpretations given by the authors.

The research carried out in the former USSR in 1960s allowed developing driven pile behavior modes; later, they were taken as the bases of pile design resistance technique by CPT data:

- Relative depth of penetration is considerably larger than that of the pile, but in most clayey soils it is not of great importance, since in the depth of 3...4 m the soil is deformed under the standard pile foundation of typical sizes (0.3 ×0.3 m) practically in the same way as under the cone.

- While drilling of a pile its side surface takes the soil along to the underlying layers; this creates the gaps and loose zones between the pile and the soil reducing pile side bearing capacity (the process was called "examination").

From similar conditions of soil behavior under the pile foundation and the cone one can conclude that in most cases pile end bearing capacity and cone resistance are not to be considerably different if the speeds of deformations are equal. The equivalence of the speeds of deformations can be attained in "relaxation-creep" CPT, i.e. in "relaxation-creep" tests ignoring pore pressure measurement developed in BashNIIstroy (section 2.3, fig. 2.12). It is evident that this refers to the limited range of conditions: rather deep penetration of the pile (usually exceeding 10 diameters of the pile), quaternary clayey soils of low strength. Table 4.7 presents the comparative results of clayey soil specific resistances under the pile foundations (R_s) and cone resistances (q_c) in "relaxation-creep" CPT.

As one can observe in the table, divergences between the values of R & q_c are not high if to consider the inevitable influence of soil heterogeneity and other random factors as well. Here, divergences to the "anticipated" side, i.e. $q_c > R_s$ correspond with low relative depths of pile penetration. In traditional (standard) mode of CPT, i.e. in penetration of cone with permanent speed, the values of q_c in clayey soils are usually higher than those in the equilibrium condition in 20...30 %. Due to this the ratios R_s/q_c are to be 20...30 % lower than those given in table 4.7 in standard CPT.

In sands where cone resistance is considerably higher than in clays, one can fail to observe any similarities between q_c & R. Among a number of reasons the principal one is probably the difference in soil deformations occurring under the cone and the pile. The immersion of the pile under static loading is always realized not only due to soil compaction but also to its projecting on the surface. Projecting is of greater importance in sands, it can prevail (in solid sands) in depths of pile penetration up to 10...15 m. At the same time in the same sands it can be observed in considerably smaller depths (not more than 2...3 m) under the cone. It is this difference that is the principal reason of divergences between R & q_c. Pile foundation resistances (R) in sands happen to be significantly lower than cone resistances (q_c), since projecting of soil requires smaller amount of energy than in compaction of soil. The stronger the sands, the more obvious it is.

Table 4.7 Comparison of clayey soil specific resistances under the pile foundation and the cone (in "relaxation-creep" CPT)

Section sides or diameters of piles (cm)	Depth of penetration (m)	Specific resistances of soil (MPa)		R_s/q_c
		Pile foundation resistance R_s	Cone resistance q_c	
30× 30	5.65	1.16	1.04	1.12
30× 30	5.65	1.11	0.64	1.75
30× 30	5.00	2.26	2.96	0.76
30× 30	5.00	2.70	2.00	1.35
Ø 80.0	5.60	0.58	0.59	0.99
Ø 32.5	2.00	0.29	0.41	0.71
Ø 32.5	4.00	0.94	1.10	0.85
Ø 32.5	6.00	1.05	1.44	0.73
Ø 10.8	2.00	0.83	0.81	1.23
Ø 10.8	4.00	1.63	1.57	1.04

Thorough research of sand behavior under the driven pile and the cone was carried out in 1970s in Belarus. G.S. Rodkevich (1989) and Yu.S. Kovalev (1977) suggested the values $\beta_1 = R_s/q_c$ for sands of the region on the basis of static treatment of over 200 CPT results of piles and parallel CPT by type II cone penetrometer. The data are presented in table 4.8.

Table 4.8 Conversion factors in transition from cone resistances (q_c) to pile foundation resistances (R_s) (Rodkevich 1989)

q_c (kPa)	2,500	5,000	7,500	10,000	15,000	20,000
$\beta_1 = R_s/q_c$	1.0	0.75	0.6	0.45	0.25	0.20

As it is seen from table 4.8, the difference between R & q_c increased with the increase of q_c and in higher values of q_c it attained fivefold value ($\beta_1 = 0.2$).

In the following years the values of conversion factors were repeatedly specified and at present, the Normative Document SP 50-102-2003 specifies conversion factors for driven piles given in table 4.9 (q_c corresponds to the standard CPT in the speed of 1.2±0.3 m/min).The values of β_1 given in table 4.9 are reliable for both clayey and sandy soils.

Reliability of pile end bearing capacity is evaluated by recording of adjacent soil layers strength located close by the end of the pile (over and under it), but not only by selection of the conversion factor β_1. The forming plasticity zone (of limit equilibrium) embraces the considerable volume of soil impossible to be characterized by q_c measured in pile end depth. In the National Norms of Russia pile end bearing capacity is admitted to be evaluated by mean value of q_c in the zone; its upper limit is $1D$ higher than the mark of the pile end, while the lower one − $4D$ lower (D − diameter or cross-section side of the pile). Design zone of q_c selection is shown in figure 4.4. In table 4.9 q_c implies this mean value evaluated in accordance with fig.4.4.

Table 4.9 Conversion factors in transition from cone resistances (q_c) to pile foundation resistances (R_s) after SP 50-102-2003

q_c(kPa)	≤ 1,000	2,500	5,000	7,500	10,000	15,000	≥ 20,000
$\beta_1 = R_s/q_c$	0.90	0.80	0.65	0.55	0.45	0.35	0.30

Pile side bearing capacity was studied in BashNIIstroy by experiments with in-situ piles (Kolesnik 1972). To estimate the condition of the adjacent soil, excavation of piles was performed and soil was examined in different parts of the pile (fig. 4.5) (Kolesnik 1972, Ryzhkov 1992). Six piles 6 m in length were immersed by diesel-hammer SP-6B, excavation was carried out in 3 months after the driving. Soil conditions were characterized by lithological homogeneous stratum of alluvial-dealluvial clays of 15 m exposed bed thickness mainly of overconsolidated consistency (Ufa, terrace II of Shugurovka river, BashNIIstroy test site).

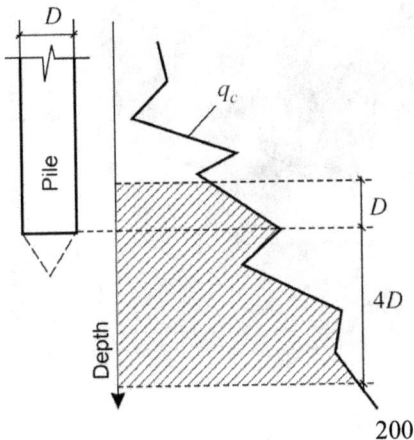

Fig. 4.4 Schematic illustration of the zone to evaluate mean value of q_c in calculation of pile end bearing capacity (R_s) after the National Norms of Russia (SNiP 2.02.03-85*, SP 50-102-2003)

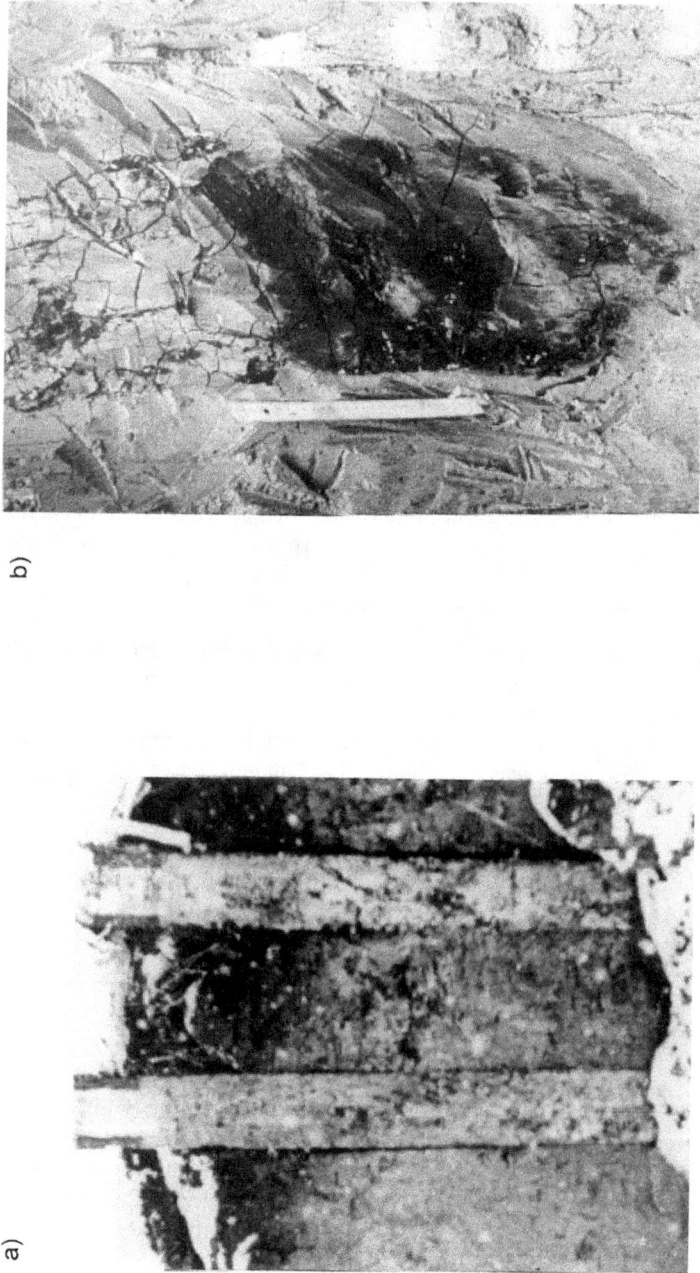

Fig.4.5 Adjacent soil after excavation of piles:

a – soil in the pile center (vertical openings between the pile and the soil are observed);

b – soil near the pile end (humus layer replaced by the pile from surface layer closely embraces the end)

Openings between pile side surface and soil were found; the largest ones were observed near the ground surface being 3...4 cm in width. In the depth of 1...2 m, the openings became broken, their width was nearly 1...3 mm decreasing in approaching to the end. In 1...3 m from the pile end they disappeared. The pile passing through various soil layers takes along the soil particles possessing the highest adhesion; in the first place, this refers to the top soil. Figure 4.5b illustrates the top soil (in black color) which remained having passed 6 m thickness of overconsolidated clays and formed dense enclosure (1...2 cm thick) around the pile end. Therefore, the pile "created" round widening of its end 1...2 cm on both sides or "reverse conicity" effecting on the contact between its surface and overlying soil.

Observations after a number of piles showed that the openings generating in the upper zone decrease when the pile is "at rest" but they do not disappear completely (at least, in clayey soils).

The analogical experiments were carried out in solid dealluvial loams (Sterlitamak) with piles 4 m in length, the results obtained being practically the same as those discussed earlier.

Type II cone penetrometers are not sensitive to the process described, as the friction sleeve is located in the minimal "examination" zone, thus one may ignore it. The results obtained correspond to the conditions of soil behavior in the pile end, i.e. near its edge. The definition of the results obtained by type II cone penetrometer distinguishes from those obtained by type I cone penetrometers where soil friction along the push rod can occur in a different way.

Figure 4.6 shows the experimental results after BashNIIstroy; friction sleeves of type II cone penetrometers were moved off the cone in distances of 0.15 cm and 64 cm (Ryzhkov 1992). Different remoteness was achieved by changing of standard cone tip on extensions of the same diameter but of different length (pos. 1, 2 & 3 in fig. 4.6b). In this case the resistances recorded by the cone sensor were not considered (they are not shown in fig. 4.6), as they reflected not "standard" cone resistance q_c but much more significant parameter $q_c + \Delta_f$, where Δ_f – incrementation conditioned by the additional friction of soil on side surfaces of extensions. Unit sleeve friction resistances were taken as usual but corresponded to different remoteness of the friction sleeve from the cone. CPT was carried out in alluvial-dealluvial overconsolidated clays (BashNIIstroy test site, Ufa). CPT locations (locations of penetration of cones with different extensions) were located randomly on a site 2×2 m; three CPT procedures were carried out with each extension, i.e. the experiment was done in accordance with the methodological principle discussed in section 2.2 (fig. 2.4).

As one can see from fig. 4.6a, unit sleeve friction resistances f_s decreased in moving of the friction sleeve away from the cone. In 64 cm moving off f_s (curve 3) decreased in comparison to the parameters corresponding to the standard location of the friction sleeve (curve 1), up to 5 times.

The issue on soil friction decreasing in its contact with metal surface moving along was described by H.K.S. Begemann (1953) for the first time. Begemann carried out a simple but obvious experiment, having measured soil layer resistance in through drilling of the soil by a long metal rod. Measurements were carried out either in drilling or in further advancement of the rod through the soil layer (the length of the rod considerably exceeded the thickness of the soil layer). The rod was moved downright, the soil was placed into the tray and its bottom was not an obstacle to moving of the metal rod.

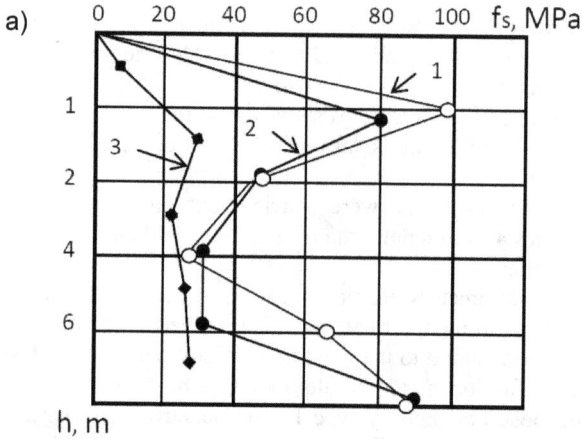

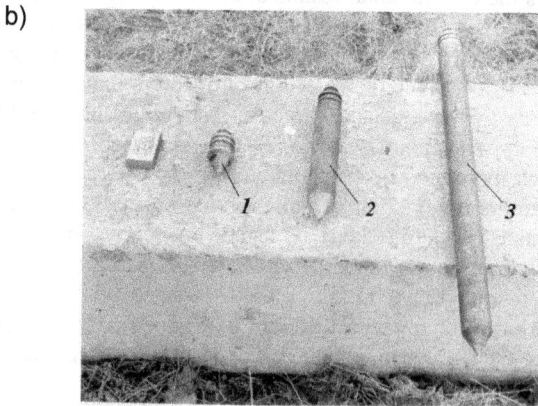

Fig.4.6 General view of penetrometers applied in experiments on effects of friction sleeve remoteness from the cone:

a – averaged graphs of f_s alteration in h depth applying the penetrometers with various extensions: 1 – in standard cone, 2 – in substitution of the cone by the extension of 15 cm, 3 – the same of 64 cm; b – general view of extensions (1, 2 and 3 – standard cone of 35.6 mm, extensions of 150 mm and 640 mm of the same diameter correspondingly)

Begemann's experiment revealed that soil resistance increased up to attainment of the lower border of soil layer by the rod. As the lower end (stabbing) of the rod drilled the tray (i.e. the soil layer was drilled) the further moving of the rod was accompanied by gradual decreasing of soil resistance (its friction upon the surface of the rod). The walls of the opening were being worn (abraded), thus decreasing the friction of the rod upon the soil. In here, soil resistance while decreasing asymptotically approached a small parameter approximate to zero.

The process of "workup" of contact between the moving solid surface and the soil always shows itself in cone or pile penetration. The upper soil layers are subjected to greater "workup", the lower ones – to smaller "workup". In piles this "workup" is intensified due to horizontal shifts during penetration and formation of "soil enclosure" at the lower end. No matter how homogeneous the soil is, when penetrating the cone, friction along the rod is distributed unevenly. The maximal "mobilization" of soil resistivity occurs nearby the cone tip, the minimal one – nearby the ground surface. Here, friction of the rod in one and the same soil layer decreases as the depth of penetration increases. When applying type I cone penetrometers measuring "lateral resistance" Q_s on the whole of the rod's length, this condition may generate additional distortions being difficult to take into account. There are situations when with increase of the depth of penetration total side friction resistance Q_s decreases; this does not mean any troubles in measuring equipment operation. It is evident that in type II cone penetrometers these troubles do not appear, since the friction sleeve is located in the zone of the closest contact with the soil where the "workup" process has not shown itself yet.

The information gives answer to the question why unit sleeve friction resistances f_s must be identified neither with side friction resistances of type I cone penetrometer nor with the soil resistances on greater part of pile side.

The experiments with tensopiles carried out in BashNIIstroy in 1960s proved the changeability of the contact between the surface of the pile and the soil (Shakhirev 1965). Reinforced concrete piles with 0.3×0.3 m in section and 6.4 m in length, equipped with membrane soil pressure cells measuring radial soil pressure on pile lateral surface (σ_r) were driven in 6 m. Soil pressure cells were placed along the pile side surface with the bay of 0.5 m. Up to 10 m soil alluvial overconsolidated clays were the soil conditions. Figure 4.7 shows distribution diagrams of side pressures drawn in accordance with soil pressure cells readings immediately after the penetration, in 6 days and in 15 days. Almost the same results were obtained by another tensopile of analogical geometry in analogical soil conditions.

As one can see from figure 4.7, pressure on pile side surface immediately after drilling was concentrated in the lower pile end; later, it increased in the upper and middle parts, while in the lower end it slightly decreased. The given information on distinction of side friction resistance and sleeve friction resistance was confirmed by G.S. Kolesnik (1972) who carried out experiments in lithological homogeneous soils of low density ($q_c < 2$ MPa). In these soils pile end bearing capacity R and cone resistance q_c (in conditional equilibrium, i.e. in "relaxation-creep" CPT) are admitted to be considered as equal ones. Pile side bearing capacity (total in the whole of the depth of drilling) evaluated by two modes (further marked as T_{test} and T_{cone}) were compared:

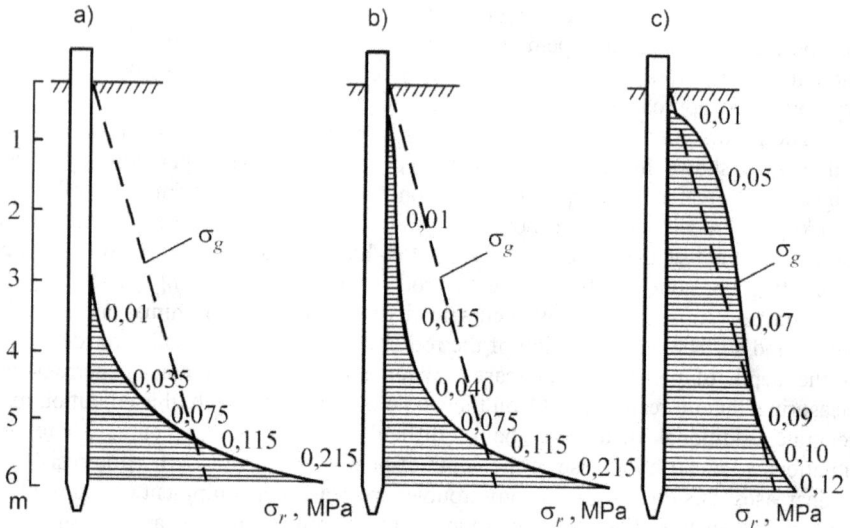

Fig. 4.7 Distribution diagrams of soil pressure on pile side surface σ_r (Shakhirev 1965):

a – immediately after drilling, b – in 6 days, c – in 15 days;
σ_g – overburden stress, σ_r – radial soil pressure on the pile

- By means of static tests of pile stamps allowing "pile side resistance" to be immediately measured or "standard" piles testing when the share was subtracted from the total pile resistance. It is the share of pile end bearing capacity evaluated by CPT from the condition $q_c = R_s(T_{test})$.
- Evaluation by CPT by total pile side bearing capacity calculation, given $\beta_2 = 1$, i.e. in $f_s = f_i(T_{cone})$.

Reinforced concrete piles and pile stamps 3...10 m in length, 0.3×0.3 m in section were used. CPT was performed by S-832 CPT rig.

Figure 4.8 illustrates the dependence of the ratio T_{test}/T_{cone} upon mean value of unit sleeve friction resistance $f_{s\ mean}$ obtained by G.S. Kolesnick. The figure illustrates the general trend of soil resistivity "degree of mobilization" decrease in its strength increase. Taking into account the data as well as publications and archives of the organizations that applied S-832 CPT rig, conversion factors β_2 were obtained; they allowed calculating unit pile side bearing capacity f_i by the value of unit sleeve friction resistance f_s by formula (4.13b).

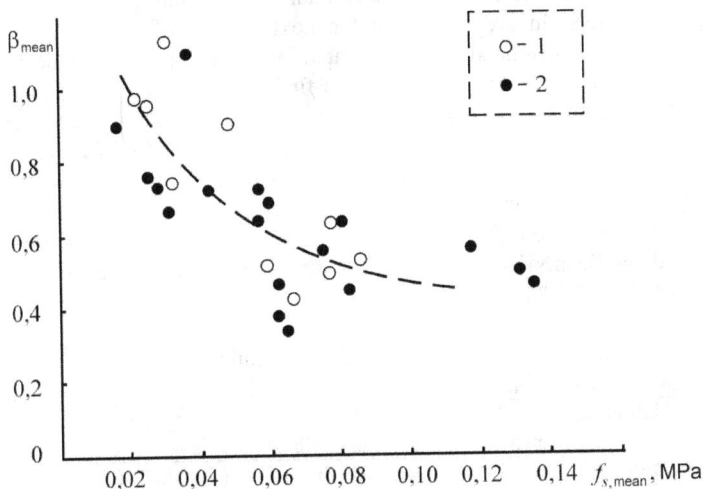

Fig. 4.8 Schematic illustration of dependence of soil resistivity "degree of mobilization" on pile side surface $\beta_{mean}=T_{test}/T_{cone}$ upon mean value of sleeve friction resistance $f_{s\,mean}$ in "relaxation-creep" CPT (Kolesnik 1972): 1 – pile stamps, 2 – "standard" piles tested ignoring division of resistance into front and side shares

Thus, the specialists of BashNIIstroy managed to develop rather reliable technique (±35 %) of driven pile bearing capacity by CPT data applying type II cone penetrometers (Temporary manual 1966). It was in 1966 for the first time in the world practice. Later, the technique was repeatedly adjusted (RSN 33-70, Temporary manual 1966, Workbook 1973, Manual 1980, Bilenko & Ryzhkov 1985, Workbook 1991) and since 1972 it has been recognized officially being included in All-Union Normative Document on Pile Foundations (SNiP) (first as the revision of chapter – SNiP II-B.5-67*, and then as the body matter of SNiP II-17-77). In the Norms on Pile Foundations Designing (Manual 1980) interpreting the requirements given in SNiP II-17-77 the tables of coefficients β_1 & β_i for two modes of CPT are presented – for traditional one (even penetration with the speed of 1.2±0.3 m/min) and "relaxation-creep" CPT reflecting soil resistances in the vanishing speeds of deformations.

It is necessary to point out that while changing the "status" of calculation (from Recommendations of a scientific organization to All-Union Normative Document compulsory for all institutions of the former USSR) its content was being simplified and adjusted to interpretations prevailing in worldwide CPT practice and designing of pile foundations. However, in such an adjustment the calculation attained not only some positive features but lost some rational elements of the preceding variants. This applied to the effect of depth of soil layer location on the coefficients β_i. In preparation of SNiP II-17-77 the columns ($h-2$ m) and ($h-1$ m) were withdrawn from the table 4.10 and it was suggested the values of β_i to be evaluated by interpolation in the interval 3 m...h in the depths of more than 3 m, where h – the depth of pile penetration in meters. In preparation of the next version of the Norms, i.e. SNiP 2.02.02-85 decreasing of pile side bearing capacity in the upper three meters "fell out " of the calculation technique. Thus, one can find the table of conversion factors β_2 with absolute ignorance of the depth of soil layer location, besides, the values of q_c & f_s correspond to just standard CPT (penetration with the speed of 1.2±0.3 m/min) in contemporary Normative Documents on pile foundations – SNiP 2.02.03-85* and SP 50-102-2003 of the beginning of the 21st century. To illustrate the mentioned changes the values of conversion factors β_i are given in the tables 4.10 and 4.11 as recommended by BashNIIstroy in 1970s (Workbook 1973) and by the Normative Document SP 50-102-2003 as well.

Table 4.10 Values of conversion factors β_i in formulae (4.12) & (4.13b) after the recommendations of BashNIIstroy of 1970s (h – depth of pile penetration, m)

Unit sleeve friction resistance (in "relaxation-creep" CPT) f_s (kPa)	Conversion factor β_i from f_{si} to f_i for type II cone penetrometers in the depth of i-soil layer:					
	1m	2m	3m ... (h–3m)	(h–2m)	(h–1m)	h
≤ 20	1	1	1	1	1	1
40	0.26	0.56	0.75	0.82	0.91	1
60	0.22	0.44	0.67	0.78	0.89	1
80	0.20	0.38	0.58	0.72	0.86	1
100	0.18	0.35	0.52	0.68	0.84	1
≥ 120	0.16	0.32	0.48	0.65	0.82	1

Table 4.11 Values of conversion factors β_2 in formulae (4.12) & (4.13b) after the Normative Document SP 50-102-2003 (SP 24.13330.2011)

Unit sleeve friction resistance (in standard CPT) f_s (kPa)	Conversion factor β_2 from f_s to f_i:	
	In sands	In clayey soils
≤ 20	0.75	1.0
40	0.60	0.75
60	0.55	0.60
80	0.50	0.45
100	0.50	0.40
≥ 120	0.50	–

If to compare the tables, one should keep in mind that table 4.10 presents "relaxation-creep" CPT data, hence, when anchoring it to standard CPT data it is necessary to reduce the values of β_i in clayey soils approximately in 30 %, in sands – in 20 %. For example, in "relaxation-creep" CPT in clays f_s = 40 kPa in the depth of 4 m is obtained; according to table 4.10 this corresponds to β_i = 0.75, but if f_s = 40 kPa is obtained in standard CPT (i.e. in even penetration of the cone), then the coefficient β_i is to be lower: β_i = 0.75/1.3 = 0.58.

207

The modified calculation became simpler, evaluation of f_s in sands was specified (before this β_i in sands and clays was equally evaluated), however, its reliability slightly decreased in clays due to the correction connected with ignorance of soil layers location relative to the pile end. At the beginning of the 21st century (since 2004) the corrected calculation technique has been given in two acting Norms – SNiP 2.02.03-85* and SP 50-102-2003, the latter (SP 50-102-2003) taking somewhat higher values of β_1 & β_i for strong soils. Thus, SNiP 2.02.03-85* gives $\beta_1 = 0.3$ in $q_c \geq$ 20,000 kPa, while in $q_c \geq$ 30,000 kPa $\beta_1 = 0.2$. At the same time, SP 50-102-2003 gives $\beta_1 = 0.3$ for $q_c \geq$ 20,000 kPa. In $f_s \geq$ 100 kPa in sands SNiP gives $\beta_i = 0.45$, while SP – $\beta_i = 0.50$. In $f_s \geq$ 120 kPa in clays SNiP gives $\beta_i = 0.3$; here, there is a lack of any values (dash). It seems reasonable to take the value of β_i "in reserve" as in $f_s = 100$ kPa. The values of β_1 & β_i admitted in SP 50-102-2003 are given in tables 4.9 and 4.11.

Testing calculations show that for most soil conditions the simplification does not significantly effect on reliability of calculation particularly in piles exceeding 10 m in length. However, in lower depths of pile penetration the "errors" may be rather significant, especially in explicit increasing or decreasing of f_s with depth. In increasing of f_s pile resistance is usually underestimated in calculations, i.e. the "errors" create the additional "safety factor". In decreasing of f_s pile resistance is usually overestimated, i.e. errors become "dangerous".

Figure 4.9 shows the results of testing calculations after by V.M. Enikeyev and S.L. Berezina (1989) corroborating the consideration given. Ultimate pile resistances obtained by static tests of full-sized piles were compared with the analogous resistances obtained by two modes of calculation after CPT data: the "new" one (in accordance with SNiP 2.02.03-85), in here, β_i was independent upon the depth and the "old" one (SNiP II-17-77), in here, the depth was taken into account. The archives of the Bashkir surveying institution (ZapuralTISIZ) were used. The institution carried out large amount of work on pile static testing and CPT by S-832 CPT rig. 40 sites with soil conditions characterized *by decreasing of soil strength with depth* were chosen. CPT data obtained by the traditional (standard) mode were applied. As it was expected, the calculation done with regard to SNiP 2.02.03-85 revealed the increase of ultimate pile resistances, i.e. "dangerous" deviations prevailed (fig. 4.7a), while the "old" calculation by SNiP II-17-77 did not reveal the deviation (fig. 4.7b).

Figure 4.10 presents the general diagram of dissipation illustrating accumulated in 1960s...1980s reliability estimation results of resistance evaluation after BashNIIstroy.

Ultimate driven reinforced pile resistances obtained by calculations after CPT data by S-832 CPT rig (type II cone penetrometer, i.e. with friction sleeve) in accordance with SNiP 2.02.03-85 (see formula (4.13) and tables 4.9 & 4.11) were compared with the ultimate resistances obtained by static tests of full-sized piles. Both CPT and pile tests (503 piles) were carried out in different regions of the former USSR including the Ukraine, the European part of Russia and Western Siberia. The driven reinforced piles 3...15 m in length, with the cross section from 0.2×0.2 m to 0.4×0.4 m were considered.

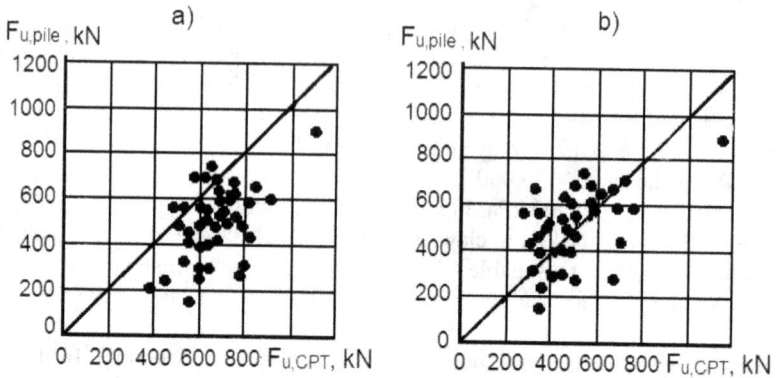

Fig. 4.9 Schematics of comparison of ultimate prismatic pile resistances evaluated by CPT data applying S-832 CPT rig ($F_{u,CPT}$) and by static tests of full-sized piles ($F_{u,pile}$) in soils losing their strength with depth:

a – in calculations of $F_{u,CPT}$ after SNiP 2.02.03-85, b – in calculations of $F_{u,CPT}$ after the former Norms of SNiP II-17-77

As one can see from fig. 4.10, most of comparison results between $F_{u,CPT}$ & $F_{u,pile}$ (75...80 % of locations) are characterized by nearly 30 % divergence, while the other ones – 20...25 % of results diverge more. Mean values of $F_{u,CPT}/F_{u,pile}$ ratios in the whole of data were 1.02; mean-square (standard) deviation – $\sigma = 0.31$. With 0.95 probability this corresponds to maximum divergence between $F_{u,CPT}$ & $F_{u,pile}$ nearly ±62 %. Nevertheless, there is no need to introduce large safety factors – $\gamma_k =1.62$ in choosing permissible loads due to the following:

- In general, when designing a pile foundation mean values of $F_{u,CPT}$ in one or the other section or the whole of the foundation play the principal role rather than particular values of pile resistances, since redistribution of loads on piles occurs in a structure; mean values present less inaccuracies than particular ones.

- The discussed standard deviation characterizes *particular* values of $F_{u,CPT}/F_{u,pile}$ ratios, standard deviation of *mean* values in definite selections is to be in $\sqrt{n}$ times lower (n – number of CPT locations at the site or its section), i.e. in determining mean value of pile resistance, say, in 6 CPT locations, standard deviation of the mean value shall be $0.31/\sqrt{6} = 0.13$, i.e. the "error" of the mean value shall be $2.01 \cdot 0.13 = 0.26$ or 26 % (2.01 – Student's coefficient in degree of freedom number 6 - 1=5 and one-sided confidence probability – 0.95 (GOST 20522-96).

- The spread of results is determined by inaccuracy of $F_{u,CPT}$ computation and heterogeneity of soil as well (see section 3.2.1); this inaccuracy considerably increases the spreading, thus, if the coefficient of correlation for the "ideally

209

accurate computation" was 0.8 in real heterogeneity of soil (section 3.2.1), the standard deviation of $F_{u,CPT}/F_{u,pile}$ would decrease from 0.31 to 0.28 without the effect of heterogeneity.

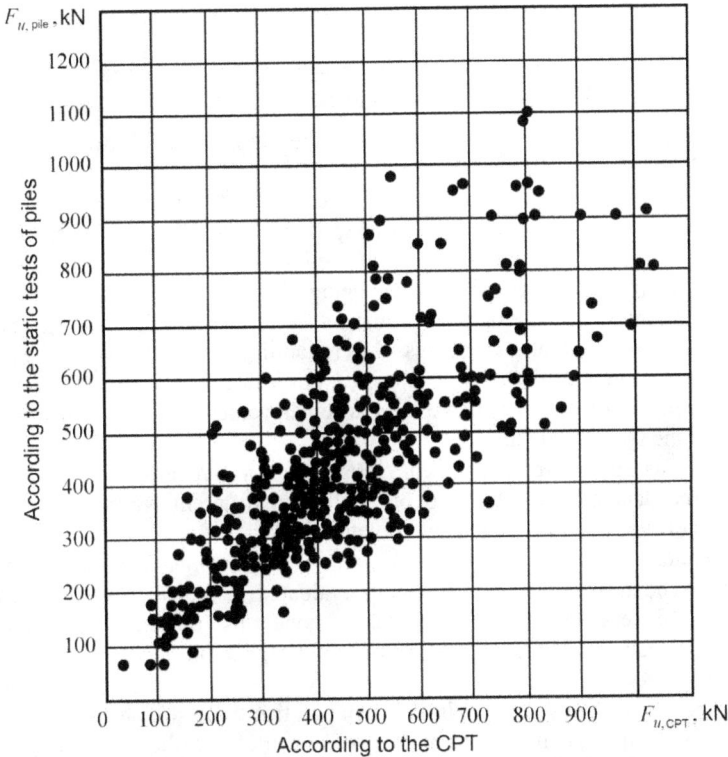

Fig. 4.10 Schematic illustration of comparison of ultimate pile resistances evaluated by CPT data in accordance with SNiP 2.02.03-85 applying S-832 CPT rig equipped with type II cone penetrometer ($F_{u,CPT}$) with ultimate resistances evaluated by static tests of full-sized piles ($F_{u,pile}$)

The given argument concerning the difference between inaccuracies of particular and mean values may be illustrated by the experiment carried out in BashNIIstroy in 1967 in alluvial-dealluvial clays (Ryzhkov 1992). 11 driven piles were tested by static load at the site of approximately 150 m^2 (after CPT in each location of driving). The piles were 6 m in length with the section of 0.3×0.3 m. The results of $F_{u,CPT}$ computation obtained and the results of static tests $F_{u,pile}$ are given in table 4.12.

210

Table 4.12 Ultimate pile resistances obtained in-situ

№ of a location	1	2	3	4	5	6	7	8	9	10	11
$F_{u,CPT}$ (кН)	241	234	193	260	234	220	219	231	229	256	247
$F_{u,pile}$ (кН)	250	250	250	230	300	250	200	325	275	250	250
$F_{u,CPT}/F_{u,pile}$	0.96	0.94	0.77	1.13	0.78	0.88	1.10	0.71	0.83	1.02	0.99

Mean values were the following: $\overline{F}_{u,зон} = 241$ kN, $\overline{F}_{u,св} = 257$ kN, i.e. $\overline{F}_{u,зон}/\overline{F}_{u,св} = 241/257 = 0.94$. Thus, in the range of particular values alteration of $F_{u,CPT}/F_{u,pile}$ from 0.71 to 1.13, mean value of the ratios deviated towards decreasing in 0.06 (6 %).

As it was mentioned earlier, reliability coefficient $\gamma_k = 1.25$ is applied in the National Norms of Russia SNiP 2.02.03-85*, SP 50-102-2003, SP 24.13330.2011 in computation of admissible loads; it is introduced not to separate (particular) values of $F_{u,CPT}$, but to pile bearing capacity F_d being the corrected mean value of pile resistances on the whole site or on its section (see formulae 4.11 & 4.12). Therefore, the reliability coefficient γ_k ensures the reliability of mean values estimation (normative resistances, pile bearing capacity), but it does not exclude the occurrence of separate points on the site where the particular values of $F_{u,CPT}$ will be lower or higher than the specified load.

The discussed features of soil behavior are displayed regardless of the cone type; however, the data obtained have to be applied in different ways due to different geometries of types I and II cone penetrometers.

Computation by data obtained by type I cone penetrometers are done by formula (4.14), pile end bearing capacity R_s being evaluated in the same way as in type II cone penetrometers. Different geometries of cone tips (presence or absence of a mantle – see section 1.1) are usually ignored in computation of pile bearing capacity. Significant difference in evaluation of q_c in types I and II cone penetrometers is displayed in weak clayey soils mainly, but in practice these soils hardly ever support the piles. As in type II cone penetrometers, when evaluating pile end bearing capacity R_s mean value of q_c in the zone located above or below the mark of the lower end is applied (in $1D$ & $4D$ correspondingly – see fig. 4.4).

In type I cone penetrometers evaluation of pile side bearing capacity has to be done in a different way, since *total friction force* Q_s is measured in all embedded part of the push rod in these cones, while in type II cone penetrometers *unit sleeve friction resistance* f_s is measured. "Workup" processes occur in contact zone between side surface of the push rod (type I cone penetrometer) and the soil, thus, friction forces are unevenly distributed along the push rod concentrating in the lower part of the push rod even in homogeneous-in-depth soils. Though one can fail to accurately determine distribution of friction along the push rod during CPT, it is similar to distribution of friction along the side surface of the pile. It is this fact that allows applying total resistance Q_s in pile bearing capacity evaluation.

The technique of pile bearing capacity evaluation by the data obtained by type I cone penetrometers as well as the analogous technique applied for type II cone penetrometers was repeatedly adjusted in 1960s...1980s. The specialists of Fundamentproject (Leonychev 1964, Trofimenkov & Mariupolsky 1975, Trofimenkov & Vorobkov 1981) were involved in this work. At first, conversion factors β_1 & β_2 were taken as constant, β_1 equaled 0.5, and $\beta_2 = 1$. However, in practice the approach did not ensure the required accuracy of results; thus, the technique was reviewed time and again aiming at differential approach to selection of coefficients β_1 & β_2. In the present-day Normative Documents of Russia on pile foundations (SNiP 2.02.03-85*, SP 50-102-2003, SP 24.13330.2011) conversion factors β_1 & β_2 are taken with CPT data obtained: β_1 is defined with q_c, and β_2 – with f_s (see table 4.13).

Table 4.13 Values of conversion factors β_1 and β_2 in formula (4.14) in type I cone penetrometers after SP 50-102-2003

q_c (kPa)	β_1 – conversion factor from q_c to R_s	f_s (kPa)	β_2 – conversion factor from f_s to f for type I cone penetrometers
$\leq 1,000$	0.90	≤ 20	1.80
2,500	0.80	40	1.30
5,000	0.65	60	1.00
7,500	0.55	80	0.80
10,000	0.45	100	0.60
15,000	0.35	≥ 120	0.50
$\geq 20,000$	0.30	-	-

The same as for type II cone penetrometers the rules of the Norms (SNiP 2.02.03-85*, SP 50-102-2003, SP 24.13330.2011) slightly differ: SNiP 2.02.03-85* determines different values of β_2 for sands and clayey soils, but in SP 50-102-2003 they are identical. The same as in type II cone penetrometers in $q_c \geq 20,000$ kPa $\beta_1 = 0.3$ after SP.

Figure 4.11 illustrates the diagram of dissipation showing reliability estimation results of driven reinforced concrete pile bearing capacity evaluation by CPT data obtained by type I cone penetrometers with S-979 and SP-59 CPT rigs during 1960s-1980s (Trofimenkov & Vorobkov 1981). As in figure 4.10, ultimate driven pile bearing capacity obtained by CPT data ($F_{u,CPT}$), was compared with resistances obtained by data after static loading ($F_{u,pile}$). Test results of 153 reinforced concrete piles from 3 to 11m in length, with the section from 0.25×0.25 to 0.35×0.35 m were applied. Comparisons were made in various regions of the former USSR (on 57 sites).

Boring pile bearing capacity computation by CPT data was developed and included in the National Norms much later than the analogous computation for driven piles. In SP 50-102-2003, SP 24.13330.2011 the formula is given for such a computation; in it ""side friction resistances", i.e. sleeve friction resistances – f_s in

212

type II cone penetrometers or the ones on the push rod – Q_s in type I cone penetrometers are not used. It is cone resistance q_c that is the initial data.

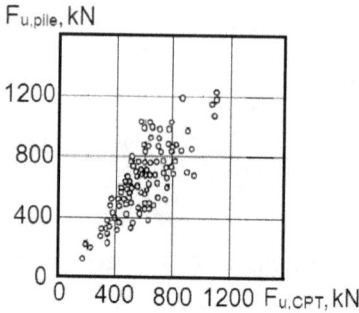

Fig.4.11 Schematic illustration of ultimate pile bearing capacity evaluated by CPT data computation by S-979 and SP-59 equipped with type I cone penetrometer($F_{u,CPT}$), and ultimate pile bearing capacity evaluated by static tests of full-sized piles ($F_{u,pile}$) comparison (Trofimenkov & Vorobkov 1981)

Unfortunately, one can notice the error in SP 50-102-2003, SP 24.13330.2011 in terminology admitted; here, "pile bearing capacity" is understood as a generalized parameter characterizing site conditions but not a separate location. Resistance of boring pile in a CPT location is called "pile bearing capacity in a CPT location F_{du}" but not a particular value of ultimate pile resistance (i.e. "bearing capacity" is again understood as a general term). The "bearing capacity" of the boring pile is evaluated by the formula

$$F_{du} = RA + u\sum \gamma_{cf} f_i h_i ,$$ (4.16)

where:

$R =$ design pile end bearing capacity admitted by table 4.14 due to mean value of cone resistance q_c in the zone located *one diameter above and up to two diameters below* the pile foundation;

$A =$ pile foundation area;

$f_i =$ mean value of design pile side bearing capacity in design pile zone h_i evaluated by CPT data after table 4.14;

Table 4.14 Pile end bearing capacity and boring pile side bearing capacity in the formula (4.16) after SP 50-102-2003, SP 24.13330.2011

Cone resistance q_c (kPa)	Design boring pile end bearing capacity R (kPa)		Mean value of design pile side bearing capacity f_i (kPa)	
	Sands	Clayey soils	Sands	Clayey soils
1,000	-	200	-	15
2,500	-	580	-	25
5,000	900	900	30	35
7,500	1,100	1,200	40	45
10,000	1,300	1,400	50	60
12,000	1,400	-	60	-
15,000	1,500	-	70	-
20,000	2,000	-	70	-

$h_i =$ thickness of i-soil layer which is to be taken up to 2 m;

$\gamma_{cf} =$ coefficient dependent upon pile manufacturing technology taken:
- In piles being concreting dry $\gamma_{cf} = 1$;
- In concreting under water, under clay mortar and in application of inventory casing pipes $\gamma_{cf} = 0.7$.

The given computation presupposes the pile diameter to be 0.6...1.2 m, embedded in not less than 5 m.

Ultimate screw pile bearing capacity in jacking and pulling loading is evaluated by the same formula as in driven piles, i.e. by the formula (4.13), but the conversion factor β_1 is different. Table 4.15 contains the values of β_1 for screw piles. Here, the operating range where mean value of q_c is evaluated in computation of R, is taken equal to *one diameter* of a pile lobe. Screw pile side bearing capacity is evaluated in the same way as in a driven pile, i.e. β_2 is taken from tables 4.11 or 4.13.

Table 4.15 Values of conversion factor β_1 in the formula (4.14) in evaluation of screw piles after SP 50-102-2003, SP 24.13330.2011

q_c (kPa)	β_1 – conversion factor from q_c to R_s for screw piles in loading	
	jacking	pulling
≤ 1,000	0.50	0.40
2,500	0.45	0.38
5,000	0.32	0.27
7,500	0.26	0.22
10,000	0.23	0.19
15,000	-	-
≥ 20,000	-	-

For screw piles embedded in water-saturated sands values of β_1 are twice reduced in table 4.15.

4.3.3 Application CPT for pile bearing capacity evaluation in foreign countries

In non-Russian practice CPT has been applied for pile bearing capacity evaluation since 1930s, the popularity of the application being typical for European countries where pile foundations are widely spread (Netherlands, Sweden, etc.). In foreign countries pile bearing capacity evaluation is usually done by Recommendations not possessing the status of State Codes. They are the recommendations of scientific institutions, scientific production associations, higher institutions or developers of geotechnical work not available to aliens. In many countries the National Norms lack in evaluation techniques of pile bearing capacity by CPT data. Nevertheless, nowadays, the calculations are being included in the Norms. This refers to the latest European "Pre-standards", i.e. International Normative Documents taken as the basis of International and National Standards' development. For example, "pre-standard" on geotechnical designing ENV-1997-3 related to Eurocod 7 system includes the ultimate ("maximum") pile bearing capacity evaluation technique by CPT data. The computation is based on conventional dependence (4.13) being added by formulae and tables for evaluation of soil resistances under the pile foundation (lower end) and on a pile side surface (driven pile, boring pile without widening, boring pile with widening, bored pile, screw pile, etc.).

Pile end bearing capacity and pile side bearing capacity are evaluated by more complex ways in the pre-standard (as compared to Russian approach); this is due to consideration of a number of factors. Thus, in the original *ultimate ("maximum") pile end bearing capacity* $p_{max, base}$ (corresponding to Russian symbol R_s) is evaluated by the formula

$$p_{max;base} = 0.5 \cdot \alpha_p \cdot \beta \cdot s \cdot \left(\frac{q_{c;I;mean} + q_{c;II;mean}}{2} + q_{c;III;mean} \right), \qquad (4.17),$$

where:

$\alpha_p =$	coefficient depending on pile and soil types is evaluated by table 4.16;
$\beta =$	coefficient depending on the relative depth of lower end and relative dimensions of pile widening is evaluated by the graph (see fig. 4.12c);
$s =$	coefficient considering lower pile end form is evaluated by the graph (see fig. 4.18d);
$q_{c;I;mean}, q_{c;II;mean}, q_{c;III;mean} =$	mean cone resistances in the zones adjacent to the pile foundation (fig. 4.12a):
$q_{c;I;mean} =$	corresponds to the zone located below the pile foundation extended to the so called "critical depth" d_{crit}, corresponding to the nearest weak soil layer roofing with the resistance of < 2 MPa, this zone being propagating below in minimum $0.8D_{eq}$ and maximum – $4D_{eq}$ (D_{eq} = diameter or cross-section side of the pile foundation);
$q_{c;II;mean} =$	corresponds to the same zone, but with the assumption that cone resistance equals to minimum parameter $q_{c;II;mean}$, equal to q_c in the critical depth – d_{crit};
$q_{c;III;mean} =$	corresponds to the zone located above the pile foundation propagating up to the level in $8D_{eq}$ higher than the pile foundation.

Limitation of maximum soil resistance under the pile foundation $p_{max, base}$ which shall not be taken above 15 MPa is the additional condition, i.e. in obtaining $p_{max, base}$ > 15 MPa, $p_{max, base}$ = 15 MPa is taken.

215

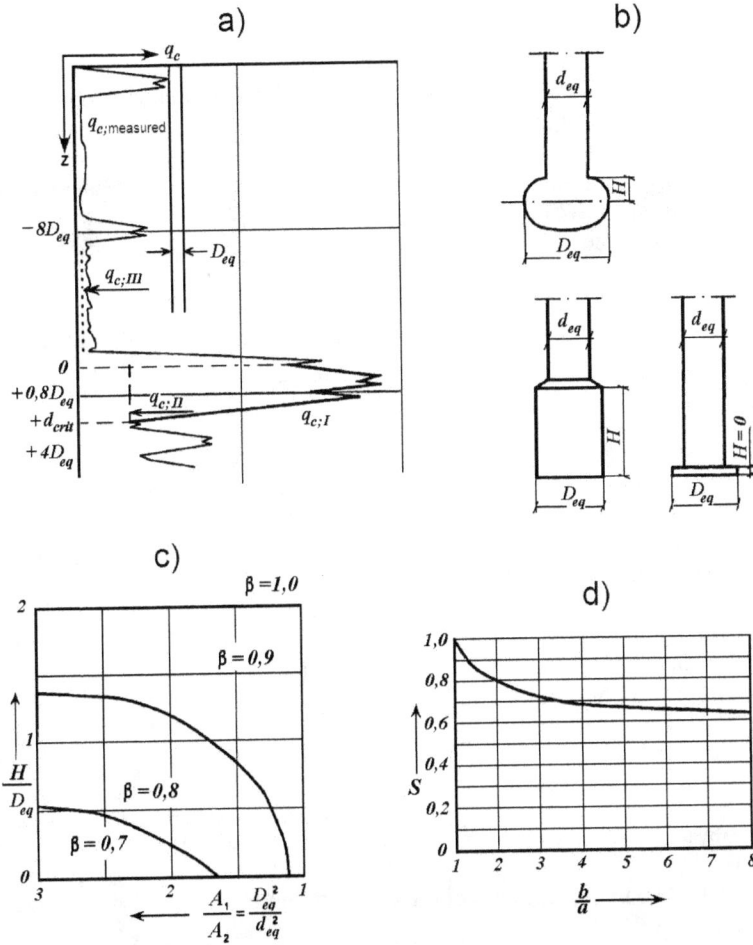

Fig. 4.12 Schematics illustrating the formula parameter 4.17 (ENV 1997-3):
a – explanation of values $q_{c;I;mean}$, $q_{c;II;mean}$, $q_{c;III;mean}$, b – dimensions of pile members (pile body and widening), c – graph of β evaluation, d – graph of coefficient s evaluation

If the pile foundation is of a rectangular form (dimensions $a \times b$), the diameter ("equivalent diameter") is taken $D_{eq} = 1.13a/(b/a)$, where a – shorter section side, b – longer side.

Ultimate ("maximum") pile side bearing capacity $p_{max,shaft}$ is evaluated by the formula

216

$$p_{max,shaft} = \alpha_s \cdot q_{c;z;a} , \qquad\qquad (4.18)$$

where:

α_s = coefficient considering types of pile and soil as well is evaluated by tables 4.16 and 4.17;

$q_{c;z;a}$ = cone resistance in the examined depth z in two additional conditions:
- If the soil layer with $q_c \geq 15$ MPa possesses the capacity of 1 m or more, $q_{c;z;a} \leq 15$ MPa is taken,
- If the soil layer with $q_c \geq 12$ MPa possesses the capacity of less than 1 m, $q_{c;z;a} \leq 12$ MPa is taken.

Table 4.16 Maximal values of coefficients α_p & α_s in formulae (4.17) and (4.18) for sands (ENV 1997-3)

Types of piles	α_p	α_s^*
Piles over 150 mm in diameter arranged by extrusion (forced release) of soil:		
- Driven piles,	1.0	0.01
- In-situ piles arranged by immersion of a metal tube with covered lower end being driven off during concreting, and the tube is extracted while concrete is discharged	1.0	0.014
Piles over 150 mm in diameter arranged by extrusion (or insignificant release) of soil and its substitution for concrete (or other material):		
- Boring piles concreted under clay mortar,	0.6	0.005
- Screw piles	0.8	0.006**

* In gravel sands α_s increases by multiplication in 0.75, in gravel – by multiplication in 0.5.

** The given value presupposes CPT application prior to pile arrangement. If CPT is carried out nearby the immersed screw pile, α_s^* is allowed to be raised up to 0.01.

Table 4.17 Maximal values of coefficient α_s in the formula (4.16) for clayey soils and peat (ENV 1997-3)

Types of bases		Relative depth of the layer z/d_{eq}*	α_s^*
Clayey soil in	$q_c \leq 1$ MPa	6...19	0.025
The same	$q_c \geq 1$ MPa	≥ 20	0.055
The same	$q_c > 1$ MPa	Regardless of depth	0.035
Peat		Regardless of depth	0

* d_{eq} – pile body diameter (see fig. 4.18b)

In general, the calculation technique on driven piles recommended by the European pre-standard is not as thoroughly developed as in Russia SNiP 2.02.03-85* & SP 50-102-2003; however, it excels in its versatility embracing practically all types of piles. The authors of the pre-standard prefer make use of neither unit sleeve friction resistance f_s (type II), nor total side friction resistance Q_s (type I); this is probably justified for piles made of mass concrete (boring piles, drilled piles, etc.) but not for driven ones. As opposed to the National Norms, in which pile end (and side) bearing capacity are not limited, in the pre-standard the capacities increase with the increase of q_c just up to *a certain value*; after that, they remain constant whatever high the values of q_c are. It seems reasonable as there is a lot of experimental data relating to CPT applications in overconsolidated soils both in Russia and worldwide. Besides, it is important to underline that pile end bearing capacity is regulated in the European Pre-standard in more complicated way than that in the National Norms of Russia.

Relative complexity of the method discussed is considerably connected with the diversity of piles mentioned above. In limited nomenclature of the applied piles the computation may be significantly simplified. In particular, for square section driven piles $\alpha_p = \alpha_s = \beta = s = 1$. It is evident that the efficiency of the suggested ways of consideration of diversity of factors is possible to be revealed in practical testing of computation in a wide range of conditions.

Improvement of the computation technique is probably to be done by specification of coefficients and expansion of the range of soil conditions. Tables 4.16 and 4.17 do not give any information on the coefficient α_p for clayey soils. Specific soils (collapsible, swelling) are not considered either; besides, it is not clear how to do calculations in availability of filled soils, etc.

As opposed to the method of ENV-1997-3 assuming improvement and connection to the local conditions, a number of computation techniques tested in various regional conditions are applied worldwide, but they miss the status of the Norms. These techniques serve as a subject of discussion; the experiments on their reliability estimation are carried out and thus, the useful data are collected.

T. Lunne et al. (2004) present the notes on non-Russian publications; in them comparison results between the known pile bearing capacity computation techniques and CPT results of the piles are given. The independent research was carried out in different countries (in general, in 1980s). As in other non-Russian publications, the research of Soviet and Russian professional engineers was ignored.

Comparative experiments were carried out in various soil conditions; CPT results of 150 driven, boring, drilled, etc. piles were applied. Pile lengths and diameters varied, thus, maximum testing loads were from 170 to 8,000 kN. CPT equipment included both mechanical and electrical penetrometers (without a friction sleeve and with it).

The general conclusion was as follows: *calculations based on CPT application exceeded other available ones in their reliability* (e.g. in soil estimation laboratory data, in dynamic tests, etc.). The most reliable methods were believed to be the following ones:

- Method of M. Bustamante and L. Gianeselli (method of LCPC).
- Method of J. De Ruiter and F.L. Beringen.

In most cases their application ensures the reliability of ±30...35 %

Method of M. Bustamante and L. Gianeselli (method of LCPC) (Bustamante & Gianeselli 1982) was developed in 1982 in France in the central laboratory of bridges and roads (Laboratoire Central des Ponts et Chaussées – LCPC); later in 1993 it was included in the National Norms of France (FOND-72) with some simplifications. The method presents the empirical generalization of CPT results of 197 piles of different structures (reinforced concrete and steel driven piles, boring piles, drilled piles, etc.) and parallel CPT. The dependency (4.8) being the bases of calculation has the following view:

$$Q_{ult} = q_p A_p + f_p A_f$$ (4.19)

In Russian signatures (SNiP 2.02.03-85*, SP 50-102-2003)

$$F_u = R_s A + f A_f,$$ (4.19a)

where:

Q_{ult} = ultimate pile bearing capacity corresponding to ultimate pile resistance F_u according to SNiP 2.02.03-85* & SP 50-102-2003,

q_p = pile end bearing capacity(in the Norms of Russia SNiP 2.02.03-85* & SP 50-102-2003 the parameter is marked as R),

$A_p (A)$ = pile cross-sectional area,

A_f = pile side surface area,

$f_p (f)$ = averaged pile side resistance.

In heterogeneous soils the side surface is divided into sections, soil resistance is evaluated in each section and the results obtained are added, i.e. instead of $f_p A_f$ the sum $\Sigma(f_{p,i} A_{f,i})$ is used. There are not any strict rules of this division.

Pile end bearing capacity q_p (in the Norms of Russia – R_s) is evaluated in agreement with the value of cone resistance q_c

$$q_p = q_{ca} \cdot k_c,$$ (4.20)

where q_{ca} = averaged (and corrected by definite rules) value of cone resistance in pile lower end zone.

q_{ca} is evaluated by the graph "$q_c \sim h$" in three stages (fig. 4.13):

- q'_{ca} = mean value of cone resistance (q_c) is evaluated in the zone located at the lower end of a pile in the depth from $L–a$ to $L+a$ (L – depth of lower pile end $a = 1.5D$, D – pile diameter);
- values of q_c in this zone are excluded, those ones beyond $0.7q'_{ca} < q_c < 1.3q'_{ca}$ (the borders of the interval are shaded in fig. 4.13);
- the desired quantity of q_{ca} is evaluated as the mean value of q_c at two-sided "truncated" curve "$q_c \sim h$" obtained in the examined zone (q_{ca} is shown as a thick dashed vertical line in fig. 4.3).

k_c = "coefficient of pile lower end bearing capacity" evaluated by table 4.18.

In computation of q_p the piles are divided into two groups.

Group I includes piles made up of in-situ concrete: boring piles, concreted "dry" piles and under clay mortar ones; drilled piles with casing pipes; drilled piles with cavities; massive micropiles constructed by intrusion of mortar in holes of a small diameter (under 250 mm) under low pressure; walls in soil, etc. These do not consolidate the adjacent soil.

Group II includes piles that consolidate the adjacent soil: prismatic driven piles, steel or reinforced concrete round hollow tubes; in-situ concrete piles made up in holes formed after forced swaging (by driving or jacking of casing pipes, etc.); micropiles constructed by intrusion of mortar in holes under high pressure, etc.

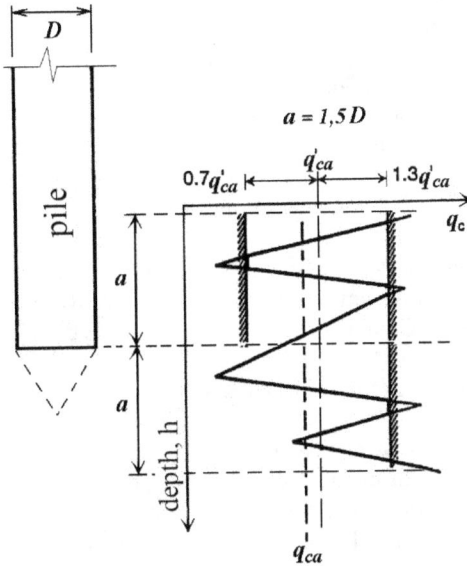

Fig. 4.13 Schematic illustration of q_{ca} evaluation in the formula (4.13), i.e. in pile resistance computation by the method of M. Busmante and L. Gianeselli (LCPC)

Table 4.18 Coefficient k_c in formula (4.20) (Bustamante & Gianeselli 1982)

Type of soil	q_c (MPa)	Coefficient k_c	
		Group I	Group II
Soft clays and muds	< 1	0.4	0.5
Clays of medium density	1...5	0.35	0.45
Silts and loose sands	≤ 5	0.4	0.5
Dense rigid clays and silts	> 5	0.45	0.55
soft chalk	≤ 5	0.2	0.3
Sands of medium density and gravels	5...12	0.4	0.5
Disintegrated chalk	> 5	0.2	0.4
Dense and very dense sands and gravels	> 12	0.3	0.4

Pile side bearing capacity f_p is also evaluated by cone resistance q_c:

$$f_p = q_c / \alpha_{LCPC},$$ (4.21)

where α_{LCPC} = "friction coefficient" evaluated by table 4.19 with regard to both soil and pile type.

The authors (Bustamante & Gianeselli 1982) thought the transition from sleeve friction resistance f_s (or side friction resistance Q_s of type I cone penetrometers) to pile side bearing capacity f_p to be rather complicated issue, thus, they decided to ignore it having admitted the values of q_c as the initial data. This approach favors the universality of the formulae applied, since they become applicable to both types I & II cone penetrometers (see section 1.1). But in this case the great demands are made to completeness and reliability of lithological identification of soil. The results prove to vary depending on whether the soil is referred to overconsolidated clay, silt or disintegrated chalk, etc. (in one and the same values of q_c). This allows applying CPT if a number of boreholes are available on the site and laboratory tests are carried out as well.

When evaluating pile side bearing capacity f_p more complex classification of piles is applied. Besides division of piles into groups I & II, each group is subdivided into subgroups – A & B. Combinations IA, IB, IIA or IIB called "categories" of piles determine the selection of the relevant column in table 4.19. Every category is characterized by the degree of the adjacent soil consolidation and the contact between pile side surface and the soil.

Category IA incorporates piles influencing minimally the adjacent soil, but ensuring close contact with the adjacent soil: boring piles made up ignoring casing pipes application ("dry concreting" or under clay mortar); boring piles with cavities; massive micropiles constructed by intrusion of mortar under low pressure; walls in soil.

Category IIA embraces driven reinforced concrete piles including pre-stressed tubular ones (round hollow); piles composed of separate sections joined together immersed by a lifting jack.

Category IB embraces piles effecting more on the adjacent soil but failing in the contact with it: boring piles constructed applying casing pipes; in-situ concrete piles arranged by driving of a casing pipe (either pulled or not) with the following filling with concrete.

Category IIB – metal driven piles; multi-sectional metal piles immersed by a lifting jack.

One of the typical features of LCPC is the determination of ultimate (maximum) values of f_p *inadmissible to be exceeded*, whatever the value of q_c / α_{LCPC} is. These ultimate values are given in table 4.19. In order to evaluate the values, piles are divided into three groups now; each group is subdivided into two subgroups – A & B, i.e. there are 6 categories – IA, IB, IIA, IIB, IIIA, IIIB. Some pile types after α_{LCPC} evaluation are referred to group II; in determination of "maximum admissible" value of $f_{p,max}$ they are related to group III (e.g. reinforced concrete driven piles).

Table 4.19 Coefficient a_{LCPC} in formula (4.21) (Bustamante & Gianeselli 1982)

Type of soil	q_c (MPa)	Coefficient a_{LCPC} Group I A	B	Group II A	B	Maximum values of f_p (MPa) Group I A	B	Group II A	B	Group III A	B
Soft clays and muds	< 1	30	90	90	30	0.015	0.015	0.015	0.015	0.035	–
Normally consolidated clays	1...5	40	80	40	80	0.035 (0.08)*	0.035 (0.08)	0.035 (0.08)	0.035	0.08	≥ 0.12
Silts and loose sands	≤ 5	60	150	60	120	0.035	0.035	0.035	0.035	0.08	–
Dense solid clays and silts	> 5	60	120	60	120	0.035 (0.08)	0.035 (0.08)	0.035 (0.08)	0.035	0.08	≥ 0.20
Soft chalk	≤ 5	100	120	100	120	0.035	0.035	0.035	0.035	0.08	–
Sands of medium density and gravels	5...12	100	200	100	200	0.08 (0.12)	0.035 (0.08)	0.08 (0.12)	0.08	0.12	≥ 0.20
Disintegrated chalk (crushed stone)	> 5	60	80	60	80	0.12 (0.15)	0.8 (0.12)	0.12 (0.15)	0.12	0.15	≥ 0.20
Dense and very dense sands and gravels	> 12	150	300	150	200	0.12 (0.15)	0.8 (0.12)	0.12 (0.15)	0.12	0.15	≥ 0.20

* Maximum values of $f_{p,max}$ are given in brackets; they are admissible in careful control after the adjacent soil condition

Category IIIA includes driven piles and in-situ concrete piles constructed alongside with casing pipe driving.

Category IIIB – piles constructed by intrusion of mortar in holes under high pressure.

It is easy to notice that values of a_{LCPC} in table 4.19 may several times vary in one and the same values of q_c depending on soil type and pile category. In here, higher values of a_{LCPC} (or simply coefficient a) are typical for denser soils, i.e. more significant decrease of f_p in relation to q_c is expected. Consolidation of the adjacent soil also causes the decrease of f_p in relation to q_c.

Method of J. De Ruiter, F.L. Beringen (Ruiter & Beringen 1979, Lunne et al. 2004) was developed in 1979 in Netherlands (Holland); here, the issues on pile bearing capacity evaluation have always been urgent. Occurrence of weak clayey soil thickness stretching under several meters of depth by dense indigenous (Neogene) sands capable to support piles is typical for engineering geological conditions of Holland. Therefore, the authors paid the greatest attention to pile end bearing capacity evaluation.

The same formula (4.19) is the basis of computation, but evaluation of q_p & f_p is done in another way. Cone resistance q_c is used as the initial data, but as opposed to LCPC method, sleeve friction resistance f_s is taken into account as an index effecting on selection of pile side bearing capacity f_p in sands (see below for details).

The method of J. De Ruiter and F.L. Beringen can hardly be referred to the "direct" ones, since it presupposes evaluation of the "intermediate" parameter s_u in clayey soils – undrained shear strength of soil (see section 3.2.3), although in other soils selection of q_p & f_p is made ignoring evaluation of such indexes. There are modifications of the method concerning different types of piles, but its basic version refers to piles consolidating the adjacent soil during the immersion (driven piles are in the first place). As opposed to the previous method, this one lacks in any division of piles into groups or categories.

As in LCPC method, pile end bearing capacity is evaluated by cone resistance q_c in a zone extending above and below the lower end. However, the zone taken is considerably larger than in LCPC method. Its upper limit is located 8 diameters (cross-section sides) higher than the lower pile end, the lower limit – 4 diameters lower (fig. 4.14). In here, the lower section of the zone is considered in the interval $0.7...4D$, but not – $0...4D$. Rather complicated correction procedure and selection of averaged cone resistance q_c are carried out; after that pile end bearing capacity is evaluated (in LCPC method the analogous parameter is marked as q_{ca}). "Envelope" curve passing through the minimal values of q_c is drawn as in figure 4.14; mean value of q_c is evaluated by typical locations of this "envelope". In here, the "envelope" curve has two branches below the lower end: descending "a - b" and ascending "b - c"; both branches are taken into account in evaluation of mean value of q_c.

Small interlayers with reduced values of q_c are ignored in sands: figure 4.14 shows the "envelope" $2a$ which passes through a - b - c - d - d''- e points, while in clays these zones are taken into account, i.e. the "envelope" 2 passes through a - b - c - d - d'- e points.

223

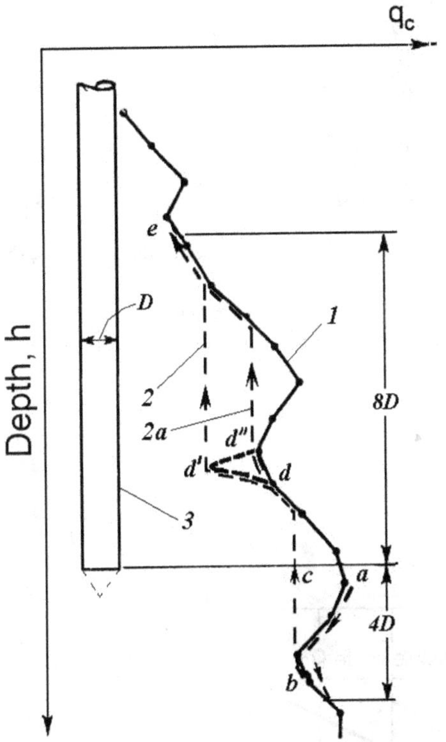

Fig. 4.14 Schematic illustration of averaged cone resistance q_c for pile end bearing capacity evaluation in J. De Ruiter and F.L. Beringen's method:

1 – curve of measured cone resistances "$q_c \sim h$", 2 – "envelope" curve of minimal cone resistances q_c in clayey soils, 2a – the same in sands, 3 – pile (with diameter D)

Pile end bearing capacity q_p is evaluated by averaged value of cone resistance obtained (in LCPC method it corresponds to q_{ca}). The procedure is carried out in different ways in clayey soils and in sands.

In sands connection between q_p and q_c is determined by coarseness of these sands and overconsolidation ratio (*OCR*) as well (see section 3.2.5). One can fail to do calculation not knowing these parameters.

Figure 4.15 shows the graph drawn by Te Campa in 1977 (Trofimenkov 1995); it is used in the method of J. De Ruiter and F.L. Beringen. As one can see from figure 4.15, unit pile end bearing capacity in normally consolidated sand (*OCR*=1) agrees with cone resistance, i.e. $q_p = q_c$ (position 1), while in overconsolidated sand (*OCR* = 6...10) it is twice lower than q_c, i.e. $q_p = 0.5q_c$ (position 3). In here, it is taken in computation that q_p of any sand cannot exceed 15 MPa. For example, in sands with *OCR* = 1 in q_c = 15, 20 or 25 MPa the value of q_p is taken one and the same – 15 MPa. It is evident that in all cases averaged (corrected) value of q_c is implied in accordance with the scheme given in fig. 4.14.

224

In clayey soils pile end bearing capacity q_p is evaluated by formula

$$q_p = N_c s_u , \qquad (4.22)$$

where s_u = undrained shear strength (see section 3.2.3)

$$s_u = q_c / N_k, \qquad (4.23)$$

N_c, N_k = "pile factor" and "cone factor" correspondingly, in calculations – N_c=9, N_k=15...20.

Value of corrected cone resistance q_c is taken in accordance with fig. 4.14.

Pile side bearing capacity in sand f_p is taken as minimal value from the following four values of f_i:

- f_1=0.12 MPa,
- f_2=f_s (f_s – sleeve friction resistance),
- f_3 = q_c/300 – in jacking load on a pile,
- f_4 = q_c/400 – in pulling load.

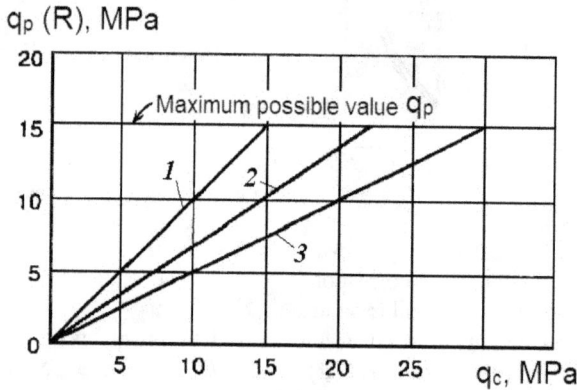

Fig. 4.15 Determination of ultimate pile end bearing capacity q_p (in Russian Norms marked as R) in the method of J. De Ruiter and F.L. Beringen (Trofimenkov 1995, Lunne et al. 2004):

1 – sand from fine to coarse in OCR = 1; 2 – gravel sand or sand from fine to coarse, bur overconsolidated with OCR = 2...4; 3 – fine gravel or very overconsolidated sand with OCR = 6...10

Pile side bearing capacity in clayey soils f_p is evaluated by formula

$$f_p = \alpha \, s_u , \qquad (4.24)$$

225

where:
  $s_u =$ the same as in formula (4.22),
  $\alpha =$ coefficient taken:
  - $\alpha = 1$ for normally consolidated clays,
  - $\alpha = 0.5$ for overconsolidated clays.

Other methods applied worldwide are based on the same dependency (4.19) but give other ways of q_p & f_p evaluation. **Method of J.H. Schmertmann** (Trofimenkov 1995) deserves particular attention; its reliability is as high as of those two discussed earlier. The method was developed in the USA in 1987. It is like the method of J. De Ruiter and F.L. Beringen developed later.

Unit pile end bearing capacity q_p is evaluated by averaged value of q_c in "work area" (in "neighborhood" of the pile lower end). The zone spreads upwards – $8D$ above the pile lower end, downwards – $3.75D$ below it (D – diameter or pile cross-section side). Minimal value of q_c is evaluated in the lower zone (but not the mean one), while in the upper zone – mean value of q_c. The values are added and, hence, total mean value of q_c is evaluated in the whole of the zone under consideration; it is this value that is taken as q_p, i.e. unit pile end bearing capacity. J.H. Schmertmann believes that q_p must never exceed 10 MPa.

Total pile side bearing capacity Q_f is evaluated in different ways in sands and clays. In sands it is evaluated by formula

$$Q_f = k\left[\frac{\left(f_s A_f''\right)_{0-8D}}{2} + \left(f_s A_f''\right)_{8D-L}\right], \qquad (4.25)$$

where:
  $k =$ relationship of reinforced concrete pile side bearing capacity to sleeve friction resistance evaluated by the graph given in fig. 4.16a,
  $(f_s A_f'')_{0-8D}$, $(f_s A_f'')_{8D-L} =$ friction of soil upon pile side surface in the zone 0-8D & 8D-L above the pile lower end; in here, D – pile diameter (cross-section side), L – pile length:
  $f_s =$ unit sleeve friction resistance obtained in the examined zones,
  $A_f'' =$ area of contact between pile side surface and the soil (in the examined zones).

In clayey soils pile side bearing capacity Q_f is evaluated by the formula

$$Q_f = \alpha' \bar{f}_s A_f, \qquad (4.26)$$

where:
  $\bar{f}_s =$ mean value of unit sleeve friction resistance (in total pile length),
  $A_f =$ total area of contact between pile side surface and the soil,

226

α' = relationship of reinforced concrete pile side bearing capacity to sleeve friction resistance evaluated by the graph given in fig. 4.16b.

In absence of the friction sleeve J.H. Schmertmann recommends application of q_c for evaluation of f_p, thus, giving the empirical formulae.

Method of J. Zhou et al. (1982) developed in China in 1982 aims at evaluation of driven pile bearing capacity applying type II cone penetrometers with the cone diameter of 5.1 *cm* and the friction sleeve - 18.7 cm in length (300 cm²). The idea of the method is practically the same as of those ones discussed earlier, but evaluation of q_p & f_p has its specific character. The averaged value of $\bar{q}_c$ to evaluate pile end (edge) bearing capacity q_p, is evaluated in the zone from the level of 4D above the edge ($\bar{q}_{c1}$) to 4D below it ($\bar{q}_{c2}$). Here, if $\bar{q}_{c1} < \bar{q}_{c2}$, then

$$\bar{q}_c = \frac{q_{c1} + q_{c2}}{2}, \tag{4.27}$$

If $\bar{q}_{c1} > \bar{q}_{c2}$, then $\bar{q}_c = \bar{q}_{c2}$.

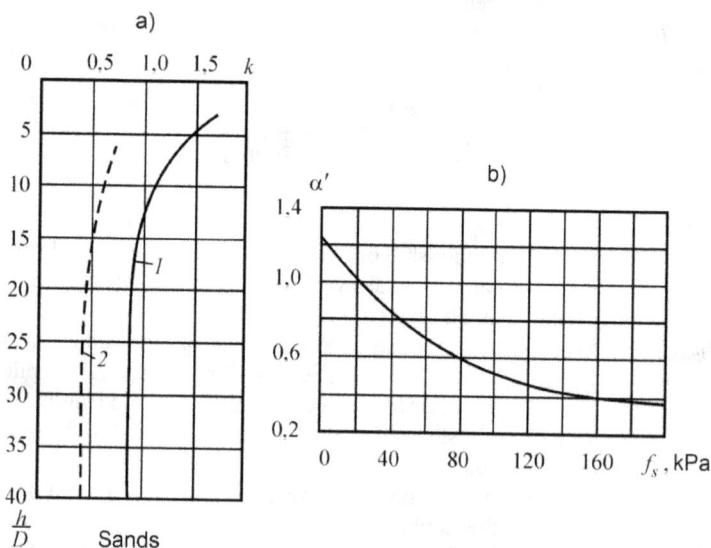

Fig. 4.16 Schematics of coefficients k & α' evaluation in the formulae of pile end (and side) bearing capacity evaluation (4.25) & (4.26) in the method of J.H. Schmertmann:

a – evaluation of coefficient k by formula (4.25), i.e. for sands: 1 – electrical type II cone penetrometer, 2 – mechanical type I cone penetrometer; *b* – evaluation of coefficient α' by formula (4.26), i.e. for clayey soils; h – depth, D – pile diameter (cross-section side), f_s – unit sleeve friction resistance

227

Pile end (and side) bearing capacity is evaluated by multiplication of $\bar{q}_c$ & f_s by the coefficients α & β correspondingly, i.e. $q_p = \alpha \bar{q}_c$ and $f_p = \beta f_s$. Here, f_p is evaluated for each layer separately, but not for the whole of pile side surface. Coefficients α & β, depending upon $\bar{q}_c$ & f_s may be evaluated by the graphs given in fig. 4.17.

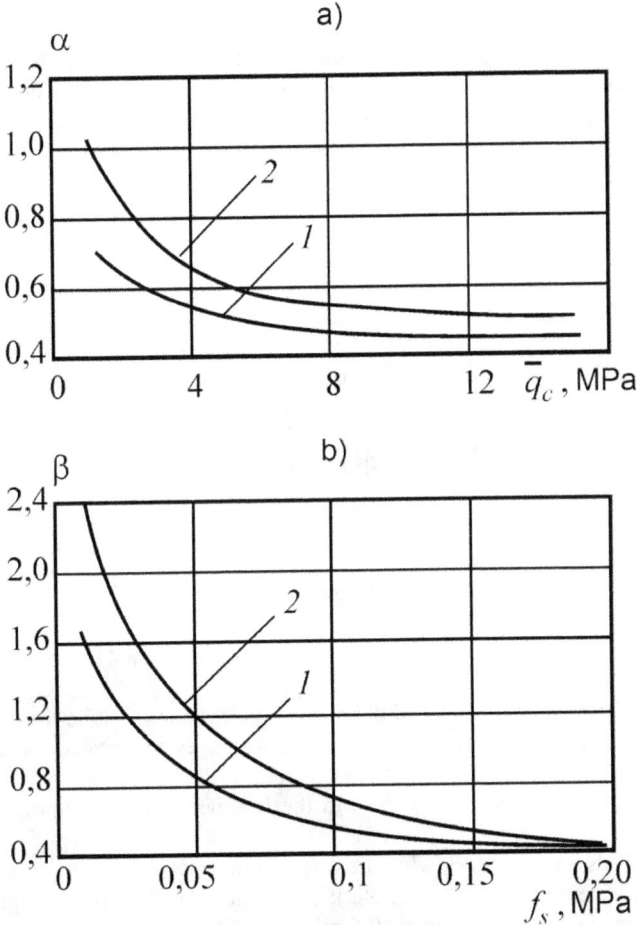

Fig.4.17 Schematics of coefficients α & β evaluation by the method of J. Zhou et al. (1982):

a – evaluation of α, b – evaluation of β;

1 – clayey soil, 2 – sandy soil;

$\bar{q}_c$ – averaged value of cone resistance in "work area" at the edge ($\pm 4D$), f_s – unit sleeve friction resistance

Conditional method of J.L. Briaud (1988) aims at driven pile bearing capacity evaluation applying type II cone penetrometers. Here, pile end bearing capacity q_p and pile side bearing capacity f_p are taken to be equal to q_c & f_s correspondingly ignoring any correction coefficients. The computation means further specification of results by means of CPT of piles or the additional tests by the same cone applying special mode. The cone is penetrated up to the expected depth of pile driving, then there is 1 hour wasting and the penetrometer is immersed with low speed – 2.5 mm/h.

There are the methods applying the data obtained by piezocones. **Method of M.M.S. Almeida et al.** (Lunne et al. 2004) developed in 1996 is an example. Pile side bearing capacity (f_p) and pile end bearing capacity (q_p) are evaluated by the formulae

$$f_p=(q_t - \sigma_{vo})/k_1 , \qquad (4.28)$$

$$q_p=(q_t - \sigma_{vo})/k_2 , \qquad (4.29)$$

where:
$q_t =$ the same as in the formula (3.3) – corrected value of cone resistance,
$\sigma_{vo} =$ the same as in the formula (3.2) – overburden stress,
$k_1 =$ coefficient evaluated by the formula (usually $k_1 = 40...45$)

$$k_1 = 12+14.9 \ log[(q_t - \sigma_{vo})/ \sigma'_{vo}], \qquad (4.30)$$

$\sigma'_{vo} =$ effective overburden stress,

$$k_2 = N_{kt}/9, \qquad (4.31)$$

$N_{kt} =$ empirical cone factor (see section 3.2.3 and formula 3.4), usually $N_{kt} =$ 10...30.

T. Lunne et al. (2004) point out that this method is less conservative than LCPC. However, one can fail to find out any significant advantages of piezocones in their application for pile bearing capacity evaluation.

If to compare the discussed methods with the Russian ones, one can conclude the following:

- Most of methods applied worldwide are directed towards various types of piles application (driven, boring, drilled, etc.); this appears to be possible ignoring new design models to be accepted: the piles are differentiated by selection of conversion factors "cone-pile".

- In methods embracing wide range of pile types cone resistance q_c is used as the initial data (i.e. values of f_s or Q_s are ignored); perhaps, it seems to be rational for in-situ piles (boring, drilled, etc.).

- In evaluation of pile end (foundation) bearing capacity "work area" where the averaged value of q_c is evaluated is considerably different in different authors, rather complicated techniques of the averaged value of q_c selection being applied.

• In most methods the highest acceptable values of pile end (and side) bearing capacity are determined; they must not be exceeded whatever the values of q_c are; it seems to be rational since not much experimental data have been collected for overconsolidated soils.

4.4 SOLUTION OF TECHNOLOGICAL PROBLEMS

4.4.1 Evaluation of available equipment capacity for pile driving

Practical application of foundations on driven piles has demonstrated that in spite of correct pile bearing capacity evaluation, the assigned foundation effectiveness is not guaranteed without detailed elaboration of pile driving technology. During construction operations there may occur situations when piles do not reach the design depth or are damaged during driving; that results in contingent delays and possible updates of the project design. In intensive construction the situations may cause noticeable economic damage. Annually nearly 0.5 mln m^3 of precast reinforced concrete was lost due to poor driving (and, hence, breaking) of piles in the former USSR.

SP 50-102-2003 gives formulae for evaluation of pile hammer capacity being used for over 50 years in Russian practice without any significant modifications. Due to the design load on piles the required *energy of hammer blow* is determined; thus, hammer brand is chosen, then the additional inspection is performed: hammer weight, pile weight and striking part of hammer head shall be in definite relationships with calculated hammer energy. Besides, SP 50-102-2003 contains the instructions on pile failure avoidance – stresses from hammer blows shall not exceed 60 % from "normal" (normative) compression strength of concrete R_{bn}.

However, in practice these instructions hardly ever consider specific features of the construction sites and the share of remaining piles proves to be huge both in Russia and worldwide. Design load does not always reflect the difficulties of driving. In occurrence of solid bearing stratum with weak soils located over it, one needs neither powerful hammers nor shock-resistant piles even if the load on the pile is heavy (800...1000 kN). On the contrary, in hard clay soils pile driving to design depth can encounter difficult problems even in the case of moderate pile loads (500...600 kN). Besides, actual hammer blow energy evaluation is no easy question requiring a number of unpredictable factors to be considered. Thus, different authors usually draw different conclusions. The issue on stresses in the pile in the moment of hammer blow is complicated and rather confusing.

The investigations of BashNIIstroy in 1970s...1980s revealed that the simplest solution of the problem of poor driving avoidance could be found provided CPT was applied (VSN 29-76, Ryzhkov & Enikeyev 1979, Enikeyev et al. 1989, Workbook 1991, Ryzhkov 1992). G.F. Novozhilov (1987) carried out a vast research on the problem; the results obtained agreed with those of BashNIIstroy.

The approach is based on the assumption that in CPT application one can easily evaluate ultimate pile resistance $F_{u,j}$ in the whole range of the "intermediate" depths of penetration ($h_j = 1, 2, 3 \ldots h$, m) that the pile passes during driving before it attains the design depth h. Then these values of $F_{u,j}$ serve as the initial data in evaluation of anticipated set per blow at these intermediate" depths with regard to one or the other hammer. The well-known "dynamic" formulae embracing ultimate pile resistances and set per blow (taking into account specific features of piles, hammers, shock-absorbing laying in a helmet, etc.) are used. After that, the calculated set per blow s_a are used as the initial data for evaluation of anticipated hammer blows, because the set inverse value is the number of blows per a unit of length (depth of driving). The calculations are usually done with regard to all "intermediate" depths with 1 m of step, the refusal being defined in meters.

$$n_i = 1/s_{a,i}, \qquad (4.32)$$

where:

$s_{a,i} =$ anticipated set per blow of pile immersed in the depth of h_i(m);

$n_i =$ anticipated number of hammer blows in one meter of driving in refusal $s_{a,i}$.

Thus, having evaluated $s_{a,i}$ – set per blow (mean) in each i-meter of driving one can easily evaluate the anticipated number of hammer blows n_i, necessary for the pile to be capable to pass each i-meter.

Summation of hammer blows number for each meter of driving n_i ($i = 1, 2, 3, \ldots h$) yields the unknown number of blows N, required for pile driving to the whole depth h in the following:

$$N = \sum_{i=1}^{i=h} n_i. \qquad (4.33)$$

It is this *total number of hammer blows* N that is used in estimation of applicability of specific hammer in driving, since it reflects both the duration of driving process and possible rupture of the pile (in material). The number of blows N obtained here is compared with the allowable values chosen with regard to two conditions:

- Pile driving shall ensure the design operation efficiency.
- Pile strength shall exclude their damage in the course of driving.

They are expressed in the following dependencies:

$$N \leq N_t, \qquad (4.34)$$

$$N \leq N_{strength}, \qquad (4.35)$$

where:

$N =$ anticipated number of hammer blows for driving in the given depth h;

$N_t, N_{strength} =$ allowable numbers of blows by pile driving ensuring of the design operation efficiency and excluding of pile damage in the course of driving as well.

Nowadays, driving is usually performed by diesel-hammers (pipe or rod). In practice, the real number of diesel-hammer blows depends upon a number of random factors: quality of hammer operation in concrete soil conditions (reduction of hammer bounce height being particularly noticeable in rod hammers), condition of shock-absorbing laying, difference between pile resistivity in the moment of driving and later on, etc. The summation of the factors usually results in design number of blows N increase in comparison to the actual ones.

Figure 4.18 illustrates the comparison results of actual and calculated numbers of hammer blows obtained in various sites by V.M. Enikeyev (1980). The refusals were evaluated by the formula given in p. 7.3.7 of SP 50-102-2003 under the number of (7.20). As one can see from fig. 4.18, calculated (N_{cal}) numbers of hammer blows 1.5-2 times exceeded the actual (N_{act}) values. When taking into account that the requirements to the reliability of the computation are lower than those of pile bearing capacity evaluation, the mentioned errors in N evaluation can easily be compensated by the correction coefficients determined during trial driving in the analogical soil conditions.

In order to comply with the requirement (4.34) it is necessary to know the allowable number of hammer blows N_t, i.e. the requirement allowing the efficiency to meet it. The parameter does depend on real conditions of driving, but in mass construction it is appropriate to apply the number of blows corresponding to driving in the course of 5...10 min. For Russian diesel-hammers the frequency of blows is 50...60 1/min, thus, the allowable value of N_t is approximately 250...500 blows.

In order to comply with the requirement (4.35) another criterion – parameter $N_{strength}$, is a must, i.e. maximum allowable number of hammer blows in which pile failure does not occur. The parameter is determined by shock resistivity of piles which depends upon a number of random factors, the quality of pile manufacture being of primary importance. Nevertheless, when analyzing the most typical situations, shock resistivity of piles is likely to be estimated with high accuracy due to their type (brand). The issue was studied by G.F. Novozhilov (1987) in detail; he carried out experimental research on pile "endurance" including destruction of hundreds of reinforced concrete piles by hammers of different power in different modes of operation. Shock resistivity was estimated by two parameters:

- Parameter of dynamic strengthening K_{ds}, showing the number of times when the dynamic strength of pile material R_d exceeds the static strength R_{st}, i.e. $K_{ds} = R_d/R_{st}$ in single-phase loading ($N = 1$).
- Parameter of durability K_d characterizing pile resistivity to repeated dynamic loading.

The classification of piles by their shock resistivity suggested by G.F. Novozhilov includes six classes. It is presented in table 4.20. The parameter of dynamic strengthening K_{ds} is given here as a fraction reflecting two stages of destruction: *starting of destruction* – numerator, *starting of cracking* – denominator. As one can notice from table 4.20, the maximum shock resistivity is typical for piles with steel-fiber-concrete helmets, i.e. the ones made up of fibrous concrete reinforced with thin steel fibers (usually wire). The occurrence of lateral reinforcement as well as the application of ceramsite concrete increase shock resistivity of piles, other conditions being equal.

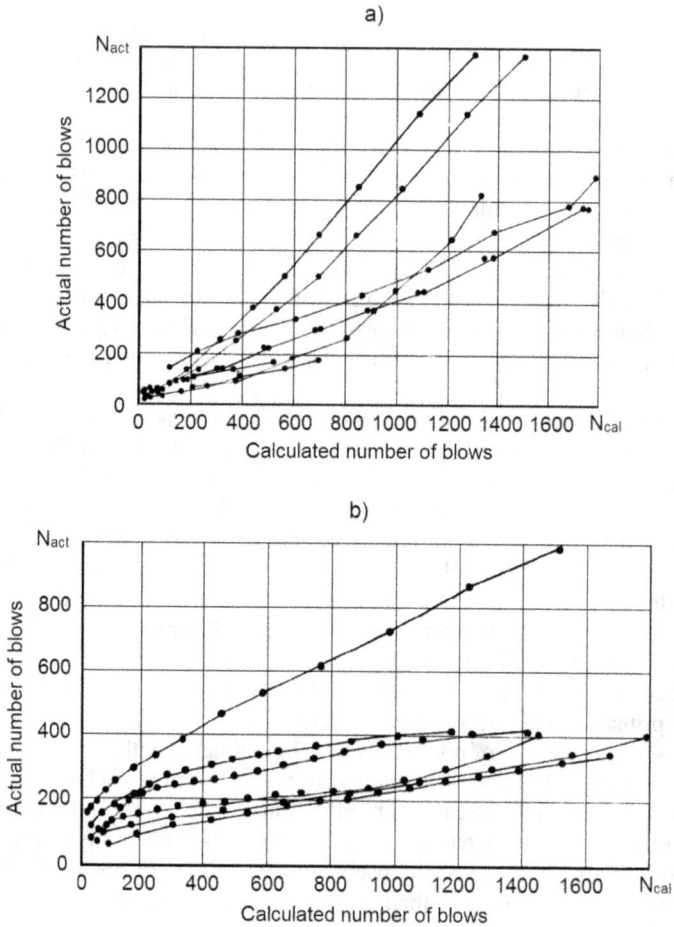

Fig. 4.18 Schematics of comparison between calculated number of hammer blows and actual number determined in pile driving (piles – prismatic reinforced concrete, soils – dealluvial loams (Enikeyev 1980)):

a – in application of pipe diesel-hammer 1,250 kg in weight dropper, b – the same 1,800 kg in weight. The lines connect the points related to one and the same pile in different depths of driving

Table 4.20 Classification of driven piles by their shock resistivity (Novozhilov 1987)

Classes	Pile brand (after Standard Specifications)	Description of piles	Parameters	
			K_{ds}	K_d
I Low shock resistant	СЦ, СПН СЦ, СЦп	Reinforced prismatic ceramsite concrete piles: Continuous ones with reinforcement located in the center of round cavity	– 2.8/2.7	– 0.8
II Minor shock resistant	С	Continuous reinforced concrete piles with lateral reinforcement (nontensional rod reinforcement)	3.3/3.2	0.8
III Medium shock resistant	СК СФ-1	Continuous piles with lateral reinforcement: Ceramsite reinforced concrete ones with steel-fiber-concrete helmet of horizontal molding (μ=1%)	3.4/3.3 3.4/3.3	0.8 0.8
IV Minor shock resistant	СН	Continuous reinforced concrete piles with lateral reinforcement (tensional rod reinforcement)	3.2/3.0	0.8
V Shock resistant	СФГ-1	Continuous reinforced concrete piles with lateral reinforcement of body with steel-fiber-concrete helmet of vertical molding in μ=1%	3.9/3.6	0.8
VI High shock resistant	СФГ-2	The same in μ=2%)	4.1/3.8	0.8

Figure 4.19 shows the graphs of pile endurance drawn by G. F. Novozhilov (1987) on the basis of the experimental research.

The number of hammer blows is evaluated due to pile brands and relative stresses σ/R_{st}, where R_{st} – brand (cube) strength of concrete, σ – stresses in the pile in hammer blow. Stresses σ are evaluated by NIIOSP formula (Novozhilov 1987):

$$\sigma = \frac{2}{3}\sqrt{\frac{6\Im_p}{\left(\dfrac{s}{E_s} + \dfrac{L}{2E_b}\right)\left(1 + \dfrac{Q}{q}\right)A}}, \qquad (4.36)$$

where:

$\Im_p =$ design energy of hammer, *kilojoule* (MN·m), taken after SP 50-102-2003 as follows:
- for tubs or single-stroke ones $\Im_p = GH,$
- for pipe diesel-hammers $\Im p = 0.9 \cdot GH,$
- for rod diesel-hammers $\Im p = 0.4 \cdot GH;$

$G, H =$ weight, MN and head, m, of striking part of hammer correspondingly;

$E_s, E_b =$ moduli of elasticity of helmet laying and pile concrete (in most cases $E_s = 400...600$ MPa, $E_b = 2.4...3.0 \cdot 10^4$ MPa);

$L, A =$ length , m, and cross-sectional area, m^2, of the pile correspondingly;

$s =$ thickness of shock absorber in the helmet, m;

Q, q = pile weight and striking part of hammer weight, MN.

Concrete B20...B25 is usually applied for driven piles, for them R_{st} = 20...25 MPa.

Thus, having calculated stress in hammer blow σ by the formula (4.36) for pile types and pile driving units and hence, having obtained the ratio σ/R_{st}, the number of hammer blows not causing pile destruction $N_{strength}$ is determined by the graph "$N_{strength} \sim \sigma/R_{st}$" (fig. 4.19). When comparing the anticipated (by CPT data) number of blows N with the obtained value $N_{strength}$, it is possible to determine driving of piles escaping their destruction, i.e. following the condition (4.35).

The calculations show that safety of pile during driving is reliably ensured by steel-fiber-concrete helmet (СФГ-1 & СФГ-2) application (table 4.20). In driving of standard piles into strong soils the destruction of piles is not always possible to be avoided. Table 4.21 presents the values of $N_{strength}$ for "minor shock resistant" piles of "C" & "CH" brands evaluated by the formula (4.36).

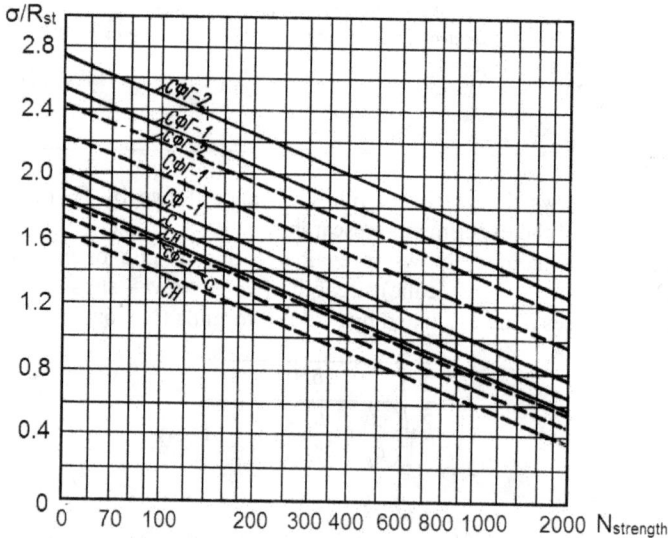

Fig. 4.19 Diagram of driven reinforced concrete pile resistivity (Novozhilov 1987): σ – stresses in pile helmet in hammer blow evaluated by the formula (4.36), R_{st} – cube strength of concrete, $N_{strength}$ – number of blows causing pile failure. Firm lines correspond to starting of failure, dashed ones – to cracking. Description of piles of the following brands (CH, C, СФ–1, etc.) is given in table 4.20

It has been taken into account in computations that piles are made up of heavy reinforced concrete of classes B20 & B25 (moduli of elasticity – $2.4 \cdot 10^4$ and $2.7 \cdot 10^4$ MPa, cube strength – R_{st} = 20 and 25 MPa correspondingly), 10 m in length, 0.3×0.3 m in section; oak laying with fibers located along the blow action, 0.15 m and 0.20 m in thickness (modulus of elasticity – 480 MPa). The head of striking part in rod hammer has been taken 1 m and 2.2 m; in pipe one – 1 m and 2.8 m, because in partial bounce of striking part the actual heads are usually in these intervals.

As one can see from table 4.21, strength of piles does depend upon thickness of laying and strength of pile concrete as well. In a number of cases, when modifying these factors, one can prevent pile failure.

Table 4.21 Numbers of blows not causing pile destruction in piles of "C" and "CH" brands 10 m in length, 0.3×0.3 m in cross section

Pile brand	Type of hammer, weight of striking part and head		Type and thickness of laying	Class of concrete	Number of blows causing:	
	Rod, 2.5 m, in head of striking part in m	Pipe, 1.8 m, in head of striking part in m			Cracking	Total pile head destruction
CH	1		Oak, s = 0.15 m	B20	500	900
The same	2.2		The same	The same	160	300
– » –		1	– » –	– » –	300	600
– » –		2.8	– » –	– » –	<50	85
– » –	1		Oak, s = 0.2 m	B25	900	>2,000
– » –	2.2		The same	The same	350	700
– » –		1	– » –	– » –	700	1,400
– » –		2.8	– » –	– » –	150	270
C	1		Oak, s = 0.15 m	B20	700	1,200
The same	2.2		The same	The same	210	380
– » –		1	– » –	– » –	400	700
– » –		2.8	– » –	– » –	<50	90
– » –	1		Oak, s = 0.2 m	B25	1,200	>2,000
– » –	2.2		The same	The same	410	800
– » –		1	– » –	– » –	700	1,300
– » –		2.8	– » –	– » –	200	370

The mode of pile driving estimation presented is not the only one. Selection of hammers does not specify CPT application (including SP 50-102-2003). But in practice this mode excels the other ones in its viability and reliability. In particular, the mentioned limitation in SP 50-102-2003, under which maximum compressive stresses in a reinforced concrete pile shall not exceed 60 % of normal compressive resistance of concrete R_{bn} in hammer blow seems to be far from reality in practical applications. It is equipotent to banning of standard piles application manufactured by Russian industry. As one can see from table 4.20, even in piles of minor shock resistivity "C" & "CH" the stresses in hammer blow are capable to exceed cube strength of concrete (not prismatic R_{bn}!) in three times without breaking.

Although the technique presupposes a great number of calculations to be done, availability of corresponding software helps solve the problem. Vice versa, designing of pile works is becoming a simple, convenient and efficient procedure. Since 1970s BashNIIstroy have been applying their own software for hammer blow computation, i.e. since the time when hardware and software were not developed at all.

In practice, the problems are to be solved before the elaboration of work production plan, i.e. in elaboration of structural part of design, namely: in selection of pile foundation geometry, and hence, pile lengths. Otherwise, serious problems may emerge connected with the necessity of redesign of the structural part of design. Such an approach to designing of pile foundations, i.e. combining of technological and structural issues opens up possibilities of effectiveness increase of pile foundations.

This is related to pile driving depth selection principle. Traditionally, it was believed that design refusal attainment ("driving up to the specified refusal") was the compulsory condition of pile foundation reliability, but not the attainment of design depth of driving. From this point of view, occurrence of remaining piles and their cutting appear to be a common practice, and vice versa, driving of the whole of piles up to the specified depth is considered to be a failure in pile foundation capabilities application and reduction of their bearing capacity. In "driving up to the specified refusal" the required resistivity of each pile is ensured as well as relative stability of this resistivity in a pile field and hence, uniformity of anticipated settings. Finally, the driving allows maximal application of foundation resistivity in the available hammer.

All these would be a sufficient argument if to ignore the loss of reinforced concrete due to underdriving and cutting of piles. Real-life soils are heterogeneous, thus, in driving up to the specified refusal the depths of driving will always be different. One has to make decision on the depth of driving taking into account the sites possessing the weakest soils; otherwise the sites will fail to attain the design refusal. As a result, in most piles the desired depth is not reached and their cutting is required.

Figure 4.20 illustrates two reference cases of pile driving. In fig. "a" the piles with h length are driven up to the "specified refusal" in the depth of $h\pm\Delta h$ (it is thought that the refusal is rather low; when attaining it the following driving is not of practical importance). Let us consider Δh to be the mean underdriving, then the maximum one $-2\Delta h$, i.e. the total volume of driving is to be $\Delta h \cdot n$, where n – number of piles. Fig. "b" illustrates the foundation taking the same total load N_{total} from piles driven by the same hammer but up to the "specified depth". The unified depth of driving is to be $h_0 = h - \Delta h$, as it would be impossible to drive the piles deeper without underdriving. It is evident that piles with depth of penetration h_0 will have lower bearing capacity and their number will have to be increased in a value Δn (in fig. "b" the additional piles are painted over black).

Thus, both schemes have similar fragments – unpainted piles and their parts. In fig. *"b"* it is volume n of piles with h_0 in length, in fig. *"a"* - part of piles from the level I–I (surface of driving procedure) to the level corresponding to the depth h_0. It is evident that material capacity of the schemes will be defined by "painted fragments". If the additional piles exceed the total volume of cutting and additional driving in fig. *"b"* (below the depth h_0), the lower material capacity will correspond to fig. *"a"* – driving up to "specified refusal". Otherwise, lower material capacity will correspond to fig. *"b"*, i.e. driving up to the "specified depth".

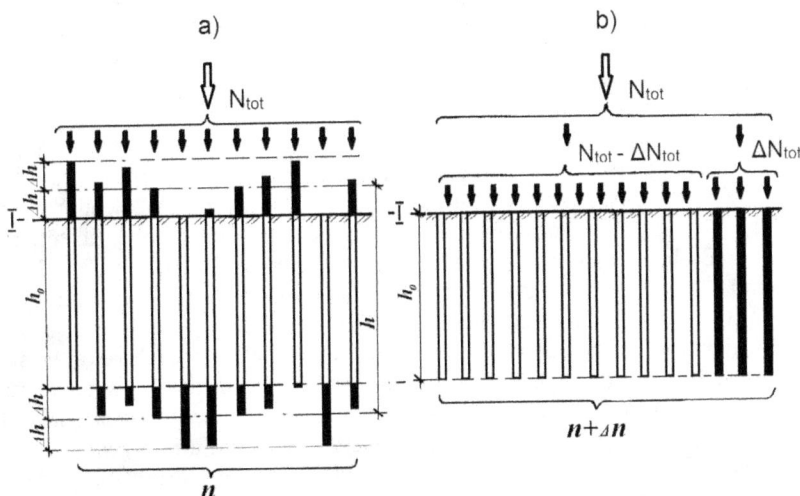

Fig. 4.20 Schematics of pile driving up to the "specified refusal" (a) and up to the "specified depth" (b)

Taking into account that the volume of cutting in fig. *"a"* equals to "additionally driven" parts, one can conclude that *pile driving up to the "specified depth" shall ensure lower material capacity on condition the volume of additional piles does not exceed double volume of underdriven piles cutting.* The criterion is approximate because foundation geometry and type of a girder are also important, and the cost of reinforced concrete of the embedded parts of piles is a little bit higher than the cost of that of the cut part. However, the restrictions are not significant; thus, one can ignore them in practice.

The book (Kolesnik & Ryzhkov 1977) gives thorough theoretical analysis of two schemes concluding that *intensity of pile bearing capacity growth from the depth* is the defining factor in selection of driving mode. Reasonability criterion of pile driving up to the "specified depth" is suggested

$$\frac{F}{h \cdot i_F} \geq \mu, \qquad (4.37)$$

where:

$F \& h$ = bearing capacity (kN) and pile driving depth (m) under study;

$i_F = \left(\dfrac{\delta F}{\delta h} \right)$ = intensity of pile bearing capacity growth from the depth growth of 1 m (kN/m);

μ = parameter, in low girder $\mu = 0.8$, in other cases – $\mu = 0.6$.

If 1 m increase of driving depth (in the interval of depths under study) increases pile bearing capacity not more than in 100 kN, then, in most cases it seems reasonable to drive piles up to the "specified depth". If more intensive growth of bearing capacity is observed, driving up to the "specified refusal" seems to be more reasonable. For friction piles this means the increase in length (in 1...3 m), i.e. acceptance of higher bearing capacity. In practice this means that in occurrence of explicit bearing layer for pile solid bearing, driving up to the "specified refusal" is necessary. In most cases, i.e. when bearing capacity increases from the depth more or less evenly, driving up to the "specified depth" seems to be more efficient. For example, in Bashkortostan these conditions are observed in 80...90 % objects (alluvial and dealluvial clays prevail; the most typical bearing capacity of a pile 10 m in length, 0.3×0.3 m in section is in the limits of 450...550 kN).

Unification of pile driving depth does not have to embrace the whole of the site. If the soil conditions at different sections of the site differ considerably and one may define the zones of stronger or weaker soils, then, in CPT data it is reasonable to divide the terrain into sections with the own *unification* depth of driving.

Refusal control in driving of piles up to the "specified depth" is possible, although it lacks its priority becoming a means of obvious errors avoidance having been made in design and exploratory work, e.g. in situations of a gap of any "weak" zone, backfilled cave, etc. when other types of piles are required. Control refusal in driving of piles up to the "specified depth" may be enhanced oriented towards the possible errors. For example, it could be 30...40 % higher than the value corresponding to the admitted pile bearing capacity (according to the "dynamic" formula).

4.4.2 Earth excavation, soil compaction and stabilization technology aspects

CPT may be applied in solution of a number of technological problems of zero cycle works not connected with piles. The problems are being resolved in designing or during construction works as well. They involve the following:

- Soil excavation assessment (in designing of excavation works).
- Current quality control on placement and compaction of soil in bulk arrangement.

- Quality control on improvement (compaction, fixing) of in-situ soil layers or the existing bulks.
- Quality control on foundation bed reinforcement in reconstruction of buildings and structures.

Soil excavation assessment by CPT data is not considered to be the most important issue, since the problem may be solved ignoring CPT application. However, quickness and cheapness of CPT make it the desired method in this sphere as well. According to the National Norms of Russia ENiR Collection E2, classification of soils by troubles of their excavation, displacement and placement is based on their engineering-geological identification regardless of presence or absence of CPT in surveying. Nevertheless, in involving CPT in the surveying program, the work on soil identification is performed quicker and more accurate, so soil excavation assessment efficiency significantly increases. Taking into account that soil lithological type identification methods by CPT data have been developed in detail (see section 3.1), their specialization with regard to soil excavation assessment appears to be quite real and reasonable.

The problem of "direct" CPT application deserved the researchers' attention in 1950s...1960s. Then, the specialists' interests turned to pile bearing capacity evaluation and estimation of soil properties, thus, publications on the issue hardly ever appeared. However, the first empirical dependencies obtained have not lost their utility and they may be used at present. One needs to consider that classifications of soils by troubles of their excavation have been specified, they slightly varying in different countries. Figure 4.21 illustrates the dependence of soil excavation upon cone resistance q_c obtained in 1950s by G. Kyun (Bondarick 1964).

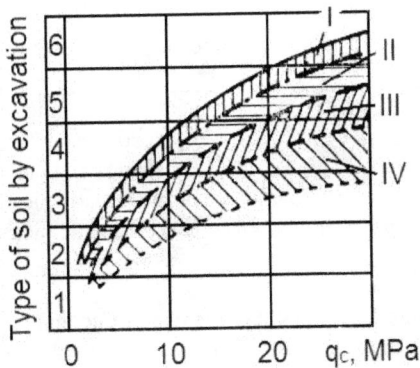

Fig.4.21 Soil excavation assessment by CPT (after G. Kyun):

I – very densely coherent soil, II – densely coherent soil, III – medium dense coherent soil, IV – poorly coherent soil; 1...6 – classification subtypes of soil by their excavation with regard to the Norms of Germany of 1950s; q_c – cone resistance

G. Kyun used the terminology different from the modern one, thus, figure 4.21 needs some explanations to be done. "Very densely coherent" soils (marked under the symbol "I") are probably clays with high degree of lithification with case-hardening links. The same refers to cemented sands. "Densely coherent" soils (II) – the same soils, but less dense. "Medium dense coherent" soils (III) – loams, sandy loams, silt sands. "Poorly coherent" soils are quaternary sands (alluvial, dealluvial, etc.) without case-hardening links.

G. Kyun divides soils into 6 classes with regard to the Norms of Germany of that period of time (table 4.22).

Table 4.22 Classification of soils by excavation after G. Kyun

Class by excavation	Soil characteristics
1	Soil is very easily excavated by a spade*
2	Soil is easily excavated by a spade
3	Soil is excavated by a spade with difficulty
4	Soil is easily excavated by a hack
5	Soil is excavated by a hack with difficulty
6	Soil is excavated by a hack with great difficulty

* G. Kyun does not characterize the soil of 1 class

The classification ENiR Collection E2 valid in Russia divides soils into *"groups"* by troubles in excavation, the division being determined for each method of excavation. Table 4.23 gives such classification for manual method of excavation.

Table 4.23 Classification of soils by troubles in manual excavation admitted in ENiR Collection E2

Soil group	Mode of opening
I	By spades
II	By spades and partially by hacks
III	By pneumatic picks or crows
IV, Ivp, Vp, V-VII and permafrost soils of all groups	By pneumatic picks or wedge bars

If to compare tables 4.22 and 4.23, one can notice that classes after G. Kyun are slightly different from the ones after EniR. Group numbers after EniR correspond with higher group numbers after G. Kyun: group I after EniR corresponds to classes 1...2 after G. Kyun, group II – to classes 3...4, group III – to classes 5...6, group IV and the other ones refer to dense soils which are not considered in the table 4.21. Nevertheless, when introducing some corrections, the dependencies given by G. Kyun (figure 4.21) can be successfully applied at present.

One factor deserves special attention, namely: according to G. Kyun, in one and the same q_c soil may be referred to different classes (and hence, to different "groups" after EniR) depending on the degree of coherence between soil particles. When the coherence decreases, the troubles in excavation decrease as well. For example, in $q_c =$ 3...5 MPa clayey soil is to be referred to class 3 (group II after ENiR), but sand – to classes 1...2 (group I after EniR), etc.

Assessment of soils by troubles in excavation is significantly simplified, if soils of concrete region characterized by the limited nomenclature are considered. Table 4.24 gives examples of assessment criteria applied in Bashkir Urals with scarcity of sands and sandy loams (found as water course deposits), alluvial and dealluvial clays and loams prevail here.

Table 4.24 Approximate assessment of troubles in manual excavation of Bashkir Urals soils

q_c (MPa)	Soil group	
	Clayey	Sandy
< 1.0	I	I
1.0...3.0	II	I
3.0...4.5	III	II
> 4.5	IV and further	III and further

Excavation works show that usefulness of CPT is appreciably observed in occurrence of strata and lens of rocks in small depth. Though the cone is not capable to penetrate them, it is CPT that allows the roof and configuration of these strata in plan to be defined quicker and more accurate than any other method.

CPT allows estimating of applicability of earth-moving machines by the amount of the anticipated power of excavation. For this purpose, one can apply the empirical dependencies worked out to be used in the conditions of the particular region. In practice, obtaining of these dependencies is not a problem and could be performed by the analysis of local excavation works.

CPT is also widely applied in *soil compaction quality control*, either in bulk construction or in compaction (in hardening by any other technique) of real-life soil layers or the constructed bulks.

Current soil compaction and placement quality control presupposes filled soil estimation in bulk construction; the small depth corresponds to thickness of excavated soil (0.2...0.5 m). In these conditions, application of heavy or medium rigs equipped with standard cones is not appropriate. Here, the simplest manual probes shown in fig. 1.19a, b or the analogous devices operating on the principle of dynamic CPT, i.e. by pushing but not pressing meet the technological requirements. The latter are the simplest, thus, they are the most popular in bulk density control. A lot of different types of these devices are applied in world practice. In the former USSR the drop point DorNII was widely used; in it penetration of the rod with the tip was carried out applying 2.5 kg weight dropping from 0.5 m (number of blows per 10 cm of penetration was taken as the criterion). Nevertheless, pushed manual probes are widely used worldwide.

The devices are simple and reliable, but their application is to be anchored to the conditions of each bulk construction. The local empirical dependencies connecting the controlled parameter (e.g. soil density) with measurement results – number of blows per 10 cm of penetration, pushing effort, etc. are needed. These dependencies are to be determined (or, at least, corrected) with regard to each particular case, i.e. with regard to the material of the bulk, measuring devices applied or conditions of measurements, etc. It is evident that the simplest devices of in-situ control shall always be applied with other reliable methods (as a rule, it is selection of monoliths followed by standard laboratory treatment). These methods serving the reference of the empirical (calibration) dependencies may be applied in minimal volumes.

Real-life soil layers and the existing bulks may require additional compaction (hardening), e.g. when they are used as bases. In here, compaction is carried out in a significant depth (deep compaction) but not layer-by-layer. It is obvious that application of manual probes is excluded in soil quality control. Only CPT rigs appear to be efficient, since they are capable to penetrate the cone in significant depths (e.g., up to 15...20 m). As a rule, CPT is used with the traditional soil estimation methods (usually, drilling followed by selection of monoliths and further laboratory treatment).

Compaction of *loessial and loess-like loams* possessing collapsible properties is typical for Russian practice, since these soils occupy most of the territory of Russia, particularly in the South of Russia. Here, impact beating is applied, and in larger thickness – soil (sandy) piles or deep-soil explosions, etc. Thus, the principal means of control is selection of monoliths followed by their laboratory treatment, as a builder is concerned with liquidation of collapsible properties of soil rather than density increase. Nevertheless, attainment of the definite density being easily determined by CPT data may serve as indirect indication of loss of compacted soil collapsible properties. Thus, CPT may be effective in drill holes and selection of monoliths decrease, as on most of the worked territory it may be carried out instead of such holes. It is evident that in all cases it is necessary to confirm correlation between CPT results and the controlled soil properties and, of course, the determination of the corresponding empirical dependencies for each concrete bulk.

Non-Russian professional engineers pay little attention to compaction or hardening of collapsible clayey soils, although the issues of *deep compaction of sands* have been worked out thoroughly. They use the methods based on vibration (vibrocompaction). Some of the professional engineers tacitly agree that compaction of soils equals to vibrocompaction of sands. In this vibration a vibrating steel shell (usually a cylindrical one) is immersed into the soil by hydraulic washout, and the forming hole is being filled with graded granular filling followed by compaction during extraction of the unit. The forming dense column is in close contact with the adjacent soil. Penetration of the vibroshell may be performed ignoring hydraulic washout ("vibroflotation").

It is evident that in solution the problems of sand compaction quality control the role of CPT becomes great, as the selection of monoliths from these soils always faces some difficulties requiring special soil samplers application. Thus, CPT is

243

widely applied either for soil compaction quality control and vibrocompaction method applicability.

Figure 4.22 illustrates the chart of T. Lunne et al. (2004) applied for recommendations during investigations. According to the chart, in friction ratio f_s/q_c > 1.5 %, as well as in q_c < 3 MPa vibrocompaction of soils is impossible. This means that most of soils are not subjected to vibrocompaction. Strong soils, in which q_c > 70 MPa before compaction, are not subjected to compaction either, but, probably, it is not needed.

It is necessary to point out that occurrence of loessial and loess-like loams is ignored in the chart 4.22, but they are subjected to compaction despite rather high friction ratios (f_s/q_c = 3...6 %). Apparently, the authors disposed of the initial data with lack of this information.

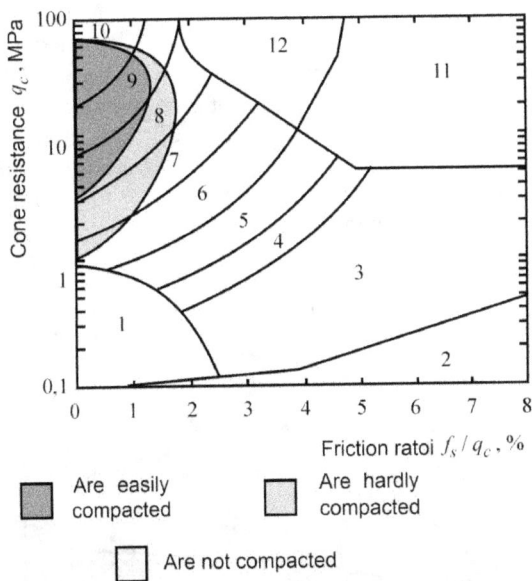

Fig. 4.22 Chart after T. Lunne et al. (2004) to determine possible application of vibrocompaction. Numbers of zones 1...12 correspond to the same soils (see fig. 3.3, section 3.1.1)

T. Lunne et al. admit application of two parameters and hence, two approaches as the criteria of the required degree of sand compaction:

- Minimal relative density of sands I_d (see section 3.2.5), i.e. "indirect" approach.
- Minimal cone resistance q_c, i.e. "direct" approach.

The second parameter – minimal value of q_c is more convenient, but its determination can be followed by a number of difficulties, as it depends upon the additional factors including initial density, depth, etc. For these reasons, T. Lunne et al. recommend to take care when considering the issue, taking into account the local experience; in here, they recommend to apply the corrected value of q_c – "normalized cone resistance q_{cl}»

$$q_{cl} = (q_c/p_a) \cdot (p_a/\sigma'_{vo})^{0,5}, \tag{4.38}$$

where

p_a = atmospheric pressure given in the same units as q_c;

σ'_{vo} = effective vertical pressure from soil net weight, given in the same units as q_c.

T. Lunne et al. pay attention to an interesting, not well-studied phenomenon – increase of soil strength in time after dynamic compaction by beating. The increase has been confirmed experimentally. Here, q_c increases without pore pressure change and bulk settings increase as well.

Figure 4.23 shows the results of research after D. Schmertmann (Lunne et al. 2004) who studied the phenomenon after beating. Silt sand stratum 10 m in thickness was compacted (the state of Florida). Increase of cone resistance q_c in the stratum was analyzed in various degrees of compaction – in different numbers of compacting blows per unit of volume. Unfortunately, D. Schmertmann marked the results obtained as the identical points on the graph (circles); this complicates spread of values estimation in the same number of compacting blows.

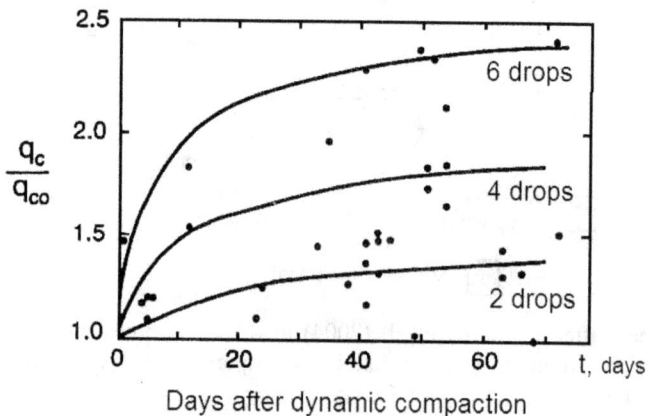

Fig. 4.23 Schematic illustration of q_c increase after dynamic deep compaction: q_c – cone resistance at the moment of time t, q_{co} – the same just after compaction

As one can see from fig. 4.23, relative cone resistance q_c/q_{co} considerably increased in 70…75 days after compaction (q_c – cone resistance in t days, q_{co} – cone resistance after compaction). The increase depended upon the degree of compaction:

given two compacting blows per unit of volume, q_c/q_{co} increased in 1.4 times, in six – in 2.4.

The analogous results have been obtained by other authors from different countries. In here, it was stated that the speed of the procedure depends upon temperature. W. A. Charlie et al. (Lunne et al. 2004) obtained the empirical formula

$$q_c/q_{co} = 1 + K \cdot log\ N, \qquad (4.39)$$

where:

$q_c =$ cone resistance after N weeks of compaction,
$q_{co} =$ the same during the first week (immediately after compaction),
$K =$ empirical coefficient dependent on temperature.

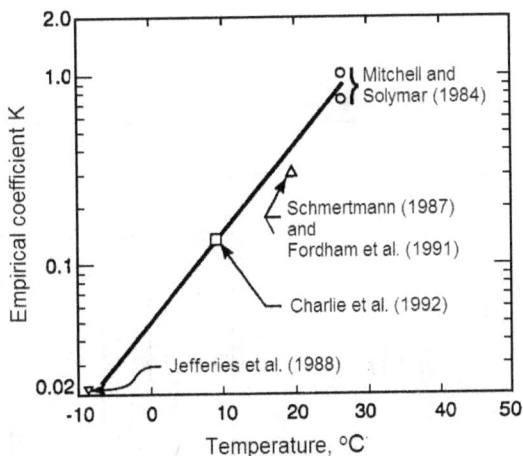

Fig. 4.24 Schematic illustration of the empirical coefficient K (in the formula (4.39) dependence upon temperature (after W.A. Charlie et al.)

Figure 4.24 shows the graph of coefficient K dependence upon temperature. As one can observe, increase of temperature from -10 to +30°C corresponds to two orders increase of K (from 0.02 to 1.0). In particular, in accordance with the formula (4.39), in 10 weeks after compaction q_c shall increase in 9 % in temperature of +5°C ($K = 1.09$), in +25°C – in 70 %.

Perhaps, the reliability of the formula (4.39) is to be verified in a wider range of conditions, but nevertheless, the results of computation make us pay greater attention to the phenomenon which it reflects. In particular, it seems to be obvious that in deep sand compaction the assessment of results applying CPT is necessary to be performed by numerous measurements (CPTs) during several weeks (8…12 weeks) after compaction (or a part of it).

Quality control of foundation strengthening is usually required in application of injection methods of foundation soil fastening (chemical or physical-chemical fastening). This is due to the fact that the process of soil fastening by injections of solutions is always "hidden", and heterogeneity of the fastened massif requires thorough control. For this, CPT rigs capable to penetrate the cone in the depth corresponding to the depth of fastening are necessary, pushing stresses are to be sufficient to "pick" the fastened layers. This appears to be possible when the fastened soil is transformed into stronger soil rather than hard rock; this soil is capable to be picked by the cone. Experience has been gained in Bashkortostan; here, for strengthening of foundations chemical fastening of water-saturated clayey soils with the alkali (NaOH) has been applied for years (Volkov et al. 1985). In the fastening given, increase of soil strength is quite large: as a rule, it corresponds to 3...10 times increase of q_c. Figure 4.25 illustrates the results of the fastened foundation testing by CPT rig (S-832).

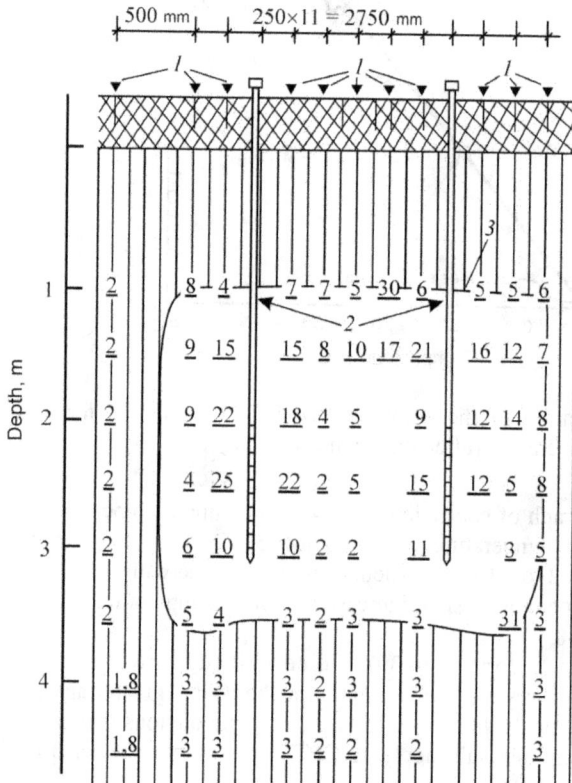

Fig.4.25 Sectional view showing cone resistances q_c in MPa in CPT of the fastened zone:

1 – CPT locations, 2 – injectors through which the alkali is delivered, 3 – border of the fastened zone. Underscored numbers show values of q_c in MPa

4.5 SAND LIQUEFACTION RISK ASSESSMENT

Under *liquefaction of soils* we mean the process of their transformation into heavy viscous liquid under certain conditions. In Russian publications the term "running sand" is usually applied; it implies the transformation of soil into liquid condition and slow yielding flow plastic deformations as well (dulling of excavation slopes, bulks, spontaneous smoothing of pitted surfaces, etc.). In most cases sands are related to the running ones (usually fine and pulverescent sands with the admixtures of clayey colloidally sized particles). The notion sometimes involves weak clayey soils showing significant thixotropic softening. The reasons of this transformation of soil into liquid condition are probably connected with static and/or dynamic (vibrational, impact) effects or change of hydrostatic/hydrodynamic pressures. As a rule, they are the dynamic effects of anthropogenic or natural (seismic) origin.

Specialists turned to forecasting of sand behavior in seismic effects after the earthquake of 1964 in Niigata (Japan). Here, soil conditions are characterized by prevalence of wet sands. The considerable part of destructions was caused by liquefaction of sandy foundations rather than "direct" seismic effects. The analogical effects were observed during another major earthquake in Koba (Japan) in 1995. The events made identification of sands capable to be liquefied under seismic effects an urgent issue.

Specialists pay attention to liquefaction of sands in vibrational effects regardless of earthquakes, e.g. vibration caused by industrial equipment (turbogenerators, compressors, etc.). In construction there may appear situations causing liquefaction of sands, e.g. shaking connected with pile driving or blasting, etc.

In order to solve the problem of sand liquefaction, one needs to understand the nature of the phenomenon. In accordance with the first ideas put forward by K. Terzagi, running sands were explained by hydrodynamic pressure effect. Thus, the idea of identification of sands capable to be liquefied was nonsense, as it implied that any sands could be liquefied under certain conditions. However, later research of Russian and non-Russian specialists revealed the connection between liquefaction of sands and their composition, properties and condition as well. Thus, only certain kinds of sands are capable to be liquefied and they should be identified during surveying. Nowadays, it is a common practice, and specialists pay attention to identification of liquefied sands efficiency increase. In Russia the problem is being resolved by *dynamic probing test* application integrated with drilling and laboratory tests SNiP 11-105-97. However, in non-Russian practice CPT is widely applied (Lunne & Keaveny 1995, Robertson & Fear 1995, Lunne et al. 2004). As it was mentioned earlier, T. Lunne et al. (2004) (see table 4.1) treat sand liquefaction risk assessment by CPT data as the problem being successfully resolved (rating "1-2"). In their general report at the symposium CPT'95 P.K. Robertson and C.E. Fear pointed out that CPT has become the *principal in-situ method* of sand liquefaction risk assessment (Robertson 1995).

At first, detailed research on the problem was undertaken by H.B. Seed (Robertson & Fear 1995, Lunne et al. 2004) in 1960s...1970s in the USA with regard to SPT. Then, the developed methods were supplemented and corrected in different countries. For instance, let us consider the experimental results obtained by Y. Suzuki et al. in 1990s (Suzuki 1995). His methods are based on sand liquefaction risk assessment with regard to the following parameters:

• Relative shear stress ratio.
• Normalized (modified) cone resistance.

In here, Y. Suzuki et al. insist on friction ratio f_s/q_c (see chapter 3) to be taken into account as well as occurrence of clayey particles in sand.

Relative shear stress ratio τ_d/σ'_{vo} is evaluated by the formula

$$\frac{\tau_d}{\sigma'_{vo}} = \tau_n \frac{a_{max}}{g} \frac{\sigma_{vo}}{\sigma'_{vo}}(1-0.015h), \qquad (4.40)$$

where:

$\tau_d =$ amplitude of cyclic shear stresses equivalent to actual shear stresses (MPa),

$\sigma_{vo}, \sigma'_{vo} =$ initial (prior to penetration) stresses in soil from the weight of overlying layers, total and effective (MPa) correspondingly;

$\tau_n = 0.1(M-1) =$ coefficient (nondimensional) dependent on magnitude:

$M =$ earthquake magnitude,

$a_{max} =$ maximum horizontal acceleration on the surface evaluated due to the earthquake magnitude (or force after MSK-64 scale), m/s,

$g =$ gravity acceleration, 9.81 m/s,

$h =$ depth of soil layer under study = depth of penetration (m).

Table 4.25 contains the data on values of a_{max} – maximum horizontal accelerations on the soil surface – given in the book of Y. Suzuki et al. (Suzuki 1995).

Table 4.25 Measured maximum horizontal accelerations a_{max} in earthquakes

Earthquakes	Magnitude M	Measured locations	Acceleration a_{max} (cm/s^2)
Kushiro-Oki	7.8	Kushiro City	150
	(6.8)*	Nemuro City	410
Togo-Oki	8.1	Kushiro City	90...290
		Nemuro City	360
Nancey-Oki	7.8	Nakodate City**	120
		Mori Town	175
		Oshmanbe Town	240
Great-Kobe	7.2	Kobe City	400
		Nishinomia City	350
		Amagasaki City	300

* corrected values (after H.B. Seed)
** including Kamiiso Town

Yu. G. Trofimenkov (1995) gives relative values of acceleration a_{max}/g as applied to MSK-64 scale:
- 7 locations $a_{max}/g = 0.1$,
- 8 locations $a_{max}/g = 0.2$,
- 9 locations $a_{max}/g = 0.4$.

Y. Suzuki et al. suggest *normalized (modified) cone resistance* q_{t1} to be evaluated by the formula

$$q_{t1} = q_t/(\sigma'_{vo} / \sigma'_a)^{0.5}, \qquad (4.41)$$

where:

$q_t =$ corrected cone resistance evaluated by the formula (3.3) implying piezocones application; for cones without piezocones $q_t \approx q_c$ is admissible,

$\sigma'_{vo} =$ initial effective stress in soil from the weight of overlying layers (the same as in the formula (4.40)) (MPa),

$\sigma'_a =$ effective pressure equal to atmospheric one $\sigma'_a = 98$ kPa.

Figure 4.26 illustrates the diagrams of dissipation obtained by Y. Suzuki et al. (1995); they show connection between τ_d/σ'_{vo} and q_t in different friction ratios f_s/q_c. The curves have been drawn on the graphs; they show connection between τ_d/σ'_{vo} and q_t in occurrence of clayey particles in sands (*FC*) obtained by K. Tokimatsu et al. (see in Suzuki 1995).

As one can see from figure 4.26, clear grouping of points related to liquefied and non-liquefied sands is observed: points related to liquefied sands are located left from the points related to non-liquefied sands and over them. Therefore, the diagrams of dissipation may be used as the charts for identification of sands in their inclination to liquefaction (e.g. in $\tau_d/\sigma'_{vo} = 0.1$ and $q_{t1} = 15$ MPa sand is not to be liquefied, and vice versa, in $\tau_d/\sigma'_{vo} = 0.3$ and $q_{t1} = 5$ *MPa* – is to).

Diagrams of dissipation related to soils with different friction ratio f_s/q_c, slightly differ. It is the most noticeable in comparison of sands in which $f_s/q_c < 1.0$ % (fig. 4.26a, b) with sands in which $f_s/q_c \geq 1.0$ % (fig. 4.26c). Percentage of clayey particles (*FC*) influence is also tangible: the lines obtained by K. Tokimatsu et al. are the division lines of points related to liquefied and non-liquefied sands.

It is important to underline that a number of other specialists including P.K. Robertson and C.E. Fear (Robertson & Fear 1995) when solving the analogical problem, apply "standard" value of cone resistance q_c (measured by penetrometer without a piezocone) instead of the corrected value of q_t; the results obtained are nearly the same as in Y. Suzuki et al. Here, the "corrected resistance" is marked as q_{c1} and given as

$$q_{c1} = q_c / (\sigma'_{vo}/\sigma'_a)^{0.5}, \qquad (4.41a)$$

where σ'_{vo} and $\sigma'_a =$ the same as in the formula (4.41).

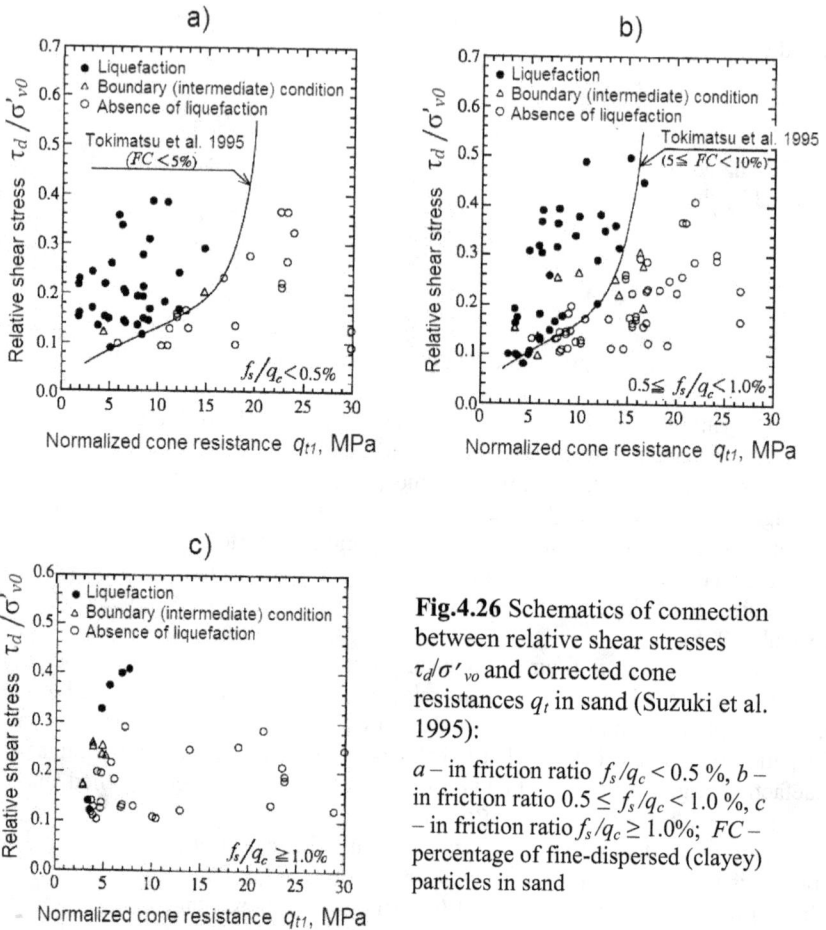

Fig.4.26 Schematics of connection between relative shear stresses τ_d/σ'_{vo} and corrected cone resistances q_t in sand (Suzuki et al. 1995):

a – in friction ratio $f_s/q_c < 0.5$ %, b – in friction ratio $0.5 \leq f_s/q_c < 1.0$ %, c – in friction ratio $f_s/q_c \geq 1.0$%; FC – percentage of fine-dispersed (clayey) particles in sand

The nondimensional value marked as q_{c1} is also used (Belyaev 2005):

$$q_{c1} = (q_c / \sigma'_a)(\sigma'_a / \sigma'_{vo})^{0,5}, \qquad (4.41b)$$

where the symbols are the same as in the formula (4.41).

Figure 4.27 illustrates the curves dividing the zones of points related to liquefied and non-liquefied sands after different authors (Lunne et al. 2004).

As one can see from figure 4.27, the results obtained by different authors in sands of different coarseness and with different clayey particles content have been the same. For instance, it seems possible to define the conditions when sand liquefaction hazards is absent according to *all* given dependencies. Thus, in $q_{c1} > 15$ MPa one should not expect any liquefaction. Vice versa, in $q_{c1} < 4$ MPa the probability of liquefaction is rather high. The reduction of the originating tangent stresses decreases probable liquefaction; this is the most noticeable in low values of τ_{max}/σ'_{vo}.

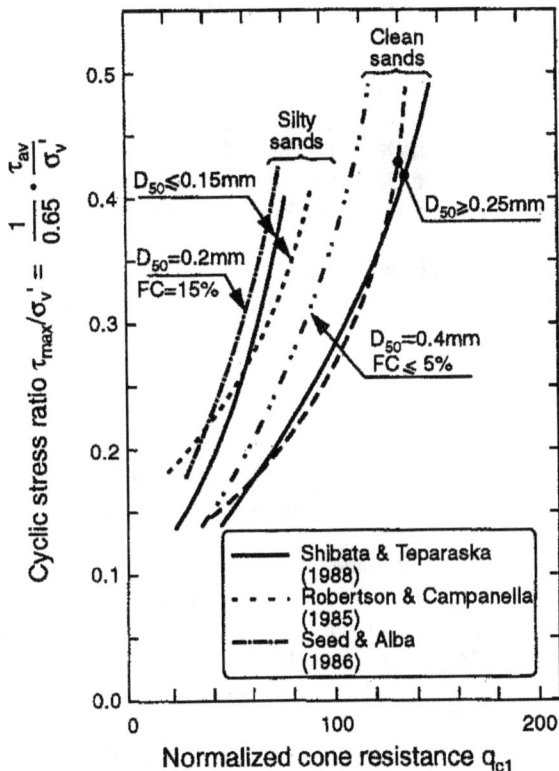

Fig. 4.27 Curves dividing the zone related to liquefied sands (on the left from each curve) from the one of non-liquefied sands (on the right from each curve) (Lunne et al. 2004):

FC – content of clayey particles in sand; D_{50} – diameter of soil particles comprising 50 % (in weight); q_{c1} – corrected cone resistances corresponding to the formula (4.41a)

Having processed a large amount of experimental data, T. Shibata and V. Teparaksa (1988) suggested the empirical formula for the normalized critical cone resistance evaluation $(q_{c1})_{cr}$; lower values of q_{c1} shall correspond to liquefied sands.

252

$$(q_{cl})_{cr} = C_2 \left(5 + 20 \frac{\frac{\tau_d}{\sigma'_{vo}} - 0,1}{\frac{\tau_d}{\sigma'_{vo}} + 0,1} \right),$$ (4.42)

where:

$\tau_d / \sigma'_{vo} =$ the same as in the formula (4.40),

$C_2 =$ coefficient of sand coarseness taken

in $D_{50} > 0.25$ mm $C_2 = 1$,

in $D_{50} < 0.25$ mm $C_2 = D_{50}/0.25$;

$D_{50} =$ diameter of soil particles comprising 50 % (in weight).

In resolving practical problems T. Shibata and V. Teparaksa suggest application of special parameter - the critical cone resistance for liquefaction $(q_c)_{cr}$, evaluated by $(q_{cl})_{cr}$:

$$(q_c)_{cr} = \left(\frac{0.07 + \sigma'_{vo}}{0.17} \right)(q_{cl})_{cr},$$ (4.43)

where:

$(q_{cl})_{cr} =$ normalized critical cone resistance (MPa) evaluated by the formula (4.42),

$\sigma'_{vo} =$ the same as in the formula (4.41) (MPa).

The value of $(q_c)_{cr}$ is the minimal value of q_c, when the sand under study may still be considered non-liquefied. If the measured values of q_c prove to be higher than $(q_c)_{cr}$, one should not be scared by any liquefaction.

The order of liquefaction hazard assessment is as follows. First of all, one has to evaluate τ_d/σ'_{vo} for different depths h with regard to magnitude M of the anticipated earthquake and hence, maximum acceleration of soil on the surface a_{max}. Initial effective stresses σ'_{vo} mean the complete consolidation of soil, thus, pore pressure may be considered as hydrostatic equal to $u = \gamma_w h$ (γ_w – specific weight of water, h – depth).

Further on, $(q_{cl})_{cr}$ is evaluated by the formula (4.42), then $(q_c)_{cr}$ is evaluated by the formula (4.43). The calculations are carried out in the whole of depths h, then the curve showing the values of *calculated* critical resistances $(q_c)_{cr}$ (i.e. the curve "$(q_c)_{cr}$ ~ h") is drawn on the graph of *measured* cone resistances q_c (i.e. the graph "q_c ~ h").

Figure 4.28 illustrates the curves obtained by M. Mimura et al. in sandy deposits of Higashi Ogashima (Japan) (Mimura et al. 1995). The calculation was done for $a_{max}/g \geq 0.2$; this approximately corresponded to 8 location after MSK-64. It is important to underline that figure 4.28 shows cone resistances corresponding to penetrometer without a piezocone, while M. Mimura et al. consider corrected cone resistances q_t measured by the piezocone. However, this correction is not of principal importance, since in absence of piezocones $q_t \approx q_c$ (see chapter 3).

Figure 4.28 shows that in the whole of depths $q_c < (q_c)_{cr}$, i.e. the sand under study *shall be liquefied* in 8 location earthquake according to MSK-64 scale ($a_{max}/g = 0.2$).

Cone resistance q_c, MPa

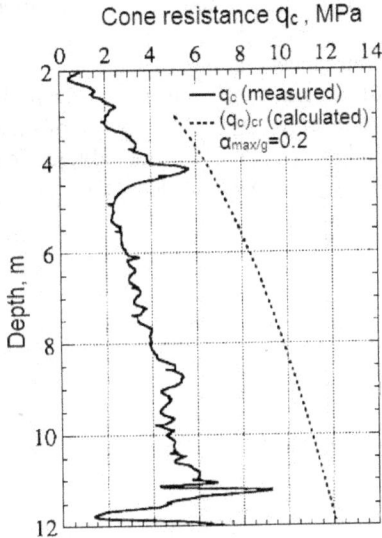

Fig.4.28 Schematic illustration of comparison between measured cone resistances q_c and calculated critical values of $(q_c)_{cr}$ (Mimura et al. 1995)

Having analyzed the results of research of Japanese authors, Yu.G. Trofimenkov (1995) drew the graphs for evaluation of critical cone resistances of sand (fig. 4.29).

$(q_c)_{cr}/C_2$, MPa

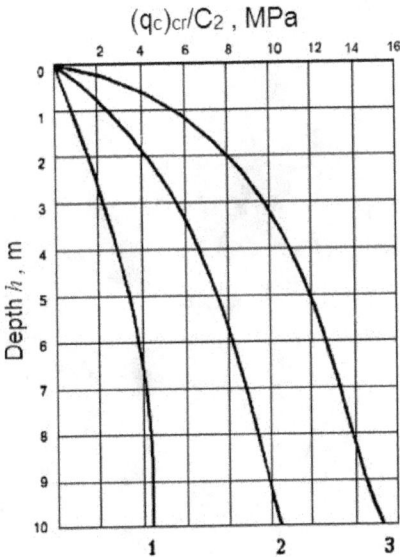

Fig. 4.29 Schematic illustration of corrected critical cone resistances $(q_c)_{cr}/C_2$ in various depths and force after MSK-64 scale (Trofimenkov 1995):

$(q_c)_{cr}$ – critical cone resistance by the formula (4.43), C_2 – coefficient of sand coarseness (see formula (4.42); 1 – curve of relative acceleration $a_{max}/g = 0.1$ (approximately 7 locations after MSK-64), 2 – the same $a_{max}/g = 0.2$ ($\approx$ 8 locations), 3 – the same $a_{max}/g = 0.4$ ($\approx$ 9 locations)

Yu.G. Trofimenkov considers corrected value $(q_c)_{cr}/C_2$, where $(q_c)_{cr}$ – critical value of cone resistance evaluated by the formulae (4.42) & (4.43), C_2 – coefficient of sand coarseness (see formula (4.42). The curves shown in fig. 4.29 have been evaluated with regard to the following conditions: magnitude $M = 7$, relative accelerations $a_{max}/g = 0.1, 0.2; 0.4$ (this corresponds to 7, 8 and 9 location earthquakes after MSK-64 scale), underground water level – 2 m in depth.

As one can see from figure 4.29, the risk of liquefaction is rather high for most of sands, especially in major earthquakes. For instance, if $q_c \leq 12$ MPa in 5 m of sand layer depth (this refers to most of sands), this sand layer is to be considered liquefied in 9 location earthquake, $q_c \leq 7.5$ – in 8 location one, $q_c \leq 3.5$ – in 7 location one.

5 CPT APPLICATIONS IN SPECIFIC SOIL ENVIRONMENT

Russian Norms on engineering investigations SNiP 11-02-96 distinguish "specific soils", such as permafrost, collapsible, expanding, organic-mineral and organic, saline, alluvial and anthropogenic. In practice, moraine soils containing boulder inclusions are often referred to these soils.

Due to space exploration, a new branch of geotechnics is being developed – extraterrestrial body soil studying, e.g. Lunar soil. It is evident that extraterrestrial soils are of specific properties, thus, other methods of studying are required as well as other equipment or calculations.

Both non-Russian and Russian specialists have been interested in CPT applications in specific conditions. The research has embraced a small part of the conditions; however, the results obtained show indisputable advantage of CPT applications here. The chapter presents the results of research on the issue.

5.1 PERMAFROST SOILS

5.1.1 General information, historical background, equipment and test methods

The most important feature of permafrost soils is the dependence of their mechanical properties on temperature. For this reason, the National Norms of Russia GOST 25100-2011 & SP 25.13330.2012 divide them with regard to their material constitution and temperature-water content conditions into three types: hard-frozen, plastic-frozen and loose-frozen.

Hard-frozen soils are dispersed soils solidly cemented by ice, they are characterized by rather brittle failure and they are incompressible under external loading possessing the coefficient of compressibility of $m_f \leq 0.01$ MPa^{-1}. The soils with the following temperatures are usually referred to these soils: below 0 ^{0}C – for large fragmental rocks, below -0.1 ^{0}C – for coarse sands and the ones of medium coarseness, below -0.3 ^{0}C – for fine and pulverescent sands, below -0.6 ^{0}C – for silty clays (sandy loams), below -1 ^{0}C – for lean clays (loams), below -1.5 ^{0}C – for clays. The properties of the soils are similar to those of semi-rocks, thus, it seems impossible to apply CPT in most of hard-frozen soils.

Plastic-frozen soils are dispersed soils cemented by ice, but they possess viscous properties and are compressible under external loading with the coefficient of compressibility of $m_f > 0.01$ MPa^{-1}. Sandy and clayey soils with the temperatures ranging from freezing of soil to the given fixed temperatures are usually referred to them. Saline soils, peat or soils with plant residues impurities are in plastic-frozen condition in lower temperatures than those given above. Both CPT and pile driving applications are possible in these soils.

256

Loose-frozen soils ("dry frozen subsoil") are large fragmental rocks and sandy soils possessing negative temperatures, but they are not cemented by ice and they are not coherent due to low water content they possess. Non-cohesive soils (sands, gravel, gruss, etc.) are referred to them in total water content $w_{tot} \leq 0.03$. Due to the fact that in negative temperatures tiny amount of ice is generated in soil pores, their properties do not depend upon temperatures and are similar to those in non-frozen condition. For this reason, immersion of the cone in loose-frozen soil seems to be possible as well as it could be done in positive temperatures of soil. One needs to take into account that immersion of the cone in large fragmental rocks, either frozen or thawed, is impossible to be performed.

Therefore, in general, CPT may be carried out in plastic-frozen and weak loose-frozen (sandy) soils.

The issues on CPT applications in permafrost soils were studied both worldwide (mainly in Canada and the USA) and in the former USSR and in Russia. The research revealed that CPT results could give geotechnical information valuable in construction on urbanized territories or in exploration of oil and gas deposits in a cryolite zone. At present, special rigs adjusted to arctic regions are produced worldwide. In them, the cones are equipped with power and temperature sensors, and besides, they may incorporate other sensors as well (e.g. inclinometer, depth sensor, piezometer for pore pressure measurement, soil electric conductivity sensor).

CPT was not used in permafrost regions for a long time. It was believed that immersion of the cone was difficult to ensure technically due to high strength of permafrost soils. It was also unclear how to consider specific properties of frozen soils, i.e. their high rheology and dependence on temperature in carrying out the tests and interpretation of the data obtained. It was thought that without special pushing rigs and cones of special geometries it was unpromising to introduce CPT in permafrost regions. However, the opinion failed when the first experiments on heavy and medium CPT rigs applications in one of the most widely-spread permafrost soils (plastic-frozen) showed positive results.

For the first time, CPT on permafrost soil was tested in 1974 by the Canadian researcher B. Ladanyi (Andersland & Anderson 1978, Ladanyi 1976, 1982) in testing of plastic-frozen strip clay -0.1...-0.3 ^{0}C in temperature. The test was carried out by means of hydraulic penetrometer "Fugro" 3.57 cm in diameter. Only cone resistance was recorded during the test. The total weight of equipment was 30 kN, that is why the machine was anchored by a pair of screw piles. In spite of this, because of freezing of soil to the push rod, CPT was difficult to be performed more than 0.6 m lower than borehole bottom.

B. Ladanyi applied two techniques in his experiments. The first one – the so-called "quasi-static" test (the term was introduced by B. Ladanyi) was applied in quite low speeds of penetration being under strict control. The second one – the test by stepwise load increasing with maintenance of permanent load on the step during the specified time (in B. Ladanyi's experiments – 15 min). The specified and observed speeds of CPT varied from 0.0025 to 2.5 cm/min.

Later, similar tests were carried out by other non-Russian specialists (Buteau et al. 2005, Fortier et al. 1993). For example, a series of analogous tests was carried out in 1999…2000 in the North of Quebec in Canada by Buteau et al. (2005) in order to study cryostratigraphy and creep of permafrost soils. In carrying out the experiments a portable rig (see fig. 5.1) was used; this allowed the stress up to 113 kN to be ensured with permanent speed of penetration to be from 0.00024 to 95 cm/min. During the tests the speed of penetration was specified from 0.006 to 6 cm/min. Standard cone penetrometer equipped with complementary sensors was applied.

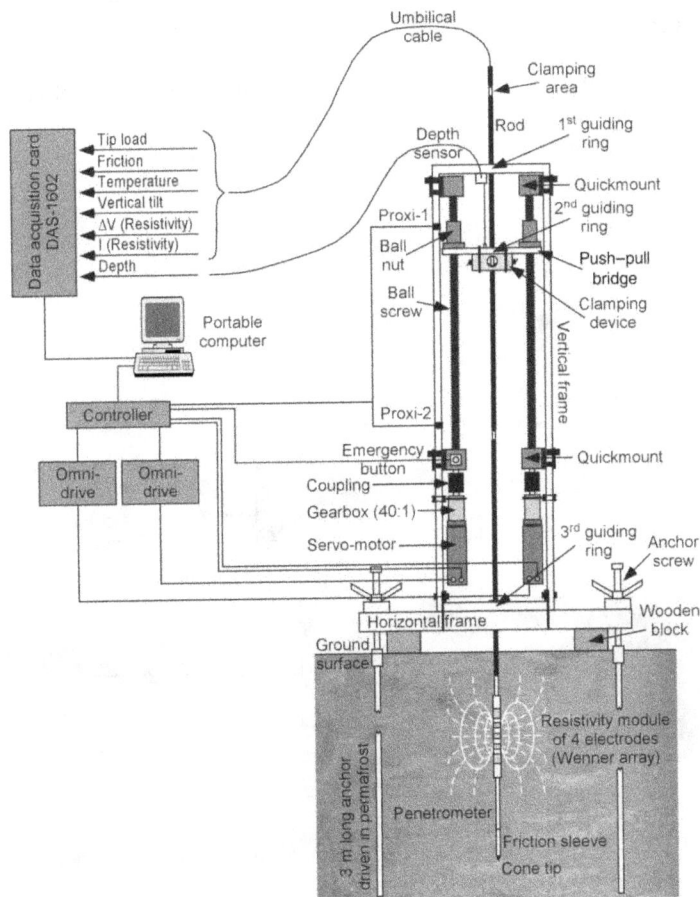

Fig. 5.1 Schematic illustration of modern portable rig with controlled speed of jacking (Buteau et al. (2005)

In 1976...1977 the specialists of US Army research laboratory on permafrost applied CPT to study frozen soil properties under the bottom of Beaufort Sea at the coast of Alaska (Blouin et al. 1979) in order to obtain engineering-geological information on sedimentation profiles and icy inclusions. Geocryological sectional view included the following: ice stratum 0.7...1.8 m in thickness (before CPT ice was drilled out), sea water layer up to 0.9...3.2 m in thickness and fine-grained sediments of sea bottom. The temperature along the sectional view varied from -1 to -5 ^{0}C. Bottom sediments were of higher salinity – the temperature of saline soils freezing was from -1.75 to -3.55 ^{0}C.

Non-standard penetrometer 6.35 cm in diameter equipped with a push rod with a tip and a casing was applied. Specially developed rig mounted in a tractor (see fig. 5.2) was used in the experiments. The equipment was placed in the cabinet installed on skis fixed to the tractor which transported the equipment. The weight of the tractor was used for jacking of the cone. The cone tip was fixed to the push rod located inside the casing. The average speed of penetration was 30 cm/min.

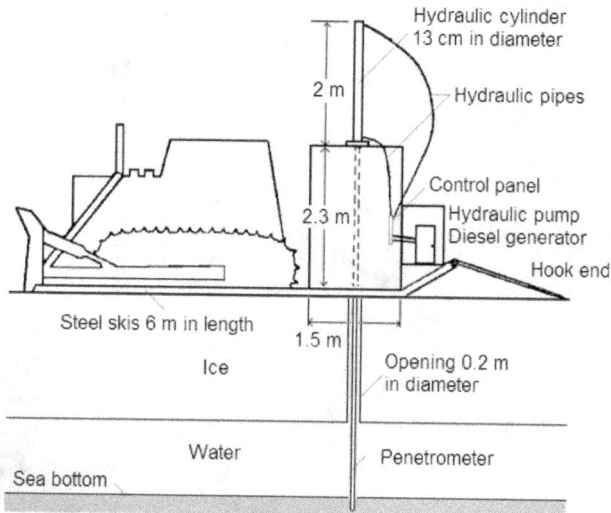

Fig. 5.2 Schematic illustration of the rig mounted in a tractor developed in US Army research laboratory on permafrost (Blouin et al. 1979)

The research resulted in the conclusion that cone resistance could provide with valuable information for lithological dissection of the ground massif. The total resistance of the cone attained 100 kN.

The cone having been penetrated up to maximum possible depth, the push rod was filled with antifreeze liquid with temperature sensor on the bottom. Measurements of sea water and bottom sediments temperatures were done bottom-up by periodical lifting (with the interval of 1.5 m) and "dwell operation" of temperature sensor. Temperature measurement technique was similar to traditional measurements in thermometric openings and, hence, it had similar disadvantages, in the first place, longer dwell operation of the opening (thermal balance was attained in 6...8 hours) due to vertical convection of antifreeze liquid located inside and subjected to temperature gradients. It is important to underline that the Norms of Russia GOST 25358-82 do not allow the openings filled with water, brine or other liquid to be used for temperature measurements. In all the disadvantages given, it was the first attempt to apply CPT for measurement of soil temperature.

When analyzing the results of non-Russian researchers and the techniques applied, it seems important to point out that in spite of absolute advantages, the approaches have certain drawbacks.

Very low speed CPT increases the duration of tests, thus, it is no longer an express-method then. Besides, application of additional special devices to measure very low speeds of jacking is required. Non-Russian authors ignore such an important parameter in CPT as sleeve friction resistance; this results in lack of information and, hence, validity of tests. They do not consider thermal interaction between the cone and frozen soil; therefore, it is difficult to understand the physics of processes occurring between the cone and frozen soil. The formulae for frozen soil geotechnical parameters assessment suggested by non-Russian authors, besides CPT data often contain a number of extra parameters – theoretical ones or those evaluated experimentally. The enlargement of parameters included in the formulae assist in more errors in calculations. The distinctive feature of non-Russian publications is the fact that there is a lack of comparisons between the suggested methods and/or formulae and the results of standard (reference) in-situ and/or laboratory tests of frozen soils. The Russian scholar of CPT tends to anchorage (local correction) of calculations by CPT results by their comparison with the reference tests. Revelation of empirical dependencies or introduction of empirical coefficients in theoretical formulae is usually the result of such a comparison.

In the former USSR the first in-situ tests of frozen soils were carried out by BashNIIstroy (the former NIIpromstroy) (Volkov & Isaev 1983, 1985): in artificially frozen soil in Ufa in 1982, in permafrost in Vorkuta in 1983 (jointly with PechorNIIproject Research Institute). CPT procedure was performed by S-832M CPT rig (see section 1.3.5); figure 5.3 illustrates the rig. Later, Fundamentproject, MIIT and NIIOSP joined the research on CPT of permafrost soils: the experiments were done in Vorkuta, Labytnangi, Yamal peninsula and other test sites (Trofimenkov et al. 1986, Isaev et al. 1987, Isaev 1988, Sadovsky et al. 1988, Isaev 1989, Isaev et al.1991, 1995, Minkin 2005).

At first, due to absence of permafrost in Ufa, the tests were carried out on artificially frozen soils (Volkov & Isaev 1983) – underconsolidated loams. Freezing procedure was done by blowing of cold air through vertical pipes placed in the ground. According to thermometric openings readings, frozen soil temperature attained -2.4 ^{0}C.

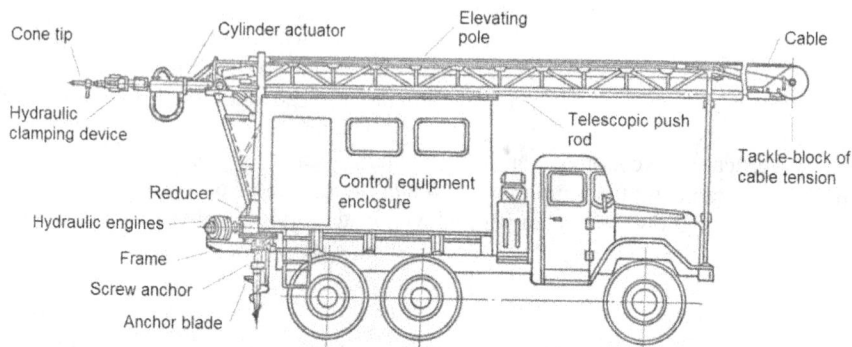

Fig. 5.3 Schematic illustration of S-832M CPT rig applied in in-situ tests in Ufa, Vorkuta and Labytnangi

Test sites in Vorkuta were characterized by presence of loams, rare sandy loams and sands with negative temperatures up to -2.1 ^{0}C, water content – from 12 to 38 %, pebbles and gravels – from 2 to 40 % and presence of boulders as well. Cryogenic texture of frozen soils was massive or stratified with 1…4 mm thickness of icy inclusions, and content of ice varied from 1 to 10 %.

In Labytnangi the tests were done in loams with interlayers of pulverescent sand possessing the negative temperatures up to -0.6 ^{0}C. When frozen, the soils were characterized by massive cryotexture with rare thin ice lenses.

In sites located in permafrost regions the soils were mainly plastic-frozen, rarer solid – frozen (interlayers), melted or thawed out.

The first research carried out in the former USSR resulted in technical feasibility of CPT in permafrost[1] (mainly in plastic-frozen soils) by Russian serialized rigs S-832 (fig. 5.3, 5.4a), SP-59B (fig. 5.4b) and SP-72.

One of the crucial issues for CPT (CPTT) in permafrost soils is the complexity of penetrometer pushing. To tackle the problem, the following must be needed:

• It is necessary to use heavy CPT rigs with pushing thrust capacity of not less than 100 kN. The rigs mounted in all-terrain vehicles are preferred in the Northern regions since they allow the specified pushing thrust to be ensured ignoring anchors (it is often quite difficult to anchor in permafrost soil) just

---

[1] Wider application of CPT in permafrost soil is considerably restricted due to a psychological factor – a persistent myth on CPT failure in permafrost. At the beginning of 1980s when the author was on an arctic expedition, he faced the absolute disbelief in its success. He agreed to bet with a CPT rig operator and promised to perform "Gypsy dance" on the rig roof in case the expedition was a success.

owing to all-terrain vehicle mass and high cross-country ability in difficult to access territories.

- It is necessary to reduce the length of unsupported push rod projecting above the ground surface because of instability hazard (in CPT in standard soil conditions the procedure usually stops by this reason). It is to be done with a hydraulic clamping device (e.g., mounted in Russian CPT rig – S-832M) or special guide casing.

- If necessary, to use "combined" penetration (the test is carried out with the help of a boring machine making it possible to drill out barely passable permafrost soil layers alternately and to push the penetrometer).

- It is necessary to mount a friction reducer above the cone tip; its diameter 10 to 20 mm larger than the penetrometer; it is placed not nearer than 300 mm from the friction sleeve to avoid extra errors in measuring unit sleeve friction resistance.

For the first time the penetrometer equipped with both electrical and temperature sensors was used by BashNIIstroy in Vorkuta in 1983 (see sections 1.3.4 & 5.3.4, fig.1.18). The cone temperature was measured both in penetration and in relaxation-creep regime. These tests as well as the following ones helped better understand the essence of thermal interaction of the cone and frozen and/or thawed soils and develop methods of their original temperature measurements (see section 5.1.4). They showed that it was extremely desirable to apply the penetrometer equipped with temperature sensor (or other additional sensors) in permafrost conditions.

In the authors' opinion, the tests by means of type II tensometric cone penetrometer equipped with temperature sensor need to be considered a separate test similarly to CPTU carried out by tensometric cone penetrometer equipped with pore pressure sensor (see section 1.3.4). In accordance with the international classification, the tests may be called "Cone Penetration Test with Temperature Measurement" or CPTT.

The tests by S-832M were carried out applying the methods of permanent speed penetration and relaxation-creep regime (the so-called "relaxation-creep" CPT means single-phase "stabilization" in every point of periodical stop of the cone in depth – see section 2.3).

Partially the tests were performed in stepwise increasing relaxation-creep regime, i.e. in repeated "stabilization" of the cone in every point of its stop (the load on the cone increased stepwise and almost instantly at the next in turn step of loading by additional delivery of oil into hydraulic jacks. The delivery having been finished, the tests became relaxation-creep). Partially the tests were carried out with stepwise increasing speeds of penetration (the load on the cone increased stepwise so that the cone was evenly penetrated at each step).

In high values the speed of penetration was evaluated by chart strips of self-recording devices. In slow speeds of penetration (including "relaxation-creep" CPT) cone movements were recorded by reference system and a flexometer.

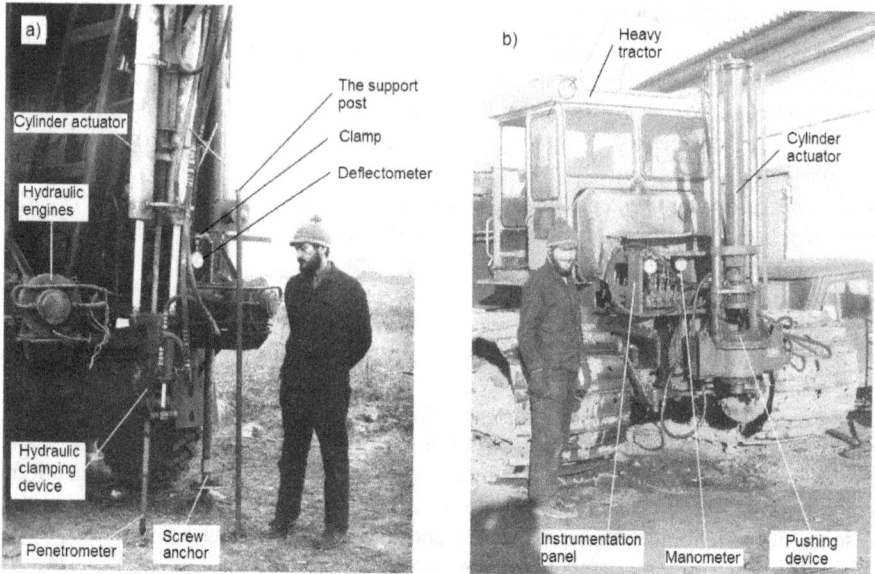

Fig. 5.4 Application of CPT rigs in permafrost soils in Vorkuta and Labytnangi:

a – cone dislocation speed measurement in tests by S-832M CPT rig in Vorkuta,
b – SP-59B CPT rig mounted in a heavy tractor in Labytnangi

During the tests cone resistances q_c and sleeve friction resistances f_s were recorded by self-recording devices. The curves of resistances (fig. 5.5) had four explicit phases showing specific features of CPT (CPTT) technique in standard regime:

"A" – even pushing of the cone with permanent speed;

"B" – stop of the cone at the specified depth by stopping of oil delivery into hydraulic jacks ("relaxation-creep" regime), transition of "cone-soil" system into relaxation-creep regime accompanied by freezing of the cone in the soil;

"C" – 0.1 m extra pushing up to maximum stroke of jacking hydraulic actuator rods;

"D" – release of the rod by hydraulic clamping device, lifting of hydraulic actuator rods and entrainment of the rod at the next level.

Three types of resistances were chosen as the most important corresponding to:

- Even jacking of the cone – q_{cv}, f_{sv} ("rapid" resistances).
- Relaxation-creep testing of the cone – q_{cs}, f_{ss} ("stabilized" resistances).
- Initial moment of extra pushing followed by relaxation-creep testing and freezing of the cone in the soil – q_{ci}, f_{si} ("peak" resistances).

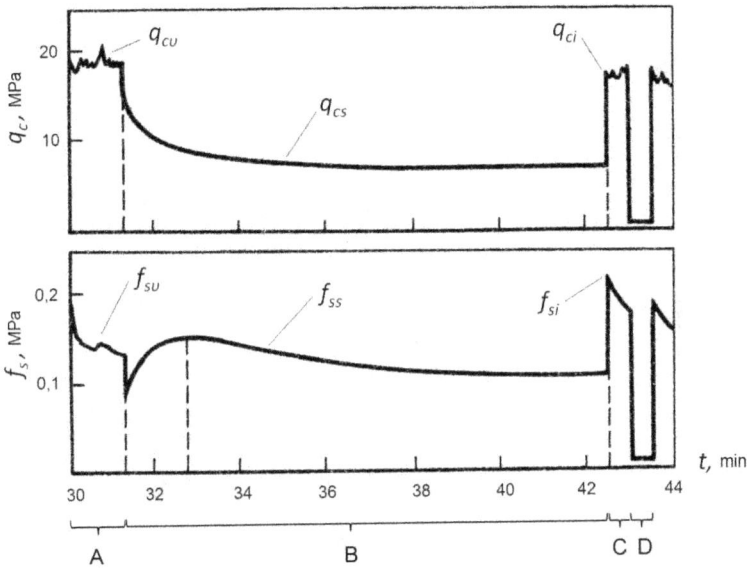

Fig.5.5 Schematics of chart strips recording cone resistances q_c and sleeve friction resistances f_s in CPT by S-832M CPT rig

5.1.2 Effects of technological factors on frozen soil CPT resistance

Effect of the speed of penetration on cone resistance in frozen soil

Velocity of penetration v greatly influences cone resistance q_{cv}. The Canadian researcher B. Ladanyi (Andersland & Anderson 1978, Ladanyi 1976, 1982) found out that "$q_{cv} - v$" diagram could be clearly described by power law (the ratio is linearized in the coordinates "$ln\ q_{cv} \sim ln\ v$"):

$$q_{cv} = q_{cv}^0 \left(\frac{v}{v^0}\right)^{\frac{1}{n}}, \qquad (5.1)$$

where:
q_{cv}^0 & $v^0 =$ coordinates of a point on the diagram;
$n =$ creep index.
In $v^0 = 1$ the formula (5.1) is of a simpler view:

264

$$q_{cv} = q_{cl}^0 \ (v)^{\frac{1}{n}} \ ,$$

(5.2)

where q_{cl}^0 = cone resistance in $v^0 = 1$.

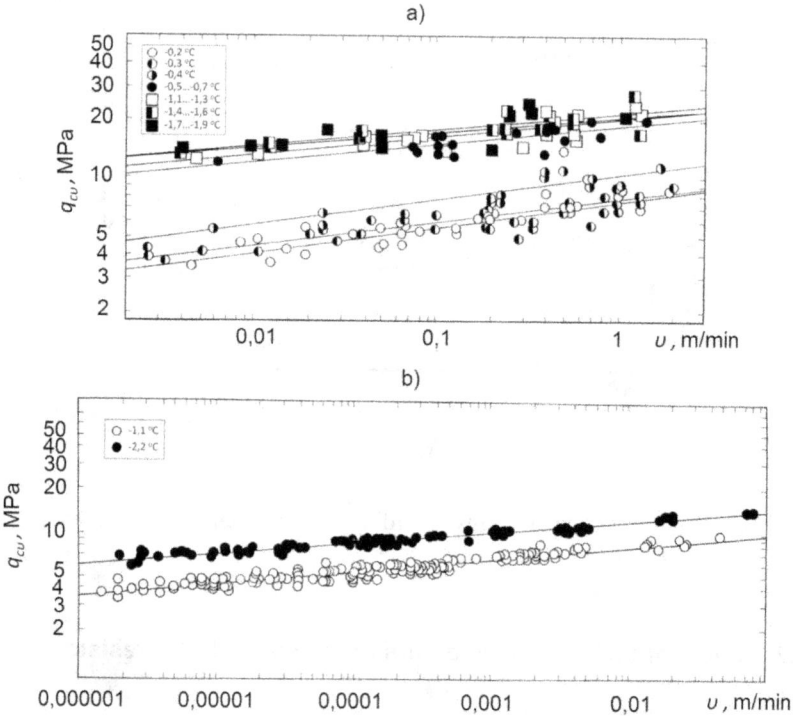

Fig.5.6 Schematics of dependence of q_{cv} in frozen soil upon CPT velocity v (Isaev 1989):

a – in-situ tests of permafrost plastic-frozen clayey soils, b – laboratory tests of artificially frozen loams in freezing chamber

In B. Ladanyi (Andersland & Anderson 1978, Ladanyi 1976, 1982) the diagram "*ln q_{cv}* ~ *ln v*" has a bending point – in the formula (5.1) in velocities lower than v = 0.025 cm/min creep index n = 3...6 (for plastic-frozen and hard-icy soils n = 3...4, for hard-frozen ones $-n$ = 5...6), in high speeds $-n$ = 10...20.

Though the experiments carried out in the former USSR and in Russia (Trofimenkov et al. 1986, Isaev 1989, Minkin 2005) proved the reliability of power formula (5.1), but a number of contradictory issues disagreeable with non-Russian research were revealed.

Specialists from Fundamentproject Research Institute (Trofimenkov et al. 1986) obtained creep coefficient to be $n = 5.7...6.4$ after CPT of plastic-frozen clayey soils with the temperature of $\theta_s = -0.6...0$ ^0C in the range of velocities $v = 0.001...100$ cm/min.

Specialists from MIIT (Isaev 1989) and NIIOSP studied the problem of effect of velocity v on q_{cv} in detail. In-situ tests (see fig. 5.6a) and laboratory ones (see fig. 5.6b) done in permafrost and artificially frozen clayey soils carried out in $v = 10^{-6}...2$ m/min showed the absence of the bending point on the diagram "$ln\ q_{cv} \sim ln\ v$", i.e. the ratio (5.2) was described by the integrated index n evaluated during the test.

Regression analysis of in-situ tests showed that the parameter $q^0_{c1} = 7...24$ MPa was significantly dependent on soil temperature θ_s. At the same time, creep index $n = 8...10$ did not significantly change and almost did not depend upon soil temperature θ_s. Laboratory tests of frozen loams in the temperatures of -1.1 and -2.2 ^0C resulted in creep index $n = 11$.

The results are of great practical importance. They indirectly substantiate the appropriateness of q_{cv} applications obtained in comparatively high velocities $v = 0.1...1$ m/min (this is extremely important for test manufacturability ensuring) for assessment of long-term values of frozen soil strength and deformability.

Effect of the speed of penetration on sleeve friction resistance in frozen soil

Velocity of penetration v effects on sleeve friction resistance f_{sv} (Isaev 1989) in different ways. In $v = 10^{-3}...2$ m/min some definite regularities can be observed in frozen soils. In high temperatures of frozen soil $\theta_s = -0.4...-0.2$ ^0C at first, f_{sv} increases and then it slightly decreases with increase of the speed of penetration. For lower temperatures $\theta_s = -1.9...-0.5$ ^0C the dependence is slightly different – f_{sv} decreases with increase of speed.

The complicated friction dependences may be explained by complex mechanical and thermal (see section 5.1.4) processes occurring at the contact zone between frozen soil and the friction sleeve.

Alteration of cone resistance in frozen soil in relaxation-creep regime testing ("relaxation-creep" CPT)

Oil delivery in jacking hydraulic actuators having been stopped, penetration of the cone slowed down; the system "cone-frozen soil" moved to relaxation-creep regime, i.e. "stabilization". The curve of alteration of cone resistance q_{cs} in frozen soil in time t_s (see fig. 5.5a, phase "B") may be divided into two sectors:

- The first sector – q_{cs} instantly falls down (in 1...2 sec) in nearly 20...40 %; the phase reflects elastic properties of the system "cone-frozen soil".
- The second sector – q_{cs} intensively falls down in 30...40 % during 1...2 min, then it gradually moves to gently sloping curve; here, rheological properties of frozen soil are clearly shown.

The diagram of q_{cs} dependence upon "stabilization" of t_s is clearly described by the formulae (5.3) & (5.4) (Isaev et al. 1987, Isaev 1989). It is evident from linearization of the graphs in the coordinates "$1 / q_{cs} \sim ln\ (t_s + t^*)$" (see fig. 5.7, *a*) and "$ln\ q_{cs} \sim ln\ (t_s + t^*)$" (see fig. 5.7b).

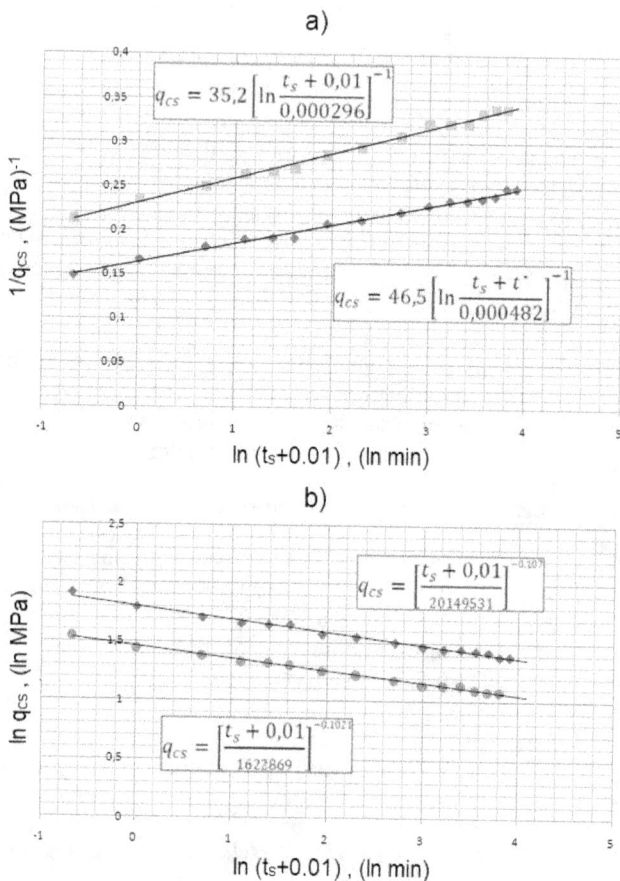

a)

b)

Fig. 5.7 Rectifying of graphic charts of q_{cs} dependence upon t_s (plastic-frozen loams, $\theta_s = -0.4...-0.6\ ^0C$) applying different formulae: *a* – formula (5.3), *b* – formula (5.4)

$$q_{cs} = \beta \left[\ln \frac{t_s + t^*}{T} \right]^{-1},$$ (5.3)

$$q_{cs} = \left[\frac{t_s + t^*}{T^\wedge} \right]^{\alpha},$$ (5.4)

where α, β, T, $T^\wedge$ – parameters evaluated by the graphs "$q_{cs} \sim t_s$"; t^* – undefined small value of time (e.g. $t^* = 0.01$ min).

A lot of geocryological factors effect on values of α, β, T, $T^\wedge$ in general case. The experiments show that in plastic-frozen soils the parameter T does not considerably depend on soil temperature θ_s (see fig.5.8a), while the parameter β – considerably depends on it (see fig. 5.8b).

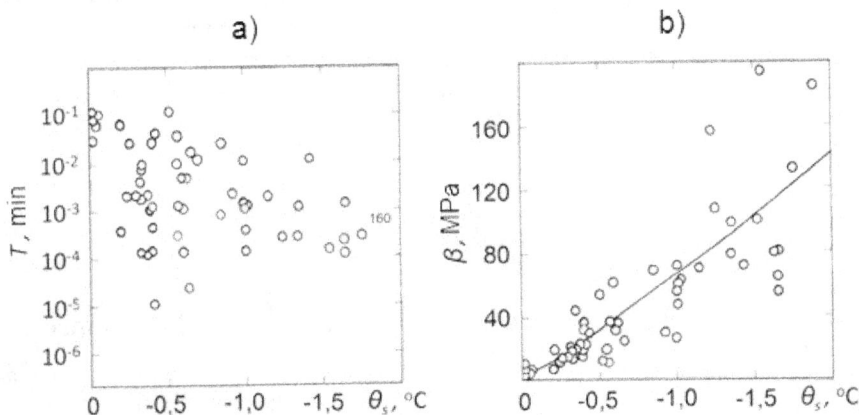

Fig. 5.8 Schematics of dependence of parameters $T(a)$ & $\beta(b)$ in the formula (5.3) upon soil temperature θ_s

Alteration of sleeve friction resistance in frozen soil in relaxation-creep regime testing ("relaxation-creep" CPT)

Curves of alteration of sleeve friction resistance in frozen soil f_{ss} in time t_s are more complex. Nevertheless, three sectors may be defined (see fig. 5.5, phase "B"):

- The first sector – f_{ss} instantly falls down (in 1...2 sec); the phase characterizes elastic properties of the system "cone-frozen soil".
- The second sector – f_{ss} increases during 0.5...5 min – the process of freezing of the cone in the soil occurs.
- The third sector – f_{ss} gradually decreases and plateaus (similar to decrease of cone resistance in frozen soil) – rheological soil properties predominate in simultaneous restoration of cone temperature and its freezing in the soil.

"Peak" cone resistances and sleeve friction resistances in frozen soil in extra pushing of the cone (after its "stabilization" and freezing)

Specific features of "peak" cone resistances q_{ci} and sleeve friction resistances f_{si} in frozen soil registered at initial time of extra pushing of the cone (see fig. 5.5, a, b, phase "C") show behavior of the cone frozen in the soil. The research revealed that "rapid" cone resistance q_{cv}, registered before "stabilization" nearly equaled the "peak" value of q_{ci}, i.e. thermal and stress-strain condition of soil corresponding to these resistances was equal. In further pushing, cone resistance slightly changed according to composition and condition of underlying soil.

Sleeve friction resistance showed different results. At initial time of extra pushing, sleeve friction resistance in frozen soil quickly attained its maximum – "peak" value of f_{si}, when the friction sleeve frozen in the soil broke. In here, "rapid" resistance f_{sv} registered before "stabilization" was lower than the "peak" value of f_{si}. In further pushing, sleeve friction resistance slightly decreased due to the generating thin lubrication layer of water and thawed soil at the contact zone of friction sleeve and the adjacent frozen soil.

As freezing of the cone in "stabilization" is quite intensive, "peak" sleeve friction resistances in frozen soil attain 80...90 % from their maximum in 10 min ; this corresponds to total freezing of the cone in the soil.

5.1.3 Effects of geocryological factors on frozen soil CPT resistance

The basic geocryological factors influencing frozen soil CPT resistance are as follows: temperature θ_s, total water content w_{tot} and plasticity index I_p (for pulverescent clayey soils). Comparative multiple-factor regression-correlation analysis of CPT results and properties of permafrost soils has been done in order to study this influence (see table 5.1).

Statistical analysis has established that soil temperature θ_s has the greatest influence on soil CPT resistance (see fig. 5.9). In particular, this refers to "peak" soil resistance f_{si}. Plasticity index I_p and total water content w_{tot} influence less in the examined soil conditions.

Table 5.1 Statistic dependence of frozen soil CPT resistance upon geocryological factors (Isaev 1989)

Geocryological factors	Frozen soil CPT resistance (MPa)				
	q_{cv}	f_{sv}	q_{cs}	f_{ss}	f_{si}
θ_s, ^{0}C	0.75	0.71	0.72	0.70	0.85
	-0.72	-0.58	-0.69	-0.34	-0.82
w_{tot}, %	0.38	0.20	0.55	0.56	0.28
	0.03	-0.09	-0.07	0.20	0.18
I_p	0.50	0.46	0.47	0.53	0.60
	-0.40	-0.16	-0.29	-0.24	-0.10

Footnotes: 1. Numerator – values of correlated ratios, denominator – values of correlation coefficients. 2. "Stabilized" values of frozen soil resistance were taken for the interval of 5 min. "Rapid" frozen soil resistances varied from 0.1 to 1.9 m/min.

269

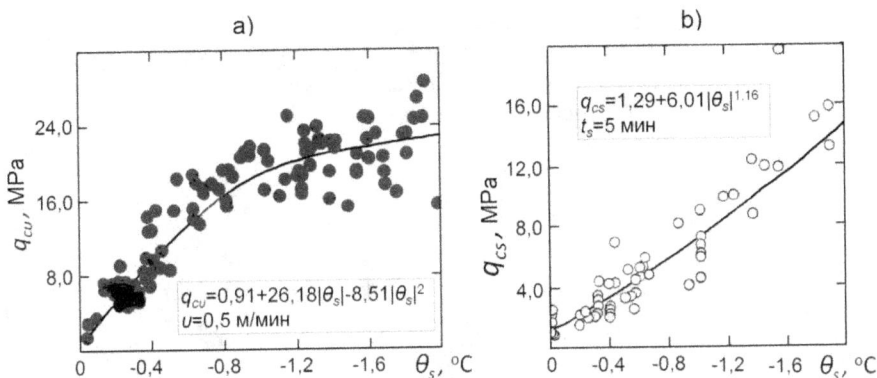

Fig. 5.9 Schematics of dependences of $q_{cv}(a)$ & $q_{cs}(b)$ in frozen soil on soil temperature θ_s (Isaev 1989)

5.1.4 Thermophysical interaction of penetrometer with thawed and frozen soils, soil temperature measurement with penetrometer

Thermophysical interaction of the penetrometer with thawed and frozen soil during its penetration

When penetrating the cone into melted or thawed soil, friction interaction between the penetrometer and the soil occurs, thus, cone temperature increases in $\Delta\theta_c = (\theta_{cv} - \theta_{cs}) > 0$, where θ_{cv} – temperature of the penetrometer, θ_{cs} – temperature of the stationary penetrometer being in "stabilization" regime. If the penetrometer attains natural temperature of soil after "stabilization", the value of its warming up is $\Delta\theta_c = (\theta_{cv} - \theta_s)$.

Warming of the penetrometer greatly depends upon mechanical and frictional properties of soil and the speed of penetration as well. Warming of the penetrometer increases with the increase of soil strength, number of coarse-dispersed particles and the speed of penetration as well. Plotted in figure 5.10 is the diagram of $\Delta\theta_c$ dependence on soil strength. In the authors' opinion, the degree of warming in thawed soils may provide with valuable information (in addition to the known methods presented in section 3.1) in identification of soil lithology by CPT data.

When penetrating the cone into frozen soil, the results are not so definite. Complex thermophysical interaction occurs at the border of "cone-frozen soil" system. Abnormal effect, as it may seem at first, (see fig. 5.11) has been discovered and studied after the experiments (Volkov & Isaev 1985, Isaev 1989) carried out applying the penetrometer equipped with temperature sensor (see fig. 1.18) as well as special thermometric penetrometer equipped with four temperature sensors (one- in the cone, three – along the side surface). If the penetrometer has been warmed in both hard-frozen and thawed soils ($\Delta\theta_c > 0$), cone temperature has slightly decreased in plastic-frozen soils ($\Delta\theta_c < 0$) (in high-temperature range of thawed soil). The authors called this unusual effect "pseudo-abnormal cooling of the penetrometer".

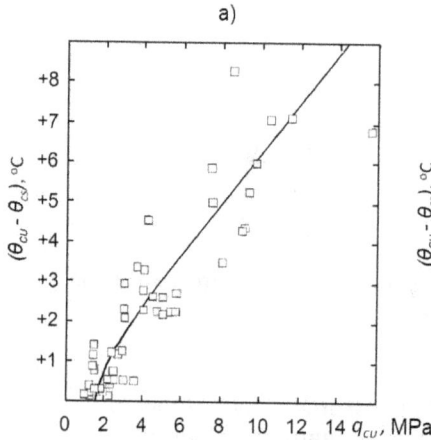

Fig. 5.10 Dependence of degree of warming upon cone resistance in melted (thawed) soil (Isaev 1989)

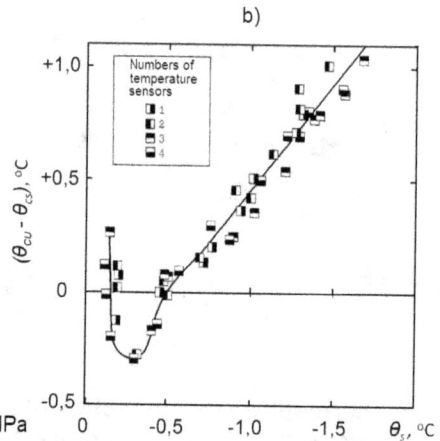

Fig.5.11 Dependence of temperature alteration of thermometric penetrometer upon the temperature of frozen soil (Isaev 1989)

Followed by theoretical research (Isaev 1989, Isaev & Ryzhkov 2010) the model of thermophysical interaction of "cone-frozen soil" system has been developed; it has been possible to explain the effect of "pseudo-abnormal cooling of the penetrometer". Two opposed thermal processes occurring at the border of "moving penetrometer-frozen soil" system have been taken as the basis of the model. The first one is conditioned by the flow of heat generating in friction and absorbed by the penetrometer; the second one – by the flow of heat generating in ice melting temperature decrease at high pressures and yielded by the penetrometer. Due to correlation of these two flows the temperature of the penetrometer may increase or decrease.

Thermophysical interaction of the penetrometer with thawed and frozen soil during "stabilization". Measurement of soil temperature by the penetrometer equipped with temperature sensor

The analysis of temperature alteration of the penetrometer during its "stabilization" in tests in thawed and frozen soils also revealed specific differences in thermal processes (Isaev 1989).

It has been stated that temperature curve is described by known task solution dependency on cooling of one-dimensional body immersed in the media with constant initial temperature in thawed soils after starting the "stabilization" regime (Manual 1982).

In frozen soils the curves of temperature alteration could be clearly described by the dependency only definite time after starting of "stabilization". Thus, the conclusion has been drawn that it seems difficult to apply the dependency in evaluation of original soil temperature. When comparing temperature curves of the penetrometer with the data obtained in standard thermometric openings (Isaev 1989), it has become possible to develop the methods of soil original temperature evaluation applying cone temperature stabilization coefficient m_θ. It is appropriate to use $m_\theta = \Delta\theta_{cs} / \Delta t_s = 0{,}01\ ^0C/min$ for practical applications.

5.1.5 Thawed and frozen soil state identification

Soil state identification as well as the border between thawed and frozen soils is one of the principal issues resolved in engineering-geocryological investigations. The issue is of principal importance in insular distribution of plastic-frozen soils. Soil state is usually identified in driving of test holes, selection of monoliths and soil samplers in engineering-geocryological investigations. The Russian researchers have revealed that soil state is possible to be identified by CPT data (Isaev 1989, Isaev et al. 1991).

As it was mentioned earlier, the penetrometer equipped with temperature sensor allows evaluating the original temperature of soil θ_s, which does show its condition. But it is not enough for accurate evaluation, i.e. it is necessary to know the initial temperature of soil freezing showing its transition from one state to another. In its turn, it depends on a number of factors: lithology, salinity, peat content, etc. This makes the approach not enough practical.

The experimental research (Isaev 1989) has showed that in application of "relaxation-creep" CPT it is possible to accurately identify soil state even in absence of temperature sensors. It has been shown that the "stabilization factor of cone resistance" $R^c_{vs} = q_{cv} / q_{cs}$ (the authors' term) significantly differs in thawed and frozen soils (fig. 5.12).

In thawed soils the stabilization factor R^c_{vs} is 1.4 at an average, while in frozen soils it is considerably higher – 3. However, this criterion is not always appropriate. For example, for weak thawed pulverscent-clayey soils the coefficient R^c_{vs} considerably increases – up to 2 times and more in $q_{cv} < 3$ MPa. This is due to the fact that weak pulverescent-clayey soils as well as frozen ones possess rheological properties to a large extent.

Therefore, two-parameter criterion (see fig. 5.13) has been suggested. It has become possible to divide soils by their state in simultaneous application of two parameters: "stabilization factor of cone resistance" R^c_{vs} and "rapid" cone resistance q_{cv}.

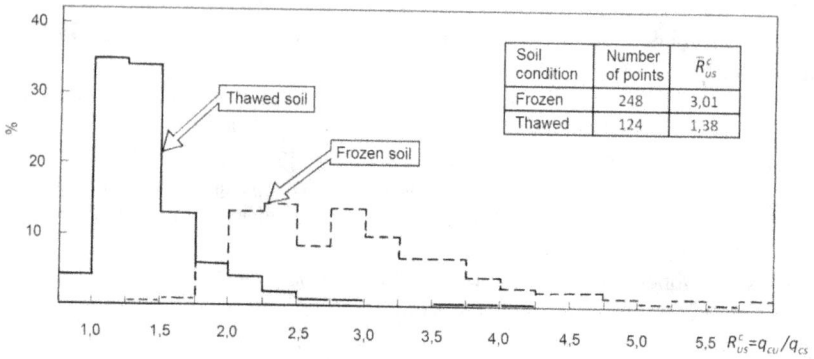

Fig.5.12 Histogram of q_{cv}/q_{cs} distribution for thawed and frozen soils

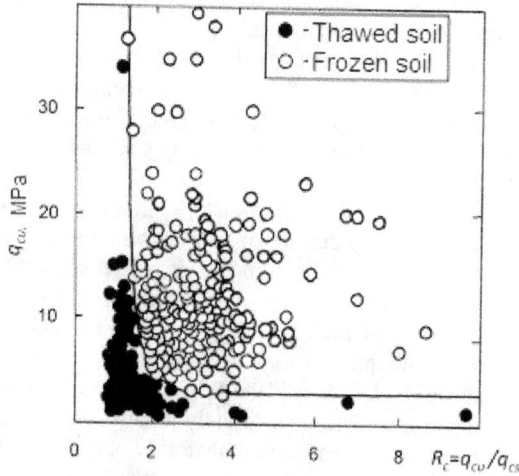

Fig.5.13 Diagram of comparison of "rapid" cone resistances q_{cv} with "stabilization factor of cone resistance" R^c_{vs} for thawed and frozen soils

5.1.6 Frozen soils mechanical properties determination

Frozen soils ultimate strength determination

Frozen soils are considered to be the ones possessing clear rheological properties. That is why their strength changes from conditionally momentary to long-termvalues due to the time of destruction. Long-term value of strength corresponds to such a stress when the deformations fade out and destruction does not occur. In exceeding of stress, sustained creep deformations occur resulting in destruction or collapse.

In accordance with the Standard Specification GOST 12248-96, in order to determine long-term strength of frozen soils, ball stamp tests, single one-dimensional shear along the frozen surface and/or uniaxial compression method are used in laboratory conditions.

Ball stamp pressure method after N.A. Tsitovich (Vyalov 2000) was widely applied in the former USSR and in Russia. Cohesion obtained during the tests includes both "pure" cohesion and friction. That is why it is called "equivalent cohesion". Long-term value of frozen soil equivalent cohesion C_{eq} is evaluated by the formula

$$C_{eq} = 0.18 N_b / (\pi d_b S_b),\qquad(5.5)$$

where:

$N_b =$ const = load imposed on the ball stamp,
$d_b =$ ball stamp diameter,
$S_b =$ depth of ball stamp immersion in the soil at the end of the test.

Correlation dependencies (5.6) and (5.7) allowing evaluation of C_{eq} (MPa) by "rapid" q_{cv} or "stabilized" q_{cs} in frozen soil have been obtained after comparative in-situ tests applying CPT and laboratory tests applying ball stamp method (Isaev 1989, Isaev et al. 1991).

$$C_{eq} = 0.003 \, (q_{cv,0.5})^{1.49},\qquad(5.6)$$

where $q_{cv,0.5}$ = cone resistance in frozen soil (MPa) in velocity of penetration $v = 0.5$ m/min.

$$C_{eq} = 0.016 \, (q_{cs,5})^{1.18},\qquad(5.7)$$

where $q_{cs,5}$ = cone resistance in frozen soil (MPa) in time of penetrometer "stabilization" $t_s = 5$ min.

According to N.A. Tsitovich (1973), in order to determine "pure" cohesion C, it is necessary to introduce the correction factor M, dependent on the angle of internal friction

$$C = M \, C_{eq},\qquad(5.8)$$

where M = correction factor; after V.G. Berezantsev: $M = 1$ (in $\varphi < 5°$), $M = 0.615$ (in $\varphi = 10°$), $M = 0.285$ (in $\varphi = 20°$).

Then, taking the formulae (5.6) & (5.7) into consideration, "pure" cohesion of frozen soils C (MPa) obtained by CPT data may be approximately evaluated by formulae

$$C = 0.0031 \, M \, (q_{cv,0.5})^{1.49} \tag{5.9}$$

$$C = 0.0016 \, M \, (q_{cs,5})^{1.18} \tag{5.10}$$

where $q_{cv,0.5}$, $q_{cs,5}$ = the same as in the formulae (5.6) & (5.7).

In Russian geotechnics SP 25.13330.2012 the relationship between long-term values of soil resistances R_c to uniaxial compression and equivalent cohesion is widely applied for frozen soils

$$R_c = 2 \, C_{eq}, \tag{5.11}$$

If to substitute dependencies (5.6) & (5.7) in the formula we obtain

$$R_c = 0.0062 \, (q_{cv,0.5})^{1.49}, \tag{5.12}$$

$$R_c = 0.032 \, (q_{cs,5})^{1.18} \tag{5.13}$$

where $q_{cv,0.5}$, $q_{cs,5}$ = the same as in the formulae (5.6) & (5.7).

Frozen soils' compressibility determination

Several types of permafrost soils possess compressive properties. Ground beds consisting of plastic-frozen soils or hard icy ones shall be evaluated by the second limit condition after the National Norms of Russia SP 25.13330.2012. Coefficient of compressibility m_f and modulus of deformation E_f, evaluated by laboratory tests in a compressive device or in-situ stamp tests are the basic parameters characterizing compressibility of frozen soils. In-situ stamp tests are rarely applied due to their labour-intensiveness, high costs and duration.

To evaluate compressive modulus of deformation E_f (MPa) of plastic-frozen soils one can use the empirical dependency (5.14) obtained after comparative in-situ CPTs and laboratory compressive tests (Isaev 1989, Isaev et al. 1991):

$$E_f = 7.39 \, (q_{cv,0.5})^{0.49}, \tag{5.14}$$

where $q_{cv,0.5}$ = the same as in the formula (5.6).

The parameters m_f and E_f are interconnected by the ratio

$$E_f = \beta \, / \, m_f, \tag{5.15}$$

where $\beta = 1 - [2v^2/(1-v)]$ = coefficient, according to GOST 12248-2011 it is allowed to take $\beta = 0.8$ for plastic-frozen soils; v = Poisson's ratio.

Taking into account the dependency (5.14), in $\beta = 0.8$, the coefficient of compressibility m_f (1/MPa) by CPT data may be evaluated by the formula

$$m_f = 0.11 \, (q_{cv,0.5})^{-0.49} \tag{5.16}$$

5.1.7 CPT application for analysis of foundations on permafrost

Application of permafrost soils as foundations of buildings and structures is put into practice following one of the principles.

- **Principle I** – preservation of the foundation frozen state during construction and operation of a building.
- **Principle II** – permafrost soils are used when thawed (thawing is allowed during construction or operation or in designed depth before erection of a structure). Application of CPT may be efficient in both principles.

Foot bearing capacity determination of a pile loaded vertically in permafrost soil applying principle I

Piles are the basic foundation type on permafrost soils applied by principle I. It was in 1980s when the issues on CPT applications for pile foundation computation on permafrost soils were started to be investigated. The issues were paid attention to both in Russia (Trofimenkov et al. 1986, Isaev et al. 1987, Isaev 1989) and worldwide (Ladanyi 1982, 1986).

Up to present time, in non-Russian investigations theoretical approaches have prevailed, while the issues on anchorage or reliability of the techniques suggested have been underestimated. At least, the authors of the book do not know any non-Russian book describing the comparison of pile bearing capacity in frozen soils obtained after static tests (being the reference method of pile bearing capacity determination) and calculated by CPT data. The suggested techniques operate on cone resistance only ignoring sleeve friction resistance.

Another feature of non-Russian approaches as opposed to the Russian ones is that CPT data are used "indirectly" here. On the contrary, in Russian approach "direct" usage of CPT data ignoring intermediate determination of soil strengthening properties is provided. It was developed taking into account semblance of pile and cone behavior under the load by determination of the empirical dependencies between long-term pile foundation resistances in frozen soils and CPT data including sleeve friction resistance.

Particular value of long-term resistance of a driven pile (bore driven pile) F_{uf} loaded vertically in a CPT location may be evaluated by the formula

$$F_{uf} = k_{\theta F} \left(R_{pf} A + \gamma_{cf} \Sigma f_{pf} A_{pf,i} \right) \tag{5.17}$$

where:

$k_{\theta F}$ = coefficient considering distinction in permafrost foundation soil state at the moment of CPT performance (in original temperature θ_s) and during operation of the designed structure (in prognostic temperature θ_{sp}) evaluated by the formula $k_{\theta F} = F_{uf,p} / F_{uf,t}$, where:

$F_{uf,p}$ & $F_{uf,t}$ = pile resistance values calculated by means of R_{pf} & f_{pf}, determined by the tables of appendix B SP 25.13330.2012 with temperature, depth of location and type of frozen soil in prognostic temperature θ_{sp} and the original one θ_s of frozen soil;

R_{pf} = unit long-term resistance of frozen soil under pile end is evaluated by the formula (5.18);

A = pile cross-section area of a pile;

γ_{cf} = coefficient of frozen soil performance along pile side surface, dependent upon pile material and side surface type; for concrete surfaces of foundations made in metal moulding as well as for wooden surfaces not treated with oily antiseptics γ_{cf} = 1; for wooden surfaces treated with oily antiseptics γ_{cf} = 0.9, for metal surfaces made of hot-rolled metal γ_{cf} = 0.7;

f_{pf} = unit long-term shear strength in frozen soil along freezing pile surface is evaluated by the formula (5.19);

$A_{pf,i}$ = freezing surface area of frozen soil i-layer with pile side surface.

In search of connection between unit long-term resistance of frozen soil under pile end R_{pf} and cone resistance in frozen soil q_c several types of resistances to CPT were studied. They corresponded to either constant velocity cone movement (q_{cv}) or "stabilization" of the cone (q_{cs}). In here, several values of technological parameters v and t_s were considered.

Correlation-regression analysis of the experimental data (Isaev 1989, Isaev et al.1991) revealed that there existed rather close correlation connection between most frozen soil resistances to CPT and R_{pf}. For practical applications it seemed acceptable to evaluate unit long-term resistance of frozen soil under pile end R_{pf} (MPa) by "rapid" cone resistance $q_{cv,0.5}$ (MPa), corresponding to CPT velocity v = 0.5 m/min

$$R_{pf} = 0.87 \, (q_{cv,0.5})^{0.54}, \qquad (5.18)$$

where $q_{cv,0.5}$ = the same as in the formula (5.6).

Pile side bearing capacity in frozen soils is conditioned by compression of the pile by the adjacent soil and fusing of soil into a frozen mass with pile side surface.

Unit long-term shear strength of frozen soil along pile side surface f_{pf} is recommended to be evaluated by means of "peak" sleeve friction resistances in frozen soil f_{si}, showing total freezing of the penetrometer into the soil.

To evaluate shear strength of frozen soils along pile frozen surface f_{pf} (MPa) one can use the formula (5.19) obtained in experimental-theoretical way (Isaev 1989, Isaev et al.1991)

$$f_{pf} = \left(1 - \frac{A_b}{A} \right)^{\frac{1-\xi}{2}} \left(f_{pf}^{(0)} - f_{pf}^{(I)} \right) + f_{pf}^{(I)}, \qquad (5.19)$$

where:

A_b = leader hole cross-section area;

277

ξ = coefficient of lateral stress of frozen soil;

$f^{(0)}{}_{pf} = 0.95 \, (f_{si})^{1.56}$ = unit long-term shear strength of frozen soil along driven pile frozen surface immersed ignoring the leader hole (MPa), $A_b = 0$;

$f^{(1)}{}_{pf} = 0.45 \, (f_{si})^{1.38}$ = unit long-term shear strength of frozen soil along drilled pile frozen surface immersed into the leader hole given (MPa), $A_b \approx A$.

Shallow foundations bearing capacity determination in permafrost soils applying principle I

Design pressure on frozen soil R_{sf} under columnar foundations is evaluated by the formula in accordance with Building Codes SNiP 2.02.04-88

$$R_{sf} = 5.7 C_n / \gamma_g + \gamma_I h, \qquad (5.20)$$

where:

C_n = normative value of long-term cohesion equal to: $C_n = C_{eq}$ in ball plate tests and $C_n = 0.5 R_c$ – in uniaxial compression tests;

γ_g = safety factor of soil;

γ_I = design value of unit weight of soil;

h = depth of foundation footing location.

When substituting the empirical dependency (5.6) in the formula (5.20) and introducing the correction factor with regard to different state of permafrost foundation soil at the moment of CPT procedure (in original temperature θ_s) and during operation of the designed structure (in prognostic temperature θ_{sp}) we obtain

$$R_{sf} = k_{\theta R} \left[0.0177 \left(q_{cv,0.5} \right)^{1,49} \Big/ \gamma_g + \gamma_I h \right], \qquad (5.21)$$

where:

$k_{\theta R}$ = coefficient evaluated by the formula $k_{\theta R} = R_{sf,p} / R_{sf,t}$, where:

$R_{sf,p}$ & $R_{sf,t}$ = values of design pressure on frozen soil determined by tables of appendix B SP 25.13330.2012 due to temperature and type of frozen soil in prognostic temperature of frozen soil θ_{sp} and the original one θ_s;

$q_{cv,0.5}$ = the same as in the formula (5.6);

γ_g, γ_I, h = the same as in the formula (5.20).

Foundations basement deformation determination in permafrost soils applying principle I

In accordance with Building Codes SP 25.13330.2012, in designing foundations in plastic-frozen, hard icy and underground ices stiffness analysis is required. Half-

278

space linear-deformable design model or linear-deformable finite thickness layer design model are used for columnar foundations and/or pile groups.

With regard to Building Codes SP 25.13330.2012, design deformation properties of plastic-frozen soils (coefficient of compressibility m_f or modulus of deformation E_f) need to be taken by compressive tests data in design (prognostic) temperature of soil.

In order to estimate deformation properties of plastic-frozen soils one can use the dependencies (5.14) & (5.16) introducing the correction factor with regard to different state of permafrost foundation soil at the moment of CPT procedure (in original temperature θ_s) and during operation of the designed structure (in prognostic temperature θ_{sp}).

Foundations basement determination in permafrost soils applying principle II

In preliminary artificial thawing of permafrost soil in given depth, CPT is capable to solve the same problems as in ordinary thawed soils, but the following specific features must be taken into account.

The first cycle of CPT is necessary to perform before thawing, i.e. at design decisions. Selection of CPT net and locations is to be set with regard to prognostic border of thawing bowl.

Ice melting results in self-consolidation of soils. If consolidation of large fragmental soils and sands occurs simultaneously with their thawing, consolidation of clayey soils may last quite long (coefficient of filtration of clayey soils after thawing is several times higher than that of thawed soils of the same composition).

Before construction, it seems reasonable to perform CPT repeatedly, with specified intervals to estimate the dynamics of their consolidation in time and define the beginning of construction works (when the foundations basement will respond to the required geotechnical properties). In order to improve the basement, consolidation or fastening of thawed soil is frequently applied, as soil index properties are worsened after melting of icy interlayers and/or particles. In here, CPT is recommended to perform just after thawing, and then during and after completeness of special work on soil improvement. Thawed soil testing is necessary to be done by the penetrometer equipped with the temperature sensor applying the methods of "stabilization" of the penetrometer.

As far as both thawed and melted soils were formed in different conditions and, thus, they are of different genesis then, in application of CPT data one should take care when using correlation dependencies obtained for definite types of thawed soils.

In authors' opinion, the following approaches of CPT applications are possible if to apply soils thawing during operation as the foundation soils. The first approach provides local melting of frozen soil in the design depth (corresponding to prognostic heat engineering computation) and its further testing by CPT. The second approach provides application of the penetrometer equipped with a heating element (the geometry of the penetrometer was suggested in BashNIIstroy and NIIOSP at the beginning of 1980s). Soil testing by special methods application followed by periodical thawing is carried out during CPT with depth-wise intervals. This approach has faced a number of technical and methodological difficulties, as it has not been thoroughly developed.

Tensometric cone penetrometers equipped with heating elements and temperature sensors prove to be the most promising, as they allow determining either temperature or mechanical properties of frozen soils with regard to their thawing, or their thermal properties as well.

5.2 COLLAPSIBLE SOILS

Application of CPT in collapsible soils has not been paid great attention to, as the majority of specialists have denied the availability of this trend. One can easily believe it if to study the Normative Documents of the Former USSR and Russia of different years of the 20th century. In them there were not any instructions concerning CPT applications in collapsible soils. The situation has been typical not only for Russia but worldwide as well. Nevertheless, the research was done time and again, particularly in the USSR, thus, a number of interesting results were obtained, but unfortunately, they were not of practical importance. Partially, this was due to the system of technical innovations implementation in the former USSR aiming at mass applications registered as obligatory requirements of National Norms and Specifications. Such a system required availability of results of rather high level; in practice, most of them did not meet the requirements. Nevertheless, the research done confirmed the rationality of many useful ideas; at present, their improvement could give a significant practical effect. For this reason, it seems reasonable to analyze some experiments comprising the ideas.

In 1960s NIIOSP (Krutov & Eiduck 1971, Krutov & Kulachkin 1974) started works on development of quantitative assessment methods of loessial soils collapsibility by CPT including development of necessary CPT equipment. The main idea of the approach consisted in application of correlation dependence between the relative collapsibility of soil ε_{sl} and "coefficient of soil strength decrease in wetting" K_w.

Relative soil collapsibility ε_{sl} at the time of the research implementation was evaluated in the same way as it is done at present time SP 50-102-2003, i.e. by compressive test results with wetting by the formula

$$\varepsilon_{sl} = \frac{h_{n,p} - h_{sat,p}}{h_{n,g}}, \qquad (5.22)$$

where:

$h_{n,p}, h_{sat,p} =$ sampler height in vertical pressure p according to natural water content and after its total water-saturation;

$h_{n,g} =$ sampler height in natural water content in overburden stress (at the depth of sampler selection).

Coefficient of soil strength decrease in wetting K_w was evaluated by the ratio of cone resistance q_c in natural state of soil to cone resistance in wetted state of soil $q_{c,wet}$:

$$K_w = q_c / q_{c,wet}, \qquad (5.23)$$

The first investigations were referred to application of **manual probes** (see section 1.3.5). V.I. Krutov and R.P. Eiduck (1971) suggested the method of soil relative collapsibility determination by CPT (pressing of the probe) from pit surface. Their method consisted in the following: the surface of the pit bottom was leveled and divided into two halves. One of them was subjected to CPT 10 cm in depth in natural water content of soil, the second one – wetting of soil 15...20 cm in depth with further CPT as well. In both cases CPT was carried out in 10 locations; mean value of soil resistance to CPT in natural water content and in water-saturation was determined. Relative collapsibility ε_{sl} was evaluated by K_w, i.e. by the ratio of mean cone resistance in natural state of soil to the analogous resistance in its wetted state (formula (5.23)). The following empirical formula was suggested

$$\varepsilon_{sl} = a\,(K_w - 1), \qquad (5.24)$$

where:

$\varepsilon_{sl} =$ relative collapsibility in pressure of 0.3 MPa;

$a =$ empirical coefficient adjusted for every region (for Dnieper loessial soils, $a = 0.021$).

According to the formula (5.24), for $a = 0.021$, soil is to be considered collapsible if $K_w \geq 1.52$, since in collapsible soil $\varepsilon_{sl} \geq 0.01$ according to the Standard GOST 25100-2011.

If it is necessary to recalculate the relative collapsibility to other (usually lower) pressures, typical dependences (for given conditions) of relative collapsibility upon pressures are applied.

Another method developed by V.I. Krutov and B.I. Kulachkin (1974) seems to be more appropriate. It does not require excavation of pits, but it requires **application of special penetrometers capable to deliver water** (the penetrometers may be installed in any rigs). CPT is carried out from earth surface. The penetrometer is pushed into the soil of natural water content, standard measurements of q_c & f_s (or only q_c) are done. CPT stops in given level of relative collapsibility determination, the push rod is raised in 10...20 mm, and then water delivery system is switched on and the soil is wetted. Wetting process lasts up to total saturation of soil. Necessary volume of water is calculated taking into account soil properties as the initial data, i.e. density, water content, total moisture, etc. The radius of the wet zone is usually taken to be equal to 20 cm. The degree of wetting is estimated by expenditure of water. Wetting having been finished, CPT continues and q_{cw} in wet soil is evaluated.

Coefficient of soil strength decrease in wetting K_w is evaluated by dividing the mean value of q_c measured before wetting (on the site of 15...20 cm located above the wetted zone) into the mean value of q_{cw} in wetted zone.

On further levels determination of resistances as well as the process of saturation are repeated in the same sequence.

V.I. Krutov and B.I. Kulachkin applied SPK and S-832 CPT rigs in their experiments (equipped with type II tensometric cone penetrometers, 62 mm and 36 mm in diameter), the penetrometers were equipped with devices for delivery of water as well. Figure 5.14 illustrates general view and geometry of the penetrometer made by re-equipment of SPK penetrometer (see section 1.3.4 and fig.1.15). The analogous penetrometer was developed for S-832 CPT rig (there was not any significant difference from SPK penetrometer).

Plotted in figure 5.14b is the penetrometer body with a tube for water delivery in borehole bottom located inside 4, and a spring valve made up of a ball 5 and a spring 6 is located in the cone. A moving-out tip 3 is mounted in the cone 2; when moved the tip, water enters the adjacent soil. Moving-out of the tip 3 occurs under water pressure being delivered by the tube 4 through the spring valve (5, 6) allowing the water to be delivered under specified pressure. The limit of valve performance is specified as well.

Cone diameter is 1 mm smaller than that of the body (friction sleeve); this excludes the generation of any openings between the friction sleeve and soil, thus, unforeseen leakage of water is excluded as well.

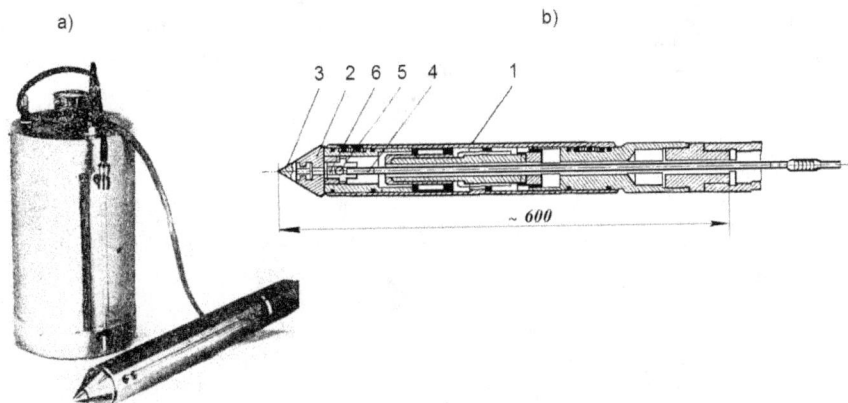

Fig. 5.14 General view and geometry of the penetrometer for soil collapsibility assessment (Krutov & Eiduck 1971):

a – general view of the penetrometer and gargets for water delivery, b – geometry of the penetrometer:1 – body, 2 – cone,3 – moving-out tip, 4 – tube for water delivery in borehole bottom, 5 – ball,6 – spring

Water delivery system (fig. 5.14a) includes a reservoir, air pressure accumulator and pressure hoses connecting the reservoir with the penetrometer and accumulator. The reservoir is equipped with a gauge tube to measure the volume of delivered water and a manometer to fix water pressure during wetting of soil.

The diagram of dissipation presented in figure 5.15 illustrates the comparative results of relative compressibility (ε_{sl}) evaluated by the standard mode (compressive tests with wetting (GOST 23161-78, GOST 25100-2011)) and the coefficient of soil strength decrease in wetting ($K_w = q_c / q_{c,wet}$) evaluated by CPT data after the method of V.I. Krutov and B.I. Kulachkin.

As one can see from fig.5.15, the connection of K_w with ε_{sl} is quite close (coefficient of correlation - 0.87).

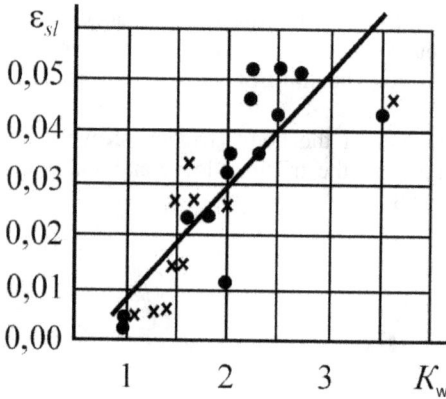

Fig. 5.15 Schematic illustration of comparison between relative collapsibility ε_{sl}, determined by standard mode (compressive tests with wetting) and coefficient of soil strength decrease in wetting K_w, evaluated by CPT data after V.I. Krutov & B.I. Kulachkin (1974):

Soils – loessial loams in Dnepropetrovsk (crosses) and Odessa region (circles)

The authors pay attention to the dependencies "$\varepsilon_{sl} \sim K_w$" which are of a regional character and are to be determined for each concrete region.

The authors give the empirical formula to evaluate relative collapsibility ε_{sl}

$$\varepsilon_{sl} = a\,K_w + b, \qquad (5.25)$$

where:

$a\ \&\ b =$ parameters determined for each examined region;

$K_w = q_c / q_{c,wet} =$ the same as in the formula (5.23), i.e. coefficient of soil strength decrease in wetting.

For loess-like loams of Dnepropetrovsk and Odessa Region the parameters in the formula (5.25) have been as follows: $a = 0.022$, $b = -0.015$. In the parameters given, 14...15 % decrease of soil resistance to CPT corresponds to collapsible soil (i.e. $\varepsilon_{sl} \geq 0.01$).

The method was successfully applied in practice being affordable, hence, one could assume that it would be quite promising in engineering-geological investigations (probably, with some improvements). The paper (Krutov & Kulachkin 1974) was published over three decades ago. Unfortunately, specialists ignored the method. Moreover, the authors of this book failed to find out any new publications describing the improvements of the method or the ones discussing its practical applications.

In 1970s in BashNIIstroy the attempt was taken to simplify the idea of soil wetting through CPT hole. Water was poured into the hole 3.6...3.7 cm in diameter, leakage of water in the soil was observed. The experiments were carried out in Ufa in macro cellular dealluvial loams possessing low collapsibility (ε_{sl} = 0.01...0.025, degree of saturation - 0.7...0.8). Positive results were not obtained, as water leaked through CPT hole very badly (after 2 hours water level in the hole was lowered in 10...15 mm only). Wetting through CPT hole might be effective only in highly porous loessial and loess-like soils.

In 1970s in BashNIIstroy V.M. Enikeyev (1980) suggested the approximate mode of soil collapsibility determination by CPT data by means of indirect characteristics **regardless of wetting** (in here, the term "indirect characteristics" is not the same as in M. Jamiolkowski (1995), but it means the methods when sagging does not occur). The method is based on comparison of sleeve friction resistances (f_s) in standard speed of penetration in equilibrium of the penetrometer, i.e. in "relaxation-creep CPT", being often called "stabilized" CPT in Russia. In water-saturated clays and loams "stabilized" values of f_{ss} are 60...70 % "from velocity ones" f_{sv} (i.e. obtained in standard speed of penetration $-$ 1.2±0.3 m/min). However, in some soils e.g. in macroporous loams (loessial, loess-like) possessing collapsible properties as well as in many sands such lowering has not been observed. On the contrary, "stabilized" values of f_{ss} may exceed "velocity" f_{sv} in these soils. This fact has not been explained theoretically, thus, one can assume that in slow speeds of deformations typical for "relaxation-creep" regime the soil of broken macroporous structure becomes denser and stronger. In sands the increase of density is probably analogous to the process described in section 4.4.2 $-$ alteration of bulks' density proceeding for some time after dynamic consolidation completion. Therefore, in f_{ss} / f_{sv} = 0.8...1.6 the soil should be considered either collapsible (loessial, loess-like) loam or sand.

As it was mentioned in chapter 3, the issue on recognition of sandy and/or clayey soils by CPT data (type II cone penetrometers) has been thoroughly studied. Though reliability of such assessments is not high, applying CPT in combination with drilling it is always possible to distinguish sands from macroporous clayey soils. Low values of f_s / q_c (< 2...3 %) and high q_c usually measured by dozens of MPa are typical for sands. In collapsible loams q_c is usually of 4...5 MPa, and the ratios f_s / q_c $-$ 3...7 %. For this reason, dependence of "equilibrium" values of sleeve friction resistance f_{ss} upon "velocity" f_{sv} appears to be an applicable criterion for soil collapsibility recognition. In particular, the recognition is facilitated due to rare occurrence of collapsible clays in combination with sands at one and the same site.

Figure 5.16 shows diagrams of dissipation after V.M. Enikeyev (1980) illustrating connection of f_{ss}/f_{sv} with known criteria of collapsibility $-$ relative collapsibility ε_{sl} (GOST 25100-2011, GOST 23161-78) and approximate criterion of collapsibility Π applied in 1960s...1980s (SNiP II-15-74, Uhov et al. 2007):

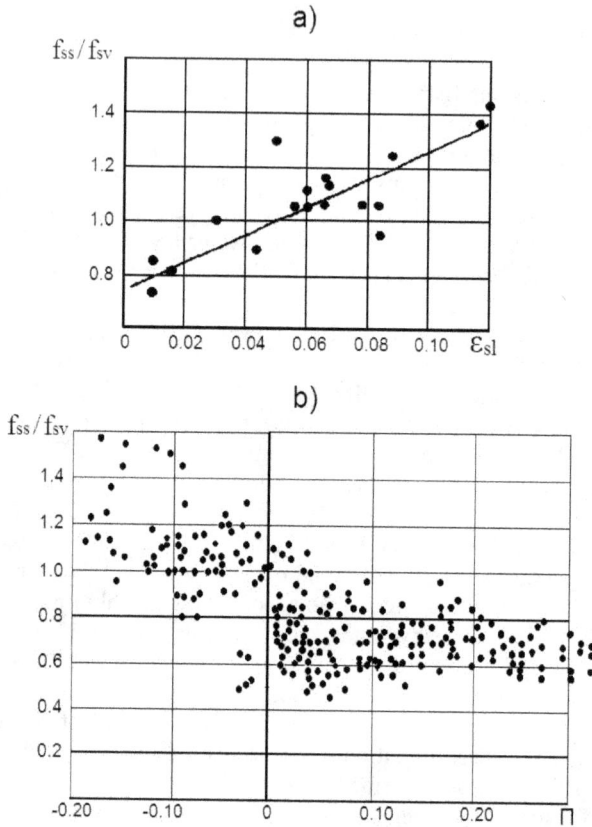

Fig.5.16 Schematics of comparison of f_{ss}/f_{sv} with traditional results of collapsibility in different years (Enikeyev 1980):

a – comparison of f_{ss}/f_{sv} with relative collapsibility ε_{sl} after SNiP 2.02.01-83*,
b – comparison of f_{ss}/f_{sv} with parameter Π after SNiP II-15-74

$$\Pi = \frac{e_L - e}{1+e}, \qquad (5.26)$$

where:

$e =$ porosity factor of soil of natural composition and water content;
$e_L =$ porosity factor corresponding to liquid limit evaluated by the formula;

$$e_L = w_L \frac{\gamma_s}{\gamma_w}, \qquad (5.26a)$$

$\gamma_s, \gamma_w =$ specific weight of solid particles of soil and water;

285

w_L = liquid limit.

The parameter Π shows alteration of porosity factor in total water-saturation. Soils are considered to be collapsible if the following conditions are observed (SNiP II-15-74, Uhov et al. 2007):

In plasticity index $\quad 0.01 \leq I_p \leq 0.10 \rightarrow \Pi \leq 0.10$;

the same $\quad 0.10 \leq I_p \leq 0.14 \rightarrow \Pi \leq 0.17$;

the same $\quad 0.14 \leq I_p \leq 0.22 \rightarrow \Pi \leq 0.24$.

With increase of collapsibility, the parameter Π tends to decrease, but the spread of such a dependency is rather large, that is why Π is not used for quantitative estimation of collapsibility; it is regarded as nomenclatural feature showing inclination of soil to collapsibility. At present, Π is withdrawn from the National Norms of Russia as being insufficiently accurate, however, in practice it is successfully applied, great attention being paid to it in the well-known Russian manual on soil mechanics and foundation engineering (Uhov et al. 2007) (a future engineer is expected to be able to apply the parameter). In other words, acceptability of collapsibility index Π for approximate assessments shall not be ignored, since it has been corroborated by years of practice in foundation engineering in the former USSR and Russia.

As one can see from figure 5.16, correlation link of parameters f_{ss} / f_{sv} with traditional criteria of collapsibility does not cause any doubts.

Thus, if the soil is not sand, and the values of f_{cm}/f_v are close to one or exceed it (f_{ss} / f_{sv} = 0.8...1.6), it is to be considered as collapsibility. It is evident that in specified conditions the assessment requires extra corroboration by more accurate methods, nevertheless, the simplicity and quickness of this mode of collapsibility recognition makes it quite useful despite its rather low accuracy.

Designing of pile foundations. Pile bearing capacity evaluation in collapsible soils by CPT data always requires additional information (experimental wetting, pile tests, laboratory tests, etc.). In accordance with the National Norms (SNiP 2.02.03-85*, SP 50-102-2003, SP 24.13330.2011) problem-solving procedures depend on the type of collapsibility and possible operating conditions of a structure.

In type I collapsible soils (i.e. being collapsed owing to their net weight up to 5 cm in foundation pit in wetting) *in impossibility of wetting* pile bearing capacity is evaluated in the same way as in "ordinary" non-collapsible soils. In here, computations on pile resistances by CPT data do not have any special features, i.e. they are done as described earlier in section 4.3. It is obvious that the confidence in impossibility of wetting (as a result of emergency wetting or elevation of ground water level) must be substantiated.

One should take into account one factor being insufficiently studied. The National Norms SNiP 2.02.03-85*, SP 50-102-2003, SP 24.13330.2011 recognize the possibility of some increase of water content in hard clayey soils in lack of their possible wetting. Such wetting may appear due to natural evaporation of water from soil pores under the erected structure after its long operation (sometimes the term "effect of blackout" is used). According to the Norms SNiP 2.02.03-85*, SP 50-102-2003, SP 24.13330.2011 when there is impossibility of collapsible soil wetting, pile end (side) bearing capacity shall be taken reduced in water content to be *lower* than the plastic limit w_p corresponding to water content *equal* to the plastic limit w_p.

However, the phenomenon would become apparent in any clayey soils of solid consistency (either collapsible or not). At the same time in the Norms SP 50-102-2003 (SP 24.13330.2011) it is required to be taken into account in non-collapsible soils just in computations of pile bearing capacity by liquidity index I_L (in water content less than 0.8 liquidity index is recalculated to total water-saturation condition). At the same time in the Norms there are indications on this account neither for static pile tests nor dynamic ones, nor for computations by CPT data, though it is thought to be logical. Moreover, years of practice has not revealed any dangerous effects of such simplifications. Perhaps, the issue needs to be carefully examined, and practical tasks are to be solved with regard to the local experience. Perhaps, it is reasonable to reduce q_c & f_s possessing lower water content than w_p, "in reserve" (e.g. in 15...20 %), thus taking into account manifestation of this phenomenon.

In *possibility of wetting* the tests or computations are to be done with regard to *total water-saturation of soil*. Three variants of pile bearing capacity evaluation by CPT data could be distinguished in the conditions given:

- Binding of computations by CPT data to the results of static pile tests carried out in conditions of foundation local wetting.
- CPT procedure before and after wetting of soil (in the pit), i.e. obtaining of q_c & f_s corresponding to natural and wetted soil.
- CPT of natural soil introducing corrections in q_c & f_s obtained due to laboratory tests of wetted soil (with regard to local experience).

In the first variant CPT is carried out either in the specified driving location (pile making) or near it (in 1...2 m) before driving (making) of a test pile and its further testing. In practice, pile testing both before and after wetting appears to be optimum as it is advantageous in the following ways:

- Comparison of ultimate pile bearing capacities obtained in testing before and after soil wetting allows evaluating the degree of pile capacities lowering in wetting with high accuracy; this can be disseminated on the whole of the site then (or its specified part).
- Comparison of ultimate pile bearing capacities obtained before soil wetting by either CPT data or pile testing allows increasing the reliability of computation by CPT, i.e. correcting it with regard to the conditions of the specified site (e.g. by specification of reliability coefficient γ_c).

In comparison the results of pile testing before and after soil wetting it seems necessary to apply the formula

$$k = \frac{F_u}{F_{u,wet}} + \Delta,$$ (5.27)

where:

k = correction factor considering the effect of wetting introduced (as a divisor) to mean pile bearing capacity evaluated by CPT data

287

obtained on the whole of the site or a part of it (with regard to its composition);

$F_u, F_{u,wet} =$ pile capacities obtained by static tests of piles before and after soil wetting;

$\Delta =$ correction considering random factors effecting on F_u and $F_{u,wet}$ ratio comparisons in different locations of the site (heterogeneity of soil, errors in computations, etc.).

The correction Δ is to be equal to 0.1...0.2 due to homogeneity of the soils at the site. In insignificant heterogeneity of soil the site is divided into sectors, here the correction of pile capacities is done with regard to each sector, i.e. pile testing is done at each sector (the issues on approximate data corrections are discussed in chapter 6).

Availability of pile testing results *before wetting* allows increasing the reliability of pile bearing capacity evaluation by CPT data, as despite their advantages the computations are approximate and they always need to be corrected. In here, correction of results appears to be possible due to obtaining of ultimate pile bearing capacities in one and the same location by means of two different methods (CPT and pile testing). Based on such a correction assuming the computation analogous to the given above (formula 5.27) correction reliability coefficient γ_c, is evaluated; it is equivalent to coefficient k in the formula. Methods of such a correction are discussed in chapter 6.

In case pile testing before wetting is not done (i.e. tests are done with soil wetting) the results of computation by CPT data are bound to the result of pile testing in wetted soil. In here, the formula analogous to the one (5.27) is used; the formula differs in that pile bearing capacity in the soil of natural water content F_u is taken after CPT data. This computation is less reliable, since distinction between F_u and $F_{u,wet}$ shows not only the influence of wetting but inaccuracy of the method as it is (drawbacks of the design model). Both factors happen to be inseparable, though they show themselves in different ways at different locations of the site. Thus, it is desirable to take the correction Δ as ($\Delta \geq 0.2$), i.e. it's maximum. From practical point of view, absence of pile testing before wetting can hardly be justified since the expenditures for test arrangements concern purchasing and manufacturing of piles and mounting of test equipment as well. Having done the preparations, it seems unwisely to restrict ourselves by tests with soil wetting only.

In the second variant test soil wetting in the pit ignoring static pile testing is assumed. CPT is performed before and after soil wetting. Then, computation of pile bearing capacity in the soil of natural water content and the one saturated with water is done based on the CPT data obtained. For soil beyond the test sector computations are done by corrected values of $q_{c,wet}$ and $f_{s,wet}$ evaluated by the formulae

$$q_{c,wet} = q_c / K_w, \quad f_{s,wet} = f_s / K_{w,f}, \qquad (5.28), (5.29)$$

where:

$q_c, q_{c,wet} =$ cone resistances before and after soil wetting;

$f_s, f_{s,wet} =$ sleeve friction resistances before and after soil wetting;

$K_w =$ coefficient of cone resistance q_c decrease in wetting evaluated by the formula (5.23);

$K_{w,f}$ = coefficient of sleeve friction f_s decrease in wetting evaluated by the formula.

$$K_{w,f} = f_s / f_{s,wet},\qquad(5.30)$$

Due to heterogeneity of soil and inevitable irregularity of wetting it is desirable to carry out CPT two-three times placing CPT locations (before and after wetting) randomly, thus comparing mean values of resistances obtained. K_w and $K_{w,f}$ are calculated for specified range of depths, usually corresponding to the engineering-geological element. In most cases $K_w \approx K_{w,f}$ (Enikeyev 1980).

In the third variant the same procedures as in the second one are supposed to be done, but K_w and $K_{w,f}$ are evaluated by laboratory tests data regardless of test wetting of the pit. There is not any integrated laboratory method of evaluation of coefficients of soil strength decrease in soil wetting. V.I. Krutov and B.I. Kulachkin (1974) suggest assessment of such a decrease by the parameter (coefficient) K_c

$$K_c = \frac{c_n}{c_{wet}} \frac{\varphi_n}{\varphi_{wet}},\qquad(5.31)$$

where:

c_n, c_{wet} = soil cohesion in natural water content and after soil wetting (up to total water-saturation);

φ_n, φ_{wet} = angle of soil internal friction in natural water content and after soil wetting.

K_c is correlated to relative collapsibility ε_{sl} similar to K_w (see formula 5.23), but K_c and K_w may significantly differ. In particular, in the diagrams of dissipation presented by the authors values of K_c nearly 2 times exceeded those of K_w for loess-like loams of Togliatti compared with those of Dnepropetrovsk (in equal ε_{sl}).

In BashNIIstroy the following parameter is used for wetted soil strength decrease assessment

$$K_\tau = \tau_{g,n} / \tau_{g,wet},\qquad(5.32)$$

where $\tau_{g,n}$, $\tau_{g,wet}$ = soil shear strength in overburden stress (see formula 2.16), in natural water content and in wetting up to total water-saturation correspondingly.

In computations one may approximately take $K_w = K_{w,f} = K_\tau$, i.e. sleeve friction resistance of wetted collapsible soil is $f_{s,wet} = f_s/K_\tau$.

In type II collapsible soils (i.e. being collapsed owing to their net weight over 5 cm in foundation pit in wetting) pile bearing capacity is usually evaluated with regard to water-saturation of the foundation, in here, additional load on the pile from setting soil layers (negative friction) is considered (fig. 5.17).

In accordance with the Norms of Russia SNiP 2.02.03-85*, SP 50-102-2003, SP 24.13330.2011 the allowable load on the pile N in this type of soils is evaluated by the formula

$$N = \frac{F_d}{\gamma_k} - \gamma_c P_n, \tag{5.33}$$

where:

$F_d =$ pile bearing capacity in the zone located under setting soil layers;

$\gamma_k =$ coefficient of reliability showing the reliability of F_d computation (after SNiP 2.02.03-85* & SP 50-102-2003 in computations by CPT data $\gamma_k = 1.25$);

$\gamma_c =$ coefficient of performance; its value depends upon the possible collapsibility of soil s_{sl}: in $s_{sl} = 5$ cm $\gamma_c = 0$, in $s_{sl} \geq 2s_u$ $\gamma_c = 0.8$, for intermediate values of s_{sl} values of γ_c are determined by interpolation;

$P_n =$ negative friction, i.e. additional load on the pile from setting soil layers.

Pile foot bearing capacity (under the setting layers) F_d is evaluated as the difference between pile bearing capacity (l in length) to pushing load and pile bearing capacity (h_{sl} in length) to extracting load (fig. 5.17). Parameter h_{sl} is the assumed depth up to which summation of side friction forces of collapsible soil layers is done. It is equal to the depth in which collapsibility of soil due to its net weight is 0.05 m in accordance with the Norms SNiP 2.02.01-83*, SP 24.13330.2011 (as a rule, the bottom of the assumed collapsible zone happens to be 2...4 m higher than the footing of the collapsible layer or groundwater level in the layer).

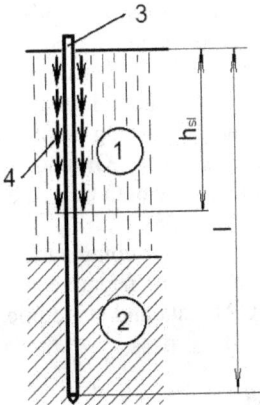

Fig. 5.17 Schematic illustration of the assumed load on the pile in type II collapsible soils:

1 – collapsible soil, 2 – non-collapsible soil, 3 – pile, 4 – zone of negative friction

Parameter F_d shows resistance of both non-collapsible layer – 2 and the bottom of the collapsible layer – 1 with the anticipated collapsibility to be up to 0.05 m, i.e. F_d characterizes "the remaining part" of foundation resistance. Computation of F_d is done by CPT data in soils of natural water content applying the formulae presented in section 4.3. Taking into account that some decrease of sleeve friction resistance f_s in wetting is to be anticipated in the bottom of the collapsible layer, it is necessary to consider the decrease in computations (introducing the corrections), however, the decrease is not significant, usually in the limits of accurate pile bearing capacity computation by CPT data. For this reason, in most cases it is allowable to apply the results of CPT in in-situ water content of soil without any corrections.

Negative friction P_n is also calculated by CPT data applying the formulae for pile resistances to extraction computation (pile length equals to h_{sl}). If wetting is possible to be performed only from below (raising of groundwater level), computation of negative friction should be done by CPT data obtained in soil of natural water content. In wetting from above, it is allowable to apply the reduced values of f_s, corresponding to water-saturated soil ($f_{s,wet}$). However, having some doubts concerning the possibility of such wetting it seems reasonable to take negative friction P_n "in reserve" as applied to the soil of natural water content, i.e. ignoring its reduction. If low non-collapsible thickness is composed of "soaked" soil capable to change its liquidity index I_L in supplying with water (rising of groundwater level), resistances of soil are to be estimated after wetting. It is necessary to underline that such modifications of non-collapsible soil properties are not significant in rising of groundwater level. To a certain extent, they are possible in any types of clays and in any foundations if changes of hydrogeological conditions are expected. This is often ignored in practice, attention being paid to these issues only in designing the most complex structures.

A.M. Dzagov (2007) gives examples of admissible load evaluation for piles of 0.35×0.35 m – 24 m in length in type II collapsible soils, on two sites of Rostov region. Resistances of the bottom non-collapsible layer F_d and negative friction P_n were determined by CPT and static pile tests data. Test results of pile extraction immersed in the depth of collapsible thickness were considered to be the accurate evaluation of negative friction. In order to evaluate the resistance of the bottom non-collapsible thickness, piles were immersed through the leader holes, Ø500 mm, drilled up to the bottom of collapsible thickness so that in pushing test the load was transferred immediately onto the low collapsible stratum. Groundwater level concurred with the foot of the collapsible stratum; that is why both CPT and pile tests were carried our ignoring wetting.

The results obtained revealed that CPT is acceptable for computation of either the resistivity of the bottom non-collapsible thickness F_d or negative friction P_n – divergences between the results of calculations by CPT data and static pile tests did not exceed 20 %.

5.3 SOILS WITH LARGE BOULDER AND BLOCK INCLUSIONS (MORAINE, ELUVIAL, ETC.)

Moraine deposits (dense loams) containing pebbles and boulder inclusions from several cm to several m in size are common in North-West of European Russia. Occurrence of large boulders significantly complicates excavation works and construction of foundations, as the boulders have to be moved away from the pit or demolished by means of special machines. Construction of driven pile foundations faces particular difficulties. Due to collisions of piles with boulders the designed depth is not achieved, thus, one has to solve the problems on remaining piles' bearing capacity or amend the project. There appears the necessity of remaining piles' cutting with further moving of reinforced concrete off the site. In a number of cases duplicating piles or other means of pile foundations strengthening may be required, and in saturation of boulders construction of pile foundation is not reasonable at all. This fact revealed with delay (during driving, but not investigations) causes serious problems of prolonged delays and more expensive construction.

These problems were a common practice in the former USSR during mass application of piles, since a lot of pile foundation structures were erected on moraine soils. These soils served a reliable basement for pile foundations in low saturation with boulders. In such soils the bearing capacity of 500...700 kN was sometimes achieved in prismatic piles only 4...6 m in length; this made pile foundations rather cost-effective.

Large boulders occur not only in moraine soils but in some marine deposits, especially in ocean shores. In Russia such soils do not occur very often, that is why a builder usually associate them with moraine soils.

In construction on eluvial soils typical for Ural region the same problems usually emerge. These soils possess increased heterogeneity and rocky inclusions of mother rock ("ribs"), such as crushed stone, blocks or cracked rocky massifs. Large blocks cause the same problems as large boulders, i.e. they complicate excavation works, construction of foundations, especially the pile ones.

As practice shows, CPT is the most convenient means of large rock inclusions disclosure (gravel or non-gravel) in the soil massif under study (Goncharov et al. 1991, Ryzhkov 1992).

In designing **structures with considerable embedded parts** (basements) on sites with large number of boulders (blocks) CPT significantly facilitates solution of many technological problems connected with excavation works. Data on presence or absence of the boulders (blocks), their size and their location allow selecting machines and machinery to excavate pits rationally. It is sometimes possible to adjust location of the designed objects in plan (if allowed by architectural-planning solutions) thus, selecting the most suitable sectors. Information on large rocky inclusions being incapable to be moved away by a bucket or a small yield bulldozer is the most important. Engineering-geological investigations are to present total information on the issues.

It is CPT that allows the mentioned problems to be more successfully resolved than traditional drilling. The number of CPT locations may be considerably larger than that of the holes (in the same time expenditures and material costs). CPT proves to be the most effective in combination with geophysical methods, e.g. geolocation.

Application of CPT in boulder (block) soils does not require any explanations when foundations on natural bed (with/without basements) are designed. It is necessary to underline that in here awareness of local conditions as well as application of local experience acquire great importance.

Application of CPT in surveying for **driven pile foundations** designing on boulder soils appears to be more complex. It is evident that in here three cases may be distinguished:

- CPT has not revealed any boulders, i.e. in all CPT locations penetration has been performed in specified depth.
- The penetrometer has run into boulders in some CPT locations, i.e. the desired depth has not been reached everywhere.
- CPT has revealed solid inclusions (boulders or blocks) just at the earth surface (in the depth from 0 to 1.5...2 m) almost in all CPT locations.

The first and the third cases do not need any explanations: in the first case piles are driven applying "standard" methods (see section 4.3), in the third one – it seems necessary to ignore it.

The second case requires the detailed consideration.

When the penetrometer attains the specified depth not in all CPT locations, then *forecasting of the number of piles incapable to be penetrated up to the specified depth* becomes the most important task. Decision on acceptability or inadmissibility of anticipated volume of remaining piles (and hence, their cutting) depends on concrete conditions, but in a number of cases cutting shall not exceed 5...8 % from the total volume of piles for the given structure.

In construction (in driving of piles) there emerges another problem as well: *whether the required pile bearing capacity is ensured when piles run into boulders and thus, they do not achieve the design depth.*

Therefore, it is necessary to consider some issues on behavior of the pile resting upon the boulder.

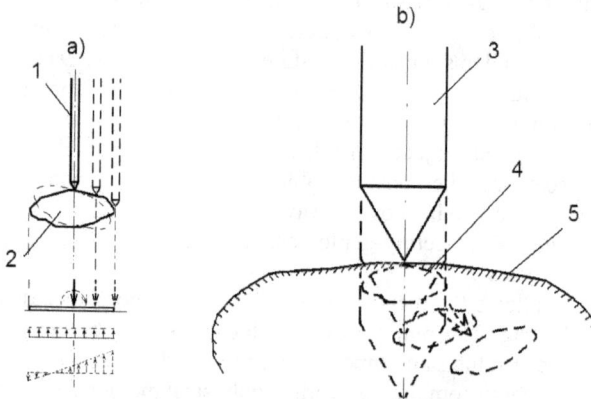

Fig.5.18 Schematics of penetrometer and pile resting upon the boulder:

a – variants of penetrometer's resting upon the boulder, possible design models, *b* – pile's resting upon the boulder: 1 – penetrometer, 2 – boulder, 3 – pile, 4 – fine boulder incapable to stop the pile (pushed aside), 5 – larger boulder capable to stop the pile

Resistance of the boulder on which the pile rests upon may be estimated as the resistance of the enlarged pile base (in case the base is constructed ignoring soil strengthening). This refers to boulders under the penetrometer as well. It is evident that there are numberless variants of pile or penetrometer resting upon the boulder. Figure 5.18 shows possible variants of such resting. As one can see from fig. 5.18a the maximal boulder resistance to moving of the penetrometer should be anticipated when the penetrometer 1 rests upon the boulder at the location above the centre of gravity of horizontal projection of the boulder 2. In inclination of the penetrometer from the center (dashed lines in fig. 5.18a), angular momentum will emerge; hence, lower loading will be required to displace the boulder with the penetrometer. This is also valid for pile resting upon the boulder (fig. 5.18b), but in order to stop the driven pile, the size of the boulder must be several times larger. Most of the boulders impassible for the penetrometer are easily pushed aside by the driven pile as it is shown in fig. 5.18b (locations of the displaced fine boulder 4 are marked with dashed line). The pile 3 having pushed aside the boulder 4, it can move forward up to the design depth or up to the next larger boulder 5 capable to stop it.

Apparently the size of the boulder capable to stop the pile or penetrometer depends on specific conditions - soil strength under the boulder, sizes of piles, hammer capacity, capacity of the pushing device, boulder shape, etc. Nevertheless, it seems interesting to discuss the most typical cases corresponding to some averaged conditions when the pile (or penetrometer) stops.

Figure 5.19 illustrates the curves showing the dependence of boulder sizes upon the depth of the boulder location in various soil consistencies (liquidity index I_L). The curves *were calculated* proceeding from the following conditions:

- Soil resistance under the boulder ("boulder resistance") equaled to enlarged boring pile end bearing capacity in this soil after of SP 50-102-2003 (SP 24.13330.2011).
- Maximum pushing thrust capacity of the penetrometer equaled to 100 kN, friction along the side surface of the push rod equaled to 50 kPa, the load was considered to be applied centrally (without concentricity).
- Maximum load from the driven pile equaled to static load of 700 kN, friction along the side surface of the pile equaled to 50 kPa in the pile of 0.3×0.3 m in section, the load was considered to be applied to the boulder centrally (without concentricity).
- The parameter $D = \sqrt{A}$, was taken as the size of the boulder, where A – area of the boulder's horizontal projection, when the pile (penetrometer) are incapable to overcome its resistance.
- Soil was clayey with the liquidity index $I_L = 0...0.6$.

Figure 5.19 shows that the boulder capable to stop the penetrometer may have linear sizes *nearly three times smaller* than that one capable to stop the pile (in here we imply the most typical pile 0.3×0.3 m in section). The stronger the soil, the smaller boulder may become impassible either for the penetrometer or the pile. Here, the deeper the boulder, the smaller sizes are necessary to stop the penetrometer or pile. In semi-solid or solid moraine loams the boulders are usually of $D = 0.2...0.4$ m to stop the penetrometer and $D = 0.6...1.2$ m – for piles.

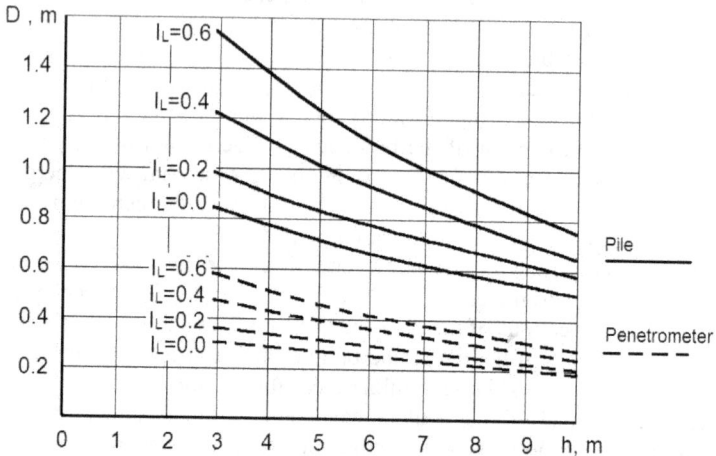

Fig.5.19 Schematic illustration of dependence of boulder capable to stop the penetrometer or pile (D) upon the depth of its location (h) in various liquidity indexes (I_L):

D – boulder size was taken as the square root of its horizontal projection area

Thus, at one and the same site "the failures" in penetration are to be met more often than underdriving of piles. The boulder stopping the pile is always impassible for the penetrometer, while the boulder stopping the penetrometer may be passed through by the pile. In here, the number of small-sized boulders appears to be considerably larger than the coarse ones. Therefore, one needs to know granulometric (fractional) composition of rocky inclusions in the soil for quantitative transition from the share (percentage) of locations where the penetrometer failed to achieve the specified depth, to the anticipated percentage of underdriving of piles.

Approximate data on the volume and fractional composition of the boulders may be obtained based on investigations done earlier at the site. Assessment of the volume and fractional composition of the boulders is significantly facilitated if to take into account one geometrical feature of two-phase media (i.e. the ones containing alien inclusions). In investigations of pore materials the regularity called "principle of Akkera-Covalery" is used. Its meaning is in the following: "volume" porosity (voids in the volume), "plane" one (area of voids in arbitrary cross-section) and "linear" one (sectors fitting the voids, in general line crossing the pore medium) are nearly the same. There are theoretical proofs of the assumption validity for separate particular cases (e.g. Babkov & Polak 1965), but in general case the "principle of Akkera-

Covalery" is usually considered as a postulate. It is evident that the larger number of voids and evener their distribution, the more valid it is.

The principle is possible to be applied for the medium incorporating hard alien inclusions instead of voids (in here – boulders). From practical point of view this is a significant simplification of boulder occurrence assessment: in excavation of a pit one can judge on either boulders' share (in total volume of ground massif) or their fractional composition by those ones found out on the bottom and the walls of the pit.

Figure 5.20 illustrates histograms of boulders' sizes of two types of soil – moraine loams of Karelia (Petrozavodsk, Drevlyanka neighborhood) and marine deposits of Singapore (after Tang & Queck 1986). The authors point out that the histogram obtained is easily approximated by gamma-distribution.

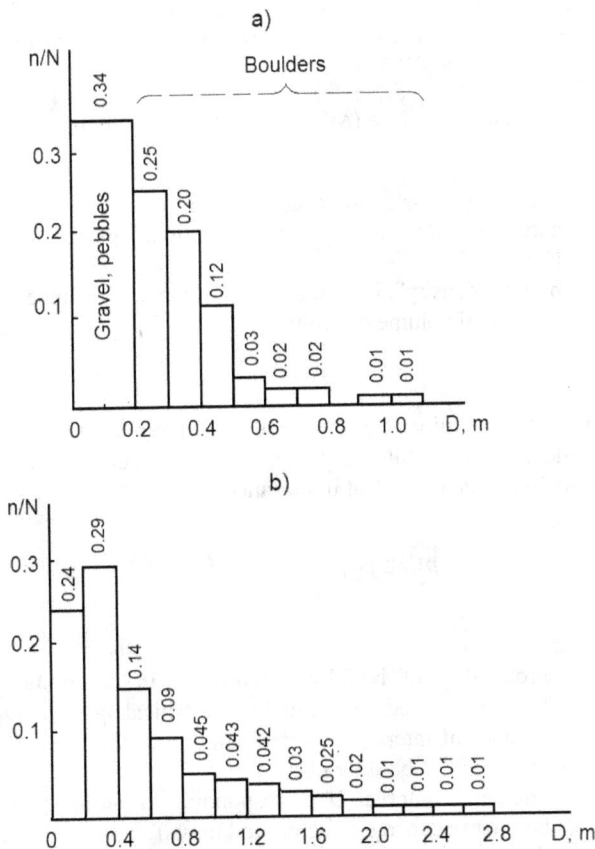

Fig. 5.20 Histograms showing fractional composition of boulders:

a – moraine loam, Karelia (Petrozavodsk), b – marine deposits, Singapore (Tang & Queck 1986); D – boulder size, n/N – relative frequencies of boulders of D in size (n & N – number of boulders of fraction D and total number of boulders in the ground massif correspondingly)

The histograms given in fig. 5.20 characterize the relationships of a *number* of boulders of each *i*-fraction n_i to the total number of boulders N in soil massif under study, i.e. $n_{0.2}/N$, $n_{0.4}/N$, $n_{0.6}/N$ ..., etc. Nevertheless, one can approximately calculate *the volume* of each fraction ΔV_i – relative "volume" frequencies $\Delta V_{0.2}/V$, $\Delta V_{0.4}/V$, $\Delta V_{0.6}/V$ ..., where V – total volume of boulders in soil massif under study. For this, it is necessary to admit one or another boulder form as the assumption (e.g. ball or pillow-like).

"Volume" frequencies $\Delta V_i/V$ may be recalculated on the *total volume of ground massif* V_o, thus, the parameters $\Delta V_i/V_o$ are obtained; they may approximately be taken as *probability* of boulder occurrence in *i*-fraction in the unit of soil volume. For this, the parameters $\Delta V_i/V$ and "boulder density" B are to be multiplied, i.e. share of boulders V in the total volume of ground massif V_o. Thus, the probability of *i*-fraction occurrence in the soil, i.e. boulders of ΔD_i in size may be estimated by the parameter $p(\Delta D_i)$

$$p\,(\Delta D_i) \approx \Delta V_i/V_o = (\Delta V_i/V) \cdot B = (\Delta V_i/V) \cdot (V/V_o\,), \qquad (5.34)$$

where:

$p\,(\Delta D_i) =$ probability of *i*-fraction occurrence in the soil;

$\Delta V_i/V_o =$ share of boulders in *i*-fraction (ΔV_i) in total volume of ground massif (V_o);

$B =$ "boulder density", i.e. relationship of the total volume of boulders V to the total volume of ground massif V_o

$$B = V/V_o\,. \qquad (5.35)$$

To estimate the possibility of collision between the pile (penetrometer) and the "impassible" boulder, the probability of "impassible" boulders occurrence in the soil, i.e. boulders up to D_k in size is of great importance

$$p\big(D \ge D_k\big) \approx \left(\sum_{i=k}^{i \to m} \Delta V_i \right) \Big/ V_o\,, \qquad (5.36)$$

where:

$p\,(D \ge D_k) =$ probability of boulders occurrence in the ground massif, i.e. boulders exceeding or equal to the critical value of D_k in size;

$k, =$ number of interval D_i corresponding to the minimum size of the "impassible" boulder D_k;

$m =$ number of interval D_i corresponding to the maximum size of the boulder occurring in the ground massif;

$\Delta V_i, V_o =$ the same as in the formula (5.34), i.e. volume of *i*-fraction and the total volume of the ground massif.

Probability of "impassible" boulders occurrence in the soil and probability of their collision with piles do not coincide. The pile may come into collision with only one boulder with further stop of its embedding, and any "impassible" boulders located below are not of any importance then. Collision of the pile (or the penetrometer) with the "impassible" boulder corresponds with another problem of theory of relativity – appearance of "at least an occurrence from n tests" where n is a number of cut soil layers with nearly the same thickness as in "impassible" boulder. When passing through n layers the pile or penetrometer have a tendency to stop (collide with *at least one* "impassible" boulder – see Gmurman V.E. (2000)) equal to

$$p\,(A) = 1 - q^n, \qquad\qquad (5.37)$$

$$q = 1 - p\,(D \geq D_k), \qquad\qquad (5.37a)$$

where:

$p\,(A) = $ probability of pile or penetrometer stop ("occurrences A"), i.e. probability of their collision with at least one "impassible" boulder;

$n = $ number of soil layers in the ground massif;

$p\,(D \geq D_k) = $ the same as in the formula (5.36).

Distinction in probabilities of pile and penetrometer stop is conditioned by distinction in critical sizes of the boulder D_k being larger for the pile than for the penetrometer.

From practical point of view the discussed problem was in percentage of pile underdriving forecasting by a number of locations (share in percent) where penetration up to the specified depth had been a failure. BashNIIstroy developed the methods based on the following:

- The anticipated share of underdriven piles equaled to probability of their collision with the "impassible" boulder, i.e. $p\,(A)$ in the formula (5.37).
- Probability of boulders occurrence "impassible" for the penetrometer was estimated by the number (share) of locations after "unsuccessful" CPT when the penetrometer had not achieved the desired depth due to collision with the boulder.
- Transition from probability of "impassible" (for the penetrometer) boulders occurrence in the ground massif to analogical probability for piles is done by means of histograms showing fractional composition of the boulders at the terrain given and the graph in figure 5.19 as well; here, liquidity index I_L was evaluated by CPT data (see section 3.2.5).

The methods are to be considered with regard to fractional composition of the boulders as in histogram 5.20.

Table 5.2 shows distribution of boulder fraction *volumes* corresponding with fig.5.20a in assumption that the whole of fractions are of spherical form with the diameter of D_i (ordinates of the histogram were not evened for simplification).

Let us consider the following conditions of pile applications in accordance with fig. 5.19 & 5.20:

- Piles of 0.3×0.3 m in section are to be embedded in a semi-solid loam ($I_L \approx 0.2$).

- Ground massif is divided into the layers of 0.5 m thick, i.e. 20 layers have been distinguished.

Table 5.2 Distribution of boulder fraction volumes corresponding with fig. 5.20a

Boulder size of the fraction under study D_i (m)	0.0...0.2 (0.10)	0.2...0.3 (0.25)	0.3...0.4 (0.35)	0.4...0.5 (0.45)	0.5...0.6 (0.55)	0.6...0.7 (0.65)	0.7...0.8 (0.75)	0.8...0.9 (0.85)	0.9...1.0 (0.95)	1.0...1.1 (1.05)	Total V (m³)
Boulder volume of the fraction under study V_i (m³)	0.02	0.20	0.45	0.57	0.26	0.29	0.44	-	0.44	0.60	3.27
V_i/V	0.01	0.06	0.14	0.18	0.08	0.09	0.13	-	0.13	0.18	1.0

It is necessary to point out that the number of layers has a low effect on the final result (in alteration of number of layers from 5 to 20 the probabilities p (A) changed in 5...7 %).

Let us assume that in 20 % of CPT locations the penetrometer has not attained the specified depth, i.e. the probability p (A) corresponding to collision of the penetrometer with, at least, one boulder of critical size may be considered to be equal to 0.2. Then, it is possible to evaluate q – probability of *absence* of the boulder of critical size in one or the other soil layer by the formula (5.37)

$$0.2 = 1 - q^{20} \rightarrow q = 0.989,$$

and, hence, the probability of presence of the boulder of critical size in the layer will be

$$p (D \geq D_k) = 1 - 0.989 = 0.011.$$

In accordance with the discussed principle of volume, plane and linear boulder occurrence equality, the probability obtained may be referred to the volume of the ground massif under study.

In depths of 3...10 m, in accordance with fig. 5.19 mean value of boulders' critical size (for the penetrometer) may equal to $D_{k,penetr} = 0.30$ m. According to table 5.2 relative frequency of fractions (V_i/V) being "impassible" for the penetrometer ($D_{k,penetr} \geq 0.3$ m) will be 0.93. The factum of this parameter and "boulder occurrence B" (see formula (5.35)), i.e. $0.93B$ may be considered as the probability of "impassible" (for the penetrometer) fractions occurrence in the ground massif. For boulders capable to stop the pile ($D_{k,pile} \geq 0.8$ m), relative frequency (V_i/V) will be 0.44 after table 5.2. Therefore, the probability of "impassible" fractions occurrence (for piles) in the ground massif will be $0.44B$.

Thus, for soils defined by the histogram (fig. 5.20a) and table 5.2 occurrence of boulders capable to stop the pile happens to be twice lower than occurrence of boulders capable to stop the penetrometer $(0.44B)/(0.93B) = 0.47 \approx 0.5$). In computation of the latter one (*penetrometer*) (0.011), the former one (*pile*) shall be $0.011 \cdot 0.47 = 0.005$ (either in a separate layer or the whole of soil). Here, probabilities of absence of the boulder being "impassible" for the pile (q) will be $1 - 0.005 = 0.995$, and thus, probability of pile stopping (its collision with, *at least, one boulder* exceeding the critical size $D_{k,pile}$) will be

$$p\,(A) = 1 - 0.995^{20} = 0.1.$$

Thus, 10 % of piles appear to be underdriven up to the design depth at the examined site.

The analogical computation may be done for each fractional composition of boulders. The initial data are as follows:

- Percentage of CPT locations in which the penetrometer stops because of its collision with boulders.
- Consistency of soil filling the voids between boulders.
- Fractional composition of boulders allowing determination of probability ratio of boulders capable to stop the pile to the analogical probability for boulders capable to stop the penetrometer, i.e. $[p\,(D \geq D_{k,pile})]\,/\,[p\,(D \geq D_{k,penetr})]$.

As one can see, knowledge of fractional composition of boulders is important for determination of probability ratio $[p(D \geq D_{k,pile})]/[p(D \geq D_{k,penetr})]$. The ratio is to be considered as a parameter $K_{pile-penetr}$, which can approximately be presented as the ratio of boulder volume being "impassible" for the pile to the analogical volume for the penetrometer

$$K_{pile-penetr} = \frac{V_{D>Dk,pile}}{V_{D>Dk,penetr}}, \qquad (5.38)$$

where:

$V_{D>Dk,pile} =$ volume of boulders exceeding *the critical parameter* $D_{k,pile}$ in size (corresponding to the right part of the formula (5.36)), evaluated by the histogram of fractional composition of boulders;

$V_{D>Dk,penetr} =$ volume of boulders exceeding *the critical parameter* $D_{k,penetr}$ evaluated in the same way as $D_{k,pile}$.

Figure 5.21 illustrates the curves showing the dependences between locations of "unsuccessful" CPT (i.e. locations in which the penetrometer failed to achieve the specified depth because of collision with the boulder) and underdriven piles. As one can see from the graph, the curves are determined by the parameter $K_{pile-penetr}$ showing the ratios of boulder volume "impassible" for the pile to the analogical volume for the penetrometer (formula (5.38)). The smaller is the difference in critical sizes of boulders "impassible" for piles ($D_{k,pile}$) and the ones "impassible" for the penetrometer ($D_{k,penet.}$), the smaller will be the difference in percentage of underdriving of piles and "unsuccessful" CPT locations.

Fig. 5.21 may be used as a nomogram saving from the computations to be done. This is complicated by the necessity to be aware of the parameter $K_{pile-penetr}$; these data obtaining is rather effortful procedure. It is evident that small difference between $D_{k,pile}$ and $D_{k,penetr.}$ is typical for large fractions of boulders. For example, for fractional composition of boulders illustrated in fig. 5.20 the parameter $K_{pile-penetr}$ is quite near to one ($V_{D>Dk\,pile} = 0.963$; $V_{D>Dk,penetr.} = 0.998$ and hence $K_{pile-penetr} = 0.965$). In soils with such a composition of boulders the percentage of "unsuccessful" CPT locations will almost agree with the percentage of underdriven piles.

In experience gained $K_{pile-penetr}$ may be evaluated approximately, proceeding from visual assessment of boulder coarseness, e.g. "prevalence if fine boulders up to 0.2 m in size" or "prevalence of coarse boulders exceeding 0.8 m in size", etc.

The approach presupposes that premature stops of the penetrometer or piles occur due to presence of boulders (or blocks). However, in high density of soil located between the boulders the stops may occur due to high resistivity of this soil to penetration of the cone or pile as well. Apparently, one cannot apply the regularities (fig. 5.21) here, i.e. the penetrometer may be immersed up to the same depth as the pile, and often – deeper. In order to avoid the errors, it is necessary to study the CPT diagrams carefully, thus, it would be easy to explain the reasons of premature stops of the penetrometer then. Collision with the boulder is usually characterized by sudden increase of cone resistance q_c and lack of such increase of sleeve friction resistance f_s. The depths where such increase occurs are usually irregular. It is obvious that the problems are to be settled applying the results of drilling, laboratory analyses of soil monoliths and local experience in construction as well.

The issue on *the required pile bearing capacity* when colliding with boulders goes beyond CPT applications. Here, a number of "non-geological" factors are important, namely: hammer capacity, design pile bearing capacity, type of pile foundation (whether the pile is a part of mat foundation, well cluster or a single bearing-pile; whether the pile shall take horizontal loads, etc.). Therefore, in every case of pile underdriving the individual approach is needed, thus there is not any common solution here. Nevertheless, CPT is always useful, as it provides with a great number of data to solve the problem.

Formally collision with the "impassible" boulder may be considered as obtaining of "zero" refusal corresponding to pile bearing capacity being the same as in piles that achieved the design depth ("control refusals" mentioned in projects are usually higher than "zero" ones). However, "zero" refusals of underdriven piles do not guarantee the same settings as in piles being immersed in the design depth; the conditions of underdriven piles' performance happen to be considerably worse in resting upon the boulder edge (breaking of the pile is possible). When underdriving occurs due to strong rocks appearance in the soil (gravel-pebble soils, dense basement clays, etc.) such dangers do not usually emerge. In practice, one may not take any special measures on foundation strengthening here. It is CPT that simplifies the solutions of the problems, as it gives possibility to take decisions on the "impassible" stratum in advance and more accurate than traditional drilling.

The paper (Goncharov et al. 1991) gives results of calculated loads on the pile causing edging of the boulder and the loads causing insignificant (admissible) settings. In the latter case resistance of soil under the boulder was considered as design resistance of the basement R after p. 2.41 of SNiP 2.02.01-83*. Computations were carried out in piles 0.3×0.3 m in section, 6 m in length. The following soils were considered (up to 6 m in depth):

- Soils, homogeneous in depth with $q_c = 1$ MPa, $f_s = 0.05$ MPa and $q_c = 3$ MPa, $f_s = 0.15$ MPa.
- Soils becoming stronger with depth up to $q_c = 0...3$ MPa, $f_s = 0...0.15$ MPa.

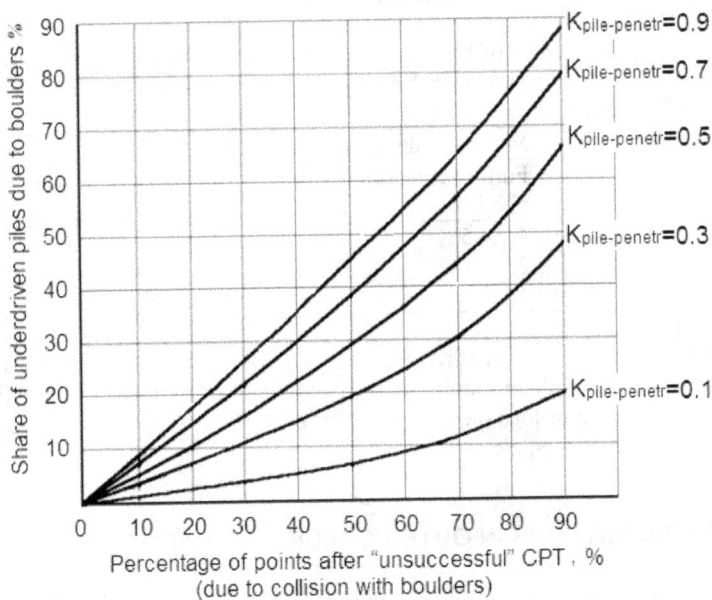

Fig. 5.21 Calculated dependences between locations of "unsuccessful" CPT (i.e. locations where the penetrometer failed to achieve the specified depth due to collision with boulders) and underdriven piles because of the boulders: $K_{pile-penetr}$ – parameter showing the ratio of "impassible" boulders for the pile and penetrometer evaluated by the formula (5.38). The shares of underdriven piles and failures of CPTs are taken in accordance with probabilities of the events [$p(A)_{pile}$, $p(A)_{penetr}$] evaluated by the formula (5.37)

Table 5.3 shows the depths of boulders' location when pile "stopping" is possible and the depths where the design bearing capacity is ensured at the "stopped" pile.

As one can see from table 5.3, the depth of boulder location is of great importance: the size of the boulder capable to stop the pile decreases with depth. This is due to two reasons. First, while embedding the pile, the load transferred by its side surface increases, thus, smaller part of the total load is left to displace the boulder. Second, soil resistance under the boulder increases with depth, as under any foundation on the natural bed. Pile bearing capacity increases due to these reasons as well, although its growth may be behind boulder resistance in driving of the pile. For example, the boulder 0.75 m in size is capable to "stop" the pile in the depth of 2 m, but pile bearing capacity may fail to attain the design parameter. At the same time, collision between the pile and the boulder at the depth of nearly 5 m will not cause any problems, since pile bearing capacity will be sufficient.

Table 5.3 Minimum depths of boulder location when pile "stopping" is possible and the depths where the design pile bearing capacity does not exceed that of the "stopped" pile (Goncharov et al. 1991)

Size ("diameter") of the boulder (m)	Minimum depth of boulder location (m) when	
	Further embedding of the pile is impossible	Design pile bearing capacity is ensured
0.5	4...4.5	6
0.75	2	4.7...5.5
1.0	1	1

In practice, 1...2 m underdriving appears to be "safe" for piles 8...12 m in length, however, the problems are to be resolved with regard to each case taking into account redistribution of load onto the adjacent piles, "reserve" pile bearing capacity, assumed sizes of boulders, etc.

5.4 CPT APPLICATIONS IN OUTER SPACE. LUNAR SOILS

Exploration of near-space is of great importance among the global problems of the 21st century that are to be resolved in future. The problem is recognized by everybody, the views differ in terms of problem-solving or material costs necessary to achieve the goals.

It is always important to examine natural conditions of space objects before taking any attempts to study them. In here, the necessity to estimate the mechanical properties of surfaces of space objects emerges.

As it is known, the surface of space objects in lack of atmosphere is composed of loose material – *regolith*. It is crushed soil (size of particles is nearly $10^{-3}...10^{-7}$ m), formed after meteorite shower on the surface of space objects and effects of cosmic rays. In occurrence of atmosphere formation of the surface layer is significantly complicated by effects of various exogenous physical or chemical processes; this results in formation of loose rocks, but they are more diversified in their structure and chemical composition as well. There are also rocky blocks several meters in size on the surfaces of the planets of Solar system. CPT seems to be one of the most promising methods of loose soil investigation, as it means *carrying out the measurements just on the examined object and it can easily be adjusted to radio transmission of information obtained in any distances.* Probably, combining CPT with some geophysical methods (radiation, geolocation, etc) appears to be the most effective.

Space CPT shall be rather economical, since selection and delivery of samplers to the Earth is the only alternative. Though study of samplers provides with relevant information, it is a complicated technical problem requiring higher material costs. Even when limiting CPT application by preliminary reconnaissance, one should recognize the exclusive value of the method, particularly in investigation of small space objects.

Performing of any soil tests on space objects is incomparably more difficult than any analogical tests carried out on the Earth, as one needs to solve rather complicated technical problems. This is related to CPT as well, despite its simplicity in "terrestrial" conditions. However, the problems are possible to be successfully resolved, because the progress has been noticed in this field in last fifty years, at least it does not yield to "terrestrial" investigations. Let us consider specific features of the problems resolved as well as the results obtained taking lunar soil as the example.

The first research of physical-mechanical properties of Lunar soil (regolith) was done by Soviet scientists in 1960s (Cherkasov et al. 1968, 1986, Cherkasov & Shvarev 1971, Cherkasov 1973). They initiated the new trend in geotechnics, i.e. "space geotechnics". Fundamentals of Lunar soil science were the first step in its creation (Cherkasov & Shvarev 1971). Russian researchers contributed a lot in Lunar soil-regolith investigations carried out applying the penetrometers of special geometry developed to work in ground-controlled lunar bases and self-propelled devices.

The first soft landing on the Moon was done on February, 3, 1966 by means of Soviet automatic lunar station "Luna-9". This successful landing disproved the hypothesis on thick layer of loose dust presence on the Moon and hence, gave possibility to instrumental determination of lunar soil mechanical properties.

"Luna-13" landed in Ocean of Storms on December, 24, 1966 (fig. 5.22); three devices were mounted in it aimed at investigations in the sphere of soil mechanics. They were the groundmeter-penetrometer, gamma-densitometer and dynamograph of "Luna-13", the first geotechnical devices delivered to the Moon.

Groundmeter-penetrometer (fig.5.23) (Cherkasov et al. 1968, 1986, Cherkasov & Shvarev 1971) was specially intended for Lunar soil mechanical strength assessment; it was the first device applied for Lunar soil study.

It was composed of plastic body; its lower part formed flat annular plate 12 cm in the outer diameter and 7.15 cm in the inner one. Movable titanic indentor was located in the upper cylindrical part; its body presented a cone with the tip angle of 103^0 and the diameter of base to be 3.5 cm. Turned upwards miniature solid fuel jet engine with a nozzle was placed above; it was intended for pushing of the indentor into the soil. Here, the thrust was 55...75 kN, duration of operation - 0.6...1.0 s. The depth of pushing was measured by means of the potentiometer equipped with an adjusting slider fixed in the device. The accuracy of measurement was nearly 0.3 mm.

Fig. 5.22 Automatic station "Luna-13"

1- leafed aerials, 2 - pin aerials, 3 - device ejection mechanism, 4 - radiation gamma-densitometer, 5 - groundmeter-penetrometer, 6 - TV camera, 7 - radiometers

a)   b)

Fig. 5.23 Groundmeter-penetrometer (*a*) and device ejection mechanism (*b*) for folded soil investigation

The initial depth of penetration happened to be 4.5 cm. Further, it varied from 4.17 to 4.33 cm, perhaps, due to temperature deformations of the ejection mechanism and station body.

Decoding of results obtained was done in two ways.

The first way involved the tests of device carried out in laboratory conditions on the Earth; in here, the device was installed in various soils and materials with subsequent determination of their properties and densities applying the conventional methods. The calibration procedure having been done, the table (table 5.4) of both indentor penetration depths and device body penetration depths into every investigated material was obtained; later, it was used for decoding the data obtained from the Moon.

Table 5.4 Calibration results of "Luna-13" groundmeter-penetrometer obtained in terrestrial conditions (Cherkasov & Shvarev 1971)

Depth of penetration, cm		Density, g/cm^3	Modeling materials	
Indentor	Body		Natural	Man-made
0	0	2.4...3.0	Hard dense rocks	Heavy concrete
0.0...1.3	0	0.25...2.0	Foamy and porous rocks; cohesive soils	Foam concrete, foamglass, resin-bonded claydite
1.3...5.0	0	1.3...1.7	Non-cohesive granular soils of medium density; quartz sand	
1.3...5.0	0...1.0	0.25...0.77	Non-cohesive granular soils of low density	Crushed foam concrete, claydite, agloporite
5.0	Above 1.0	0.16 and below	Very loose pulverescent soils	Heaved perlite sand

Special experiments were done to reveal the effect of gravity force acceleration on indentor penetration depth in pushing into granular soils. The latter were carried out in the airplane cab; in its flight path the acceleration equaled to the lunar gravity force acceleration, i.e. 1.62 m/sec^2. The indentor was pushed down by means of a spiral spring; the depth was recorded by a mechanical recorder.

The comparison of penetration depth with that of obtained in the same way on the Earth allowed determining that 6 time decrease of gravity force acceleration as compared with the terrestrial one resulted in nearly 70 % of indentor penetration increase.

The second way of groundmeter-penetrometer readings decoding involved application of soil mechanics, i.e. solutions of round plate pushing. As far as there were not any solutions for the plate of such a complicated geometry as that in "Luna-13", the solution for the flat plate after V.G. Berezantsev (1955) was applied. Evaluation by this formula gave critical pressure on Lunar soil equal to 66.7 kPa.

The analysis of readings as well as examination of the panoramas obtained allowed concluding that granular, weakly cohesive loose soil occurred at the place of "Luna-13" landing. Dust layer was not found. Density ρ, angle of internal friction φ and cohesion C of Lunar soil were determined during first approaching.

Scientific research carried out with two self-propelled apparatuses delivered to the Moon by automatic stations "Lunohod-1" and "Lunohod-2" was the zenith of Soviet cosmonautics. "Lunohod-1" was operating in the Sea of Rains from November, 17, 1970 till February, 19, 1971."Lunohod-2" was operating in the Crater of Le Montier from January, 16 till April, 22, 1973. The apparatuses were controlled from the Earth.

Measurements done on "Lunohods" with PROP were directly related to Lunar regolith mechanics (fig. 5.24). The operating body of PROP was composed of a conical plate with the tip angle of 30^0, a base – 5 cm in diameter and two vertical blades – 7 cm in width and 4.4 cm in height. The apparatus was a combined one in its geometry and function – it combined the elements of the penetrometer and vane.

a)   b)

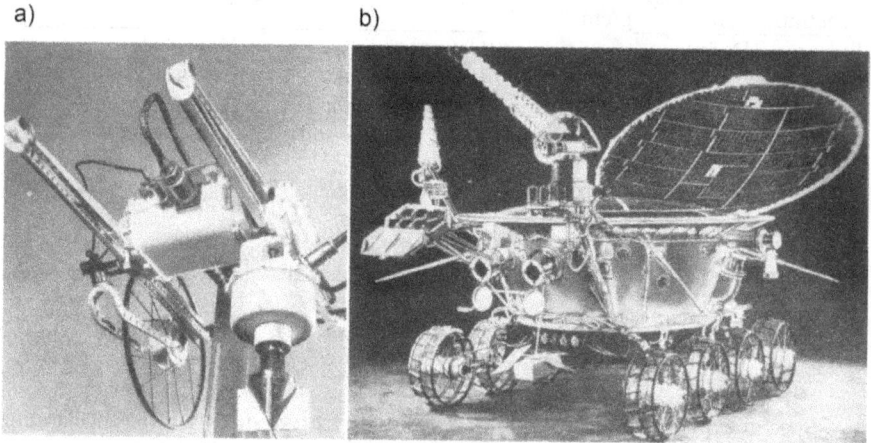

Fig. 5.24 General view of PROP apparatus (a) and self-propelled apparatus "Lunohod-1" equipped with it (b)

By the order from the Earth the cone was dropped up to the surface and then pushed into the soil. Maximum load was taken to be 200 N. The cone having been penetrated 5 cm in depth, it was rotated and the soil was cut by the blades along the cylinder side surface – 7.7 cm in diameter and its base as well. The maximum rotary angle was 90^0, the maximum running torque – 5 Nm. Thus, data on soil bearing capacity in pushing of the cone and/or ultimate shear stress of the soil were obtained.

CPT as a geotechnical method of lunar soil study was used worldwide as well. In the research program "Apollo" (Mitchell & Houston 1974) the US NASA specialists applied manual probes for CPT of Lunar soil. Here, the maximum depth of

CPT was 0.76 m. Cores selected from the same site where CPT had been carried out showed that Lunar soil was composed of pulverscent sand and sandy pulverescent soil. When analyzing the data obtained it was taken into account that lunar gravity force was six times lower than the one on the Earth. Figure 5.25 illustrates the results obtained after the tests carried out during the 16th flight of "Apollo" to the Moon. As one can see, cone resistance q_c in regolith is approximately the same as that of "terrestrial" pulverescent sand (of medium density or loose). Having analyzed CPT data obtained, the authors (Mitchell & Houston 1974) drew the conclusion that relative density of Lunar soil varied from 65 to 95 % in the depth of 0.6 m.

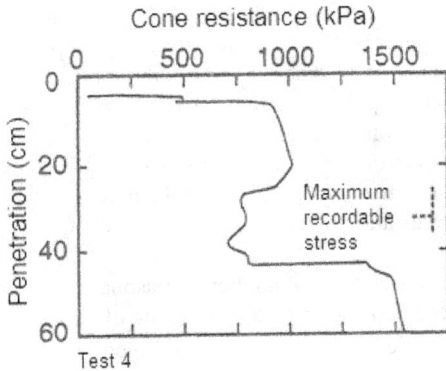

Fig.5.25 CPT profile of Lunar soil (Mitchell & Houston 1974)

6 CPT AND OTHER SOIL TESTS JOINT APPLICATIONS

6.1 JOINT TESTS APPLICATION ISSUES

Both in non-Russian and Russian practice CPT is used jointly with other soil tests, namely: engineering-geological and geotechnical (engineering) soil assessment including drilling, laboratory analyses, in-situ pile tests, plate and/or pressure meter tests, etc. There is an opinion that at present the most effective investigations involve optimization of soil study technology rather than improvement of separate surveys. Although optimization of investigations appears to be an independent complicated problem being not studied carefully, it is no doubt that clarity of joint application of CPT and other geotechnical tests is of principal importance in CPT improvement. Different techniques applied in soil tests at one and the same site always result in irrelevant information on mechanical properties of soil and its assumed behavior under the foundation. Rational choice of estimate indicators in such a diversity of the initial data requires highly-skilled employees and realization of the results obtained as well.

Such contradiction of data could be explained by a number of reasons.

First, *conditionality* of generally accepted mechanical properties of soil which are just the parameters of mathematical models of soil (φ, c – loose medium, E, v – linear deformable medium, etc.) has its impact on the results. Unfortunately, the real behavior of soil happens to be more complex and invariance of the properties is far from the anticipated one in various tests.

Second, in estimation of any parameter *heterogeneity of soils* always shows itself. Evaluation of one or the other property of soil in different locations of the site applying one and the same technique does result in different data. As it was mentioned earlier (see section 3.2.1), any two CPT locations moved away from each other (in plan) just in 1...2 m are incapable to show the same values of q_c & f_s, distinction increasing 30...40 % (see fig. 3.10). The same is referred to pile bearing capacity evaluation in different locations of the site. CPT tests with in-situ piles carried out in BashNIIstroy show that the identical piles embedded into a homogeneous stratum of clayey soil in locations spaced away from each other just in 3...4 m may differ in their ultimate bearing capacities up to 20...30 % (after CPT data).

Third, *errors in measurements* somehow effect on the data obtained. Occurrence of faulty results conditioned by *gross errors* of a researcher or *faultiness* of measuring equipment cannot be excluded. Distortions caused by these factors are not possible to be always found out, especially when they are not significant and according to the order of magnitudes they are close to deviations caused by heterogeneity of soil.

In practice, there may appear the situations when the researcher having done a great number of tests obtains extensive but contradictory information, thus neither he nor the planner are capable to sort it out. As a result, in planning the considerable part of this information happens to be unused or partially used then.

Generalization of test results of *different accuracy*, their relevant integration with each other is often the more complicated problem than interpretation of results of each test. The issue is a matter of discussion, since there is not any effective generally accepted solution either in Russia or worldwide. In practice, the *most reliable* results are preferred corresponding to *"the weakest"* sectors of the site. Other data obtained are usually applied for *qualitative* assessments – identification of soil types, their stratification, selection of locations for "accurate" tests, etc. According to this, determination of one or the other design parameter is done by the results of only one test mode ("accurate"), other data are the additional means ensuring the efficiency of such an "accurate" test. The approach is not given in the Norms, but in absence of clear rules of different reliability data generalization it is taken by the majority of researchers as the only possible ensuring the necessary safety of solutions.

Nevertheless, selection of the design parameters by the results of just one test mode (type) has one significant drawback. The possibilities of other tests are hardly ever used, as they are the means of security, selection of locations for "accurate" tests, etc. The accuracy of results of such tests decreases, as the design parameters are determined by the "accurate" tests. Negative points of the approach are particularly noticeable in absence of reliable tests (plate tests, static tests of piles, etc.). The situations occur in joint applications of CPT and laboratory tests, CPT and geophysical tests, etc. Here, selection of the "accurate" test becomes rather conditional and the loss from incomplete use of the rest ("additional") methods proves to be the most noticeable.

Besides, conducting of "accurate" tests in "the weakest" sector of the site and further introducing of various lower "reliability coefficients" in the data obtained result in unjustified "reserves" in the design parameters. To a certain extent, the focus on the weak site disagrees with the principle of foundation computation admitted in the National Normative Documents. In accordance with the National Norms SNiP 2.02.01-83*, SNiP 2.02.02-85*, SNiP 2.02.03-85*, SP 24.13339.2011, GOST 20522-96, GOST 27751-88 the design parameter of soil properties (or the design resistivity of foundation) is understood as *the mean-value* function (not minimal!).

In order to apply the whole of quantitative results effectively, other approach to their generalization is required taking into account the actual reliability of every type (method) of tests. The issue was studied in BashNIIstroy in 1980s...1990s, thus alternates solutions of the problem were developed; they were based on the results of each test taken as random variables with their own dispensing (Ryzhkov 1988, 1992). In here, data acquisition was accompanied by application of corresponding tables or software.

Dispensing of "errors" is usually carried out based on the data obtained earlier, i.e. by static analysis of comparisons between CPT and "accurate" methods, i.e. the "reference" ones (plate tests, static tests of piles, sometimes laboratory tests, etc.). As to CPT, there have been a lot of data obtained. Final results of their analysis (empirical dependencies) are well-known; they are given in the Normative Documents. However, the initial data which help assess the reliability of the dependencies are usually not well-known, as they are given in scientific reports and publications. Diagrams of dissipation are the examples of the initial data of various empirical dependencies (see chapters 3 & 4) (fig.3.15, 3.17, 3.18, 3.19, 3.21, 3.22, 3.23, 4.10, 4.11, etc.). It will be shown later that it is necessary to draw histograms (tables) for specific computations, thus it will be possible to reveal the frequencies of "errors". Their relative values may be taken as probabilities after "leveling".

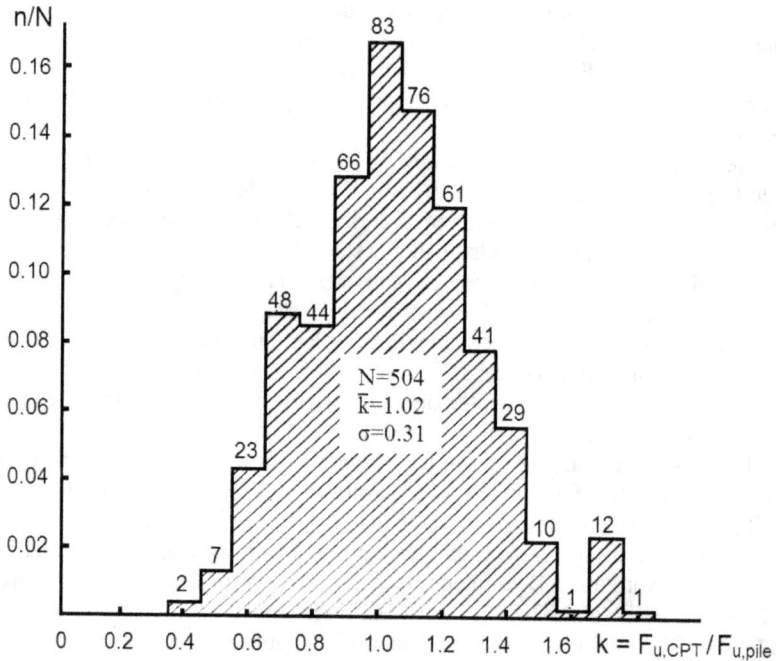

Fig. 6.1 Histogram of relative "errors" $k = F_{u,CPT}/F_{u,pile}$ drawn after the diagram of dissipation given in fig. 4.10.

$F_{u,CPT}$ – pile ultimate bearing capacity by CPT data, $F_{u, pile}$ – the same by static tests of in-situ piles, N – total number of tested piles, $\bar{k}$ – mean value of k, σ – mean square deviation of k; numbers at each interval ("steps of histogram") show the number of compared pairs "CPT – pile test" occurred in the interval

The histogram of relative "errors" (fig. 6.1) shows $k = F_{u,CPT}/F_{u,pile}$ ($F_{u,CPT}$ – ultimate pile bearing capacity by CPT data, $F_{u,pile}$ – the same by static tests of piles)

drawn after the diagram of dissipation given in figure 4.10. The term "errors" is used here in a wide sense as a generalized parameter of the whole of errors (instrumental, methodological, conditionality of design models, etc.). In here, we understand not only differences between the accurate (x) and the approximate (a) values, i.e. $(x - a)$, but the ratios of the approximate values to the accurate ones a/x (as in fig. 6.1). To prevent the ambiguity, the term "errors" is given in inverted commas.

When considering different conditions of different accuracy test results generalization it seems important to point out two cases:

- Accuracy of one of the tests is quite high; the results obtained may be considered as the "reference" to correct the rest ones (approximate results).
- The whole of the tests are approximate, being not very different in their accuracy.

In the first case when the "accurate" values of the required index are known for separate locations of the site (e.g. pile bearing capacities by their static tests, modulus of deformation by "plate" tests, etc.), the approximate values of this index obtained, for example, by CPT are corrected by introducing the corrections reflecting specific features of the site. The corrections are determined by comparisons of separate approximate values with the "accurate" ones ("reference"). It is evident that distribution of the "accurate" and approximate tests on the site shall agree with each other (be placed 1.5...2 m from each other). Based on the comparison, correction factors are introduced to the approximate results, i.e. the *specific* approximate method is anchored to the *specific* site.

In occurrence of heterogeneous soils the site may be divided into two-three sectors; correction is done with regard to each sector. Hence, the distinguished sectors become "the correction zones": "accurate" tests are to be carried out here, and comparisons between the "accurate" values and the approximate ones are to be done independently from each other. Further on, the term "mean value at the site" will be given without reservation meaning averaging at the whole of the site or at its sectors.

Definition of the approximate data by more precise tests is not new in engineering investigations. The corrections were done earlier, but the methods were not clear depending upon personal point of view of the executor. Strict correction is a complicated task followed by solutions of a number of theoretical problems and definition of some general ideas. The issues are discussed in section 6.2 in detail; however, it is important to underline three essential conditions.

First, correction is to be understood as definition of *mean values* of the required indexes, i.e. as revelation and compensation of "systematic errors" occurring in the site.

Second, the comparative results of the accurate and approximate values of the required index will not be equal in different locations of the site, since heterogeneity of soil effects on "accurate" and approximate values of the required indexes in different ways.

Third, reference properties shall correspond to operating conditions of the designed foundation. In different "references" the results may significantly differ, in failure to choose the "reference" the correction is not of any importance then. Thus, in determining the modulus of deformation E by CPT data, either "compressive" ("odometric") E_c or "stamp" E_{st} values of the index may be used as the "reference", since tightness of correlation "$q_c \sim E_c$" and "$q_c \sim E_{st.}$" is nearly the same (see fig. 3.17[1] & 3.18). Nevertheless, the results will be absolutely different showing different conditions of soil behavior. They must be used in accordance with this. For example, to calculate the strip foundation or the isolated one (bearing) it is necessary to use $E_{st.}$, to evaluate the deformations of the basement of larger area (in alluvion of soil, etc.) one can use E_c, etc.

In the second case when the applied tests are approximate, i.e. "accurate" ("reference") ones are absent; one can do without the discussed correction. Here, one should look for the most appropriate value of the required index with regard to distribution of the possible "errors" in every method and take the required index with "reserve" corresponding to the specified confidence probability.

6.2   CPT DATA UPDATING

As it was mentioned earlier, the principal advantage of CPT is the possibility to examine vast terrains very quickly revealing the capacity, extent and configuration of deposited soil strata, identification of "weak" sectors' borders and quantitative estimation of various geotechnical indexes (properties of soil, pile bearing capacities, etc.). Nevertheless, geotechnical parameters determined by CPT are always approximate, necessary to be additionally adjusted or, at least, compensated by introduction of rational "reserves". Correction with regard to the specified site becomes possible when combining CPT with any "accurate" method with the results being regarded as the "reference" one (plate test of soil, static tests of piles, sometimes laboratory tests, etc.). Otherwise the computations shall contain "reserves" aimed at the least beneficial cases, i.e. at maximum "errors" possible in a wide range of conditions.

It seems reasonable to consider the most typical example, i.e. determination of the allowable load on the pile by CPT data in conducting static tests of piles at the site (method of "key sectors"). In here, CPT locations are located more or less evenly along the territory of the designed structure, and at some (the most typical) plots – "key sectors" static tests of piles are carried out (fig.6.2).

Ultimate pile bearing capacity is determined in every CPT location $F_{u,CPT,1}$, $F_{u,CPT,2}$, $F_{u,CPT3}$, ... $F_{u,CPT,n}$, and at every (k) "key sector" comparisons between pile bearing capacities evaluated by CPT data $F_{u,CPT,k}$ and the capacities obtained by static tests of in-situ piles $F_{u,pile,k}$ are made, i.e. $k'_k = F_{u,CPT,k}/F_{u,pile,k}$ are evaluated (figure 6.2 illustrates two key sectors, i.e. two particular values of k'_1 & k'_2 have been obtained). The obtained particular values of k'_k help evaluate the correction factor to approximate value of average pile bearing capacity $\overline{F}_{u,30H}$ calculated for this site. The mode of the correction factor obtaining is presented below.

---

[1] In fig. 3.17 the compressive modulus of deformation is marked as the original one – "M" (not E_c)

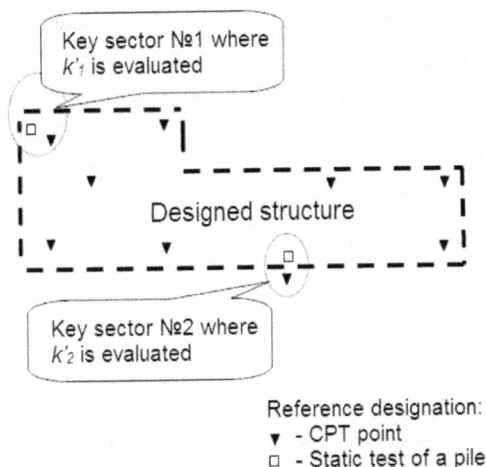

Fig. 6.2 Possible CPT locations and adjusting static tests of piles (in "key sectors" method)

Reference designation:
▼ - CPT point
□ - Static test of a pile

If divergences between the accurate and the approximate values of pile bearing capacities ($F_{u,CPT}$ & $F_{u,pile}$) were stable at the specified site, not any problems would occur: k'_k would be the same (only one "key sector" would be needed) and the correction factor to the average approximate value of $\overline{F}_{u,CPT}$ would equal to k'_k. Unfortunately, in practice the ratios $F_{u,CPT}/F_{u,pile}$ may greatly vary at one site. Apparently, the range is narrower than that one presented in the histogram of k-values (see fig. 6.1), but variability of $F_{u,CPT}/F_{u,pile}$ at the local plots is quite high, thus one must not ignore it.

Figure 6.3 illustrates the results of relative "errors" changeability assessment $k = F_{u,CPT}/F_{u,pile}$ at sites of different sizes ($F_{u,CPT}$ – ultimate pile bearing capacity by CPT data, $F_{u,pile}$ – the same by static tests of piles) (Ryzhkov 1992). The data obtained in Bashkir Urals and Tyumen Region are used. Soil conditions were characterized by alluvial, lacustrine-alluvial and dealluvial deposits composed mainly of clayey soils (sometimes with sand layers). Clayey soils were from underconsolidated to semi-solid, sands – of medium density from medium coarseness to fine ones. From 6 to 12 driven piles of the same length and sections were tested at each site; CPT was carried out nearby (in the distance of 1...1.5 m). The piles were of 0.3×0.3 m in section, 6...12 m – in length. CPT was carried out by S-832 CPT rig, computations of $F_{u,CPT}$ were done in accordance with the National Building Code SNiP 2.02.03-85*.

As one can see from fig. 6.3, variations of relative "errors" are less considerable than those of the histogram in fig. 6.1. If standard deviation of such "errors" in the general histogram is $\sigma = 0.31$, it is $\sigma_k \leq 0.25$ in the whole of cases of the examined sites (fig. 6.3). In here, the larger the site, the higher is σ_k (approximate curve "$\sigma_k \sim A$" is shown in dashed line in fig. 6.3). For sites of the most typical sizes – 1000...5000 $m^2 - \sigma_k \approx 0.2$.

314

Mean values of $\bar{k}$ are also different from the mean value of the histogram in fig. 6.1, i.e. from $\bar{k}_{tot} = 1.02$. Here, one can fail to observe any dependence of $\bar{k}$ on the sizes of the sites in the number of data available.

If to analyze σ_k, i.e. averaged dispersion of k in relation to mean local $\bar{k}$ (in separate sites) as *intragroup*, dispersion of the mean ones – as *intergroup* $\sigma_{\bar{k}}$, and σ, i.e. dispersion of the histogram in fig. 6.1 – as *total*, one can approximately take (Gmurman 2000)

$$\sigma^2 = \sigma_k^2 + \sigma_{\bar{k}}^2 . \qquad (6.1)$$

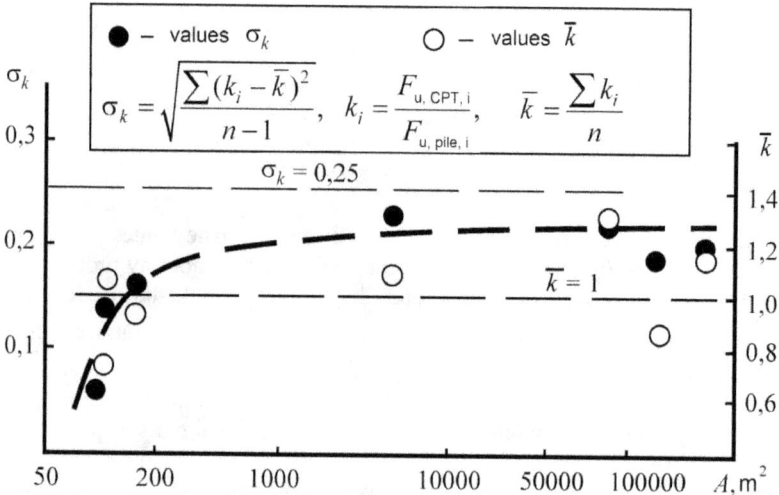

Fig. 6.3 Schematic illustration of changeability (standard deviations) and mean values of relative "errors" in pile bearing capacity evaluation by CPT data at sites of different sizes:

A – area of the site, k_i, $\bar{k}$ – instant and mean values of relative "errors" in the limits of the site; σ_k – standard deviation of relative "errors" k_i в in the limits of the site; $F_{u,CPT}$ – ultimate pile bearing capacity by CPT data; $F_{u,pile}$ – the same by static tests of piles; n – values of k_i, obtained at the site (i.e. the number of comparisons between $F_{u,CPT}$ and $F_{u,pile}$)

If to consider total standard deviation of "errors" to be $\sigma = 0.31$ (fig. 6.1), intragroup deviation to be $\sigma_k = 0.2$ (fig. 6.3), intergroup deviation will be $\sigma_{\bar{k}} = \sqrt{\sigma^2 - \sigma_k^2} = \sqrt{0.31^2 - 0.2^2} = 0.24$ then. It means that at the sites of 1000...5000 m² in size the mean local "error" $\bar{k}$ may differ from the total one $\bar{k}_{tot} = 1.02$, but the

315

range of its possible values will be 1.3 times lower than that of particular "errors" in the histogram in fig. 6.1 (0.31/0.24 = 1.3). At the separate sites the range of particular "errors" k alterations in relation to their (local) mean values of $\bar{k}$ will be 1.55 times lower than "total" range of their alteration (fig. 6.1) (0.31/0.20 = 1.55).

This fact allows obtaining of distribution (histograms) of mean "errors" at such sites approximately by general redistribution of particular values of k, given in figure 6.1. The redistribution is easily carried out if to assume that all the mentioned distributions agree with the same law being different just in their dispersions. In here, the traditional approach may be applied; it is based on acceptance of one or the other distribution law (here, asymmetrical one, e.g. lognormal) and making the histograms with new dispersions, i.e. σ_k or $\sigma_{\bar{k}}$ instead of σ. However, the solution of the problem is possible ignoring the approximation of distribution by the analytical law. Having reduced the width of histogram intervals in σ_k/σ or $\sigma_{\bar{k}}/\sigma$ times, it seems possible to regroup these intervals in the former width keeping the same center of distribution and, thus to obtain new histograms with the parameters of dissipation corresponding to σ_k or $\sigma_{\bar{k}}$ (Ryzhkov 2000).

It is important to underline that

$$\bar{F}_{u,CPT} \Big/ \bar{F}_{u,pile} \approx \overline{(F_{u,CPT}\big/F_{u,pile})} \quad , \tag{6.2}$$

i.e. the relation of mean approximate pile bearing capacity $\bar{F}_{u,CPT}$ to mean "accurate" $\bar{F}_{u,pile}$ nearly equals to mean value of particular ratios $(F_{u,CPT}/F_{u,pile})_i$, since mathematical expectation of the function nearly equals to the function from mathematical expectations of arguments. Therefore, mean value of ratios $(F_{u,CPT}/F_{u,pile})_i$, i.e. mean "error" $\bar{k}$ may be used as the correction coefficient to mean value of $\bar{F}_{u,CPT}$ at the site ("correction zones"). The correction may be considered as the specified reliability coefficient γ_k, which takes into account the accuracy of pile bearing capacity evaluation method in accordance with p. 3.10 of the Building Code SNiP 2.02.03-85*

$$\bar{F}_{u,CPT} = \frac{\bar{F}_{u,CPT}}{\bar{k}} = \frac{\bar{F}_{u,CPT}}{\gamma_k}, \tag{6.3}$$

If to divide $\bar{F}_{u,pile}$ into the reliability coefficient of soil γ_g (SNiP 2.02.03-85*, GOST 20522-96), the parameter obtained will be the admissible design load on a pile which shall not be lower than the actual design load N.

$$F = \frac{\bar{F}_{u,CPT}}{\gamma_k \cdot \gamma_g} \geq N, \tag{6.4}$$

Thus, the problem on data acquisition of pile bearing capacity evaluation by CPT data (in availability of static tests of piles) may be formulated in the following way. Probabilities of *possible* mean "errors" at the given site $\bar{k}$ are known, probabilities of *possible* deviations of particular values of k_i from the mean $\bar{k}$ either by one or several "accurate" values of k_i are known as well (at "key" sectors). It is necessary to choose the value of $\bar{k}$ which can be used as the correction coefficient to the calculated pile bearing capacity $\overline{F}_{u,CPT}$ obtained by CPT. It is the correction coefficient that will be the reliability coefficient γ_k in the formula (6.4) with regard to the given site (i.e. $\bar{k} = \gamma_k$).

In 1980s BashNIIstroy developed the technique of this correction coefficient selection applying Bayes formula (rule) as a mathematical model (Ryzhkov 1988, 1995). The formula allows revaluing the probabilities of various assumptions – hypotheses $p(H_i)$ after instant completion α (Gmurman 2000). For them, their conditional probabilities are to be known $p(\alpha/H_i)$, i.e. probabilities of their completion given validity of every "n" in alternate hypotheses H_i ($i = 1 \dots n$).

$$p(H_i / \alpha) = \frac{p(H_i)p(\alpha / H_i)}{\sum\limits_{j=1}^{j=n} p(H_j)p(\alpha / H_j)} \,, \tag{6.5}$$

where:

$p(H_i), p(H_i/\alpha) =$ validity probabilities of i-hypothesis before and after instant completion α ("a priori" and "a posteriori" probabilities);

$p(\alpha/H_i) =$ probability of instant completion α given the validity of i-hypothesis.

In computation of the admissible loads on piles, possible values of *mean* "errors" of the required parameter assessment at the given site ($\bar{k}$) are considered as "the hypotheses H_i "in the formula (6.5). Table 6.1 illustrates these hypotheses and their probabilities obtained by mentioned transformation of the total distribution (histograms) of "errors" k (fig.6.1), i.e. by compression of the mentioned histogram in 1.3 times.

Table 6.1 Hypotheses and their probabilities in "errors" of pile bearing capacity evaluation by CPT data

Hypotheses H_i	H_1	H_2	H_3	H_4	H_5	H_6	H_7	H_8
Values $\bar{k}$	0.7	0.8	0.9	1.0	1.1	1.2	1.3	1.4
Validity probabilities of hypotheses, H_i, i.e. $p(H_i)$	0.1	0.17	0.20	0.18	0.15	0.10	0.06	0.04

The results obtained at the "key" sectors, i.e. values of k'_k are considered as the "instant α ". Table 6.2 shows possible deviations of k'_k от from the mean ones of $\bar{k}$ (at the site) obtained by the same transformation (by compression of the mentioned histogram in 1.55 times). The difference $k'_k - \bar{k}$ corresponds to the situation when "error" k'_k, is obtained at k-key sector, while the true mean "error" is $\bar{k}$ at this site. Thus, probabilities $p(k'_k - \bar{k})$ correspond to the conditional probabilities $p(\alpha/H_i)$ where H_i – hypothesis that mean "error" equals to $\bar{k}$. For example, if we obtain $k'_k = 0.9$ at k-key sector, then in validity of the hypothesis H_1 (i.e. $\bar{k} = 0.7$)

$$k'_k - \bar{k} = 0.7 - 0.9 = -0.2$$

corresponds to this k'_k obtaining.
According to table 6.2 the conditional probability $p(\alpha/H_i) = p(k = 0.9/\bar{k} = 0.7)$ will be 0.17. In validity of the hypothesis H_2 (i.e. $\bar{k} = 0.8$) it will be 0.23, etc.

Table 6.2 Possible deviations of "errors" k'_k from the mean values of $\bar{k}$ at the sites

Deviations $k'_k - \bar{k}$	-0.3	-0.2	-0.1	± 0.0	$+0.1$	$+0.2$	$+0.3$
Deviation probabilities $p(k'_k - \bar{k})$ corresponding to $p(\alpha/H_i)$	0.09	0.17	0.23	0.21	0.15	0.09	0.06

Thus, tables 6.1 and 6.2 contain all the necessary data to do calculations by the formula (6.1) in obtaining any value of k'_k at the key sector. These tables show quite low accuracy of calculations (up to 0.1), but if higher accuracy is required, one needn't do any principal changes: the number of hypotheses increases, and hence computation is done applying software. Table 6.3 presents reliability coefficients γ_k, calculated with the accuracy of 0.01 applying "BAYES". As in other tables of this chapter, they presuppose calculations of driven pile bearing capacity by CPT data according to the Building Code SNiP 2.02.03-85 or SP 50-102-2003.

Mean value of pile bearing capacities is adjusted being evaluated by CPT data (see formula 6.4). Other parameters used for selection of design (admissible) load on the pile are evaluated in the same way as in lack of correction: normative pile bearing capacity $\bar{F}_{u,CPT}$ is evaluated as mean value of pile bearing capacities F_{CPT1}, F_{CPT2}, ...etc. obtained by CPT data; reliability coefficient of soil γ_g is determined after Standard GOST 20522-96, i.e. by static treatment of the same values of F_{CPT1}, F_{CPT2}, ... etc.

The approach may be applied in adjustment of either pile bearing capacities or any soil property capable to be determined by CPT data. Here, only initial distributions (histograms) showing the "errors" of the required soil property determination alter. Selection of CPT locations, location of "key" sectors and calculation technique are invariable. Table 6.4 gives reliability coefficients γ_k, to evaluate the modulus of soil deformation E by CPT data with adjustment by stamp test (in one "key" sector). It is believed that moduli of deformation obtained by CPT data acquisition in accordance with SP 11-105-97 are adjusted. Reliability coefficient γ_k is introduced as a divisor in relation to mean value of E evaluated by CPT data.

Table 6.3 Reliability coefficients γ_k determined with regard to values of $k'_k = F_{CPT}/F_{test}$ obtained at "key" sectors

a) given one key sector (one static test of a pile):

k'	≤ 0.7	0.8	0.9	1.0	1.1	1.2	1.3	1.4
γ_k	0.94	1.03	1.11	1.19	1.26	1.30	1.32	1.33

б) given two key sectors (two static tests of piles):

k'_1	Reliability coefficient γ_k in k'_2, equal to:							
	≤ 0.7	0.8	0.9	1.0	1.1	1.2	1.3	1.4
0.7	0.92	0.93	0.93	0.95	0.96	0.96		
0.8	0.94	1.00	1.02	1.04	1.05	1.06	1.15	
0.9	0.95	1.02	1.07	1.09	1.13	1.15	1.15	1.16
1.0	0.95	1.04	1.12	1.15	1.18	1.20	1.24	1.25
1.1	0.96	1.05	1.13	1.18	1.23	1.26	1.28	1.26
1.2	0.96	1.06	1.15	1.20	1.24	1.28	1.30	1.30
1.3		1.06	1.15	1.24	1.28	1.30	1.31	1.32
1.4			1.16	1.25	1.30	1.32	1.32	1.33
1.5				1.26	1.30	1.32	1.33	1.34

Table 6.4 Reliability coefficients γ_k recommended for evaluation of modulus of soil deformation by CPT data (E_{CPT}) after SP 11-105-97 with adjustment by stamp test (E_{st})

"Error" at the key sector $k' = E_{CPT}/E_{st}$	≤ 0.6	0.7	0.8	0.9	1.0	1.1	1.2	1.3	1.4
Reliability coefficient γ_k	0.78	0.84	0.90	1.02	1.14	1.18	1.27	1.32	1.36

As opposed to the traditional approach, when minimal or average result of the "accurate" test of soil (stamp or pile) is taken as the estimate indicator with the corresponding reliability coefficients, in the analyzed approach the results of the "accurate" tests serve as the *adjustment medium* of approximate estimates characterizing the site as it is. The estimate indicator is chosen after generalization and adjustment of the approximate indexes, i.e. the results obtained by CPT data.

Nevertheless, the role of the "accurate" tests does not decrease, since their effect happens to be much stronger than the accuracy of the applied empirical formulae. It could be seen from the following example.

Let us assume that stamp soil test has been carried out at the "key" sector and the accurate value of the modulus of deformation has been obtained: $E_{st} = 18$ MPa. In here, cone resistance has been equal to $q_c = 2$ MPa. Average resistance of soil to

CPT has been $\bar{q}_c = 3.1$ MPa (in the same engineering-geological element where the stamp test has been carried out).

Let us consider the result of modulus of deformation E estimated value determination (it is assumed that $E = \bar{E}_{CPT}/\gamma_k$) in application of three different formulae: $E = 5q_c$, $E = 7q_c$ and $E = 10q_c$.

Table 6.5 presents the reliability coefficients γ_k corresponding to each of the formulae. Values of γ_k have been calculated by the mode described earlier. The results of the moduli of deformation E obtained are given in table 6.6.

As one can see from table 6.5, module of deformation $\bar{E}_{CPT}$ significantly differ when ignoring the adjustment dependent on the formulae applied (15.5...31.0 MPa). The adjustment has reduced the effect of the formulae so much that the whole of module of deformation E proved to be almost the same (23.1...24.4 MPa).

Table 6.5 Reliability coefficients γ_k corresponding to different formulae $E = aq_c$ to evaluate the modulus of soil deformation

"Error" at the key sector $k' = E_{CPT}/E_{st}$	0.4	0.6	0.7	0.8	0.9	1.0	1.1	1.2	1.3	1.4	1.6
Reliability coefficient γ_k in application of the formulae											
$E = 5q_c$	0.54	0.72	0.80	0.87	0.88	0.89					
$E = 7q_c$		0.78	0.84	0.9	1.02	1.14	1.18	1.27	1.32	1.36	
$E = 10q_c$						1.2	1.27	1.38	1.45	1.52	1.52

Table 6.6 Calculations of modulus of deformation in the presented example

Formula	E_{CPTk} at the "key" sector (MPa)	"Error" at the key sector $k' = E_{CPT}/E_{st}$	Reliability coefficient γ_k (by table 6.4)	Average resistance $\bar{q}_c$ (MPa)	Average modulus of deformation ignoring the adjustment $\bar{E}_{3OH}$ (MPa)	Calculated modulus of deformation (after the adjustment) E (MPa)
$E = 5q_c$	10	0.56	0.67	3.1	15.5	23.1
$E = 7q_c$	14	0.78	0.89		21.7	24.4
$E = 10q_c$	20	1.11	1.27		31.0	24.4

The experimental test of the approach applicability (method of "key" sectors) was carried out in Ufa in 1980s. BashNIIstroy carried out a number of static tests of driven piles parallel with CPT (both in the locations of driving and in the remaining terrain as well).

On BashNIIstroy experimental site mentioned in section 4.3.2, 11 piles 0.3×0.3 m in section, 6 m in length were tested at the sector of nearly 150 m² in area, CPT being carried out applying S-832 CPT rig at every location. The piles were located in staggered order, the distance between them being 2...3.2 m. Up to the depth of 15 m the soils were composed of alluvial-dealluvial clays ranging from overconsolidated to semi-solid ones without any alien interlayers or lentils. The plan of the test sector is shown in fig. 6.4. The results of tests were given in section 4.3.2 (table 4.12).

The admissible loads on the pile were evaluated by means of the method of key sectors, all possible variants being considered:

- Adjustment after the result of one pile test only (the results of each pile test were considered by turn, i.e. it was thought that only one pile had been tested in each variant) – eleven variants.
- Adjustment after the results of two pile tests, i.e. in two key sectors (the results of each "pair" of piles tests were considered by turn in all possible combinations) – 55 variants.
- adjustment after the results of three tests, i.e. in three key sectors (the results of each "triplet" of piles tests were considered by turn in all possible combinations) – 165 variants.
- ... etc. (up to 11 tests).

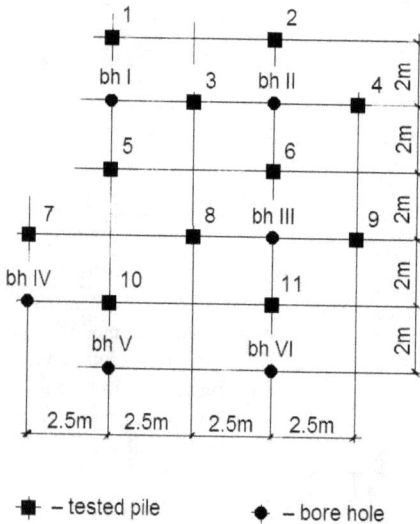

Fig.6.4 Schematic illustration of tested piles and bore holes in experiment of BashNIIstroy

Figure 6.5 illustrates the results of calculations of the admissible loads on a pile in different number of adjustments. Here, one can also see the results of calculations of these loads in traditional approach (Building Code SNiP 2.02.03-85*), i.e. when pile bearing capacity F_d is

- The *least particular value* of pile bearing capacity obtained in tests $F_d = F_{u,pile,min}$ in less than six static tests (i.e. reliability coefficient of soil γ_g equals to 1).

321

- The *least anticipated average* pile bearing capacity in six or more tests $F_d = \overline{F}_{u,pile,min} = \overline{F}_{u,pile}/\gamma_g$.

In both cases the reliability coefficient is $\gamma_k = 1.2$.

As one can notice from fig. 6.5, method of key sectors provide with both reliable and cost-effective solutions. It has not been any test when the actual mean value $\overline{F}_{u,pile} = 257$ kN has exceeded even the anticipated minimum value of average resistance $\overline{F}_{u,pile,min} = \overline{F}_{u,pile}/\gamma_g = 257/1,08 = 238$ kN. This is related to such cases when only one test of piles is being carried out.

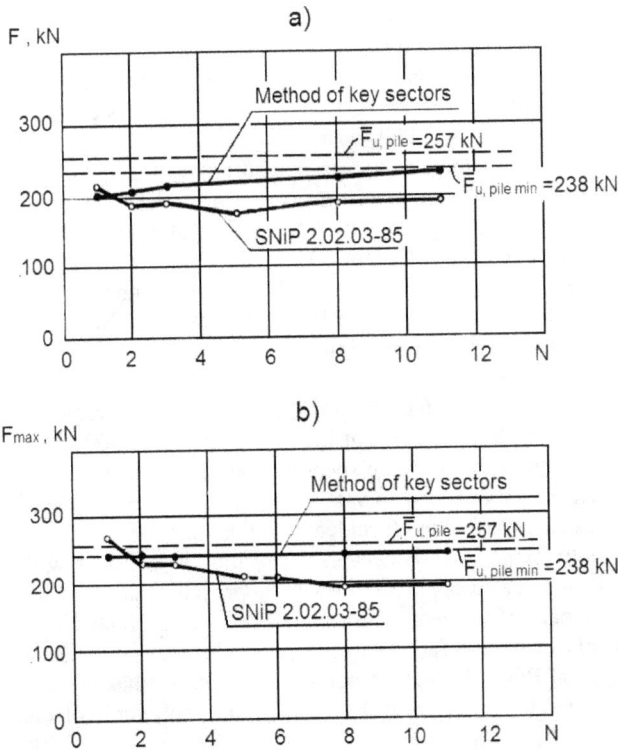

Fig. 6.5 Schematics of dependence of specified load on a pile F upon the number of static tests of piles (key sectors) N, which helped the adjustments to be done (BashNIIstroy experimental site, 150 m²):

a – mean values of specified admissible load on piles F according to combinations of the tested piles (one, two, three piles, etc., i.e. in $N = 1, 2, 3...11$), b – maximum values of pile bearing capacities F after the same variants

According to the Building Codes SNiP 2.02.03-85* and SP 50-102-2003 two static tests of piles (if there are any) are to be performed at each specified depth on the territory of the erected structure. In practice, the Code is often broken due to lack of logic concerning the prohibition to perform one static test, while the absence of these tests is considered to be allowable. For this reason, figure 6.5 illustrates the bearing capacity evaluated after a single test. As one can see from fig. 6.5b, in the single test pile bearing capacity proves to be higher than that of the average one obtained $\bar{F}_{u,pile}$ (and even the minimum average $\bar{F}_{u,pile,min}$). It is the reliability coefficient $\gamma_k = 1.2$ that can compensate this excess; it is used in transition from the bearing capacity to specified load on the pile (SNiP 2.02.03-85*). In the method of key sectors the coefficient is specified differentially for every site due to the parameter $k'_k = F_{u,CPT,k}/F_{u,pile,k}$ obtained at the key sector.

The analogical computation was done for the site of much larger sizes (150×200 m); here, 9 piles 0.3×0.3 m in section, 12 m in length were tested (Ufa, neighborhood "L"). The site was located on high-water bed of the river Belaya characterized by plane relief with quite homogeneous (in spreading) soils. Up to 3...8 m, the soils were composed of alluvial clays and loams (overconsolidated and underconsolidated) with interlayers and lentils of pulverescent water-saturated sands. Then, up to 12...15 m alluvial pulverescent sands with lentils of underconsolidated clays were deposited, lower – gravel deposits with sandy filling. The depth of ground waters was 3...5 m.

The calculations of the admissible loads were done in the same way as for the experimental site of BashNIIstroy. Figure 6.6 shows the results of calculations as given in fig. 6.5.

As one can see from fig. 6.6, the principal regularities revealed at the small site (150 m^2) (fig. 6.5) show themselves at the larger site (150×200 m). Apparently, the example is a little bit artificial, since specification of the unified bearing capacity F_d (admissible load F_d/γ_k) on such a vast territory is not appropriate due to either economic or technical reasons (accidental deviations of pile bearing capacities become extremely dangerous). However, the results obtained show that there is not any principal difference in application of the approach at large or small sites. It is necessary to pay particular attention to rational division of the territory of the erected structure into zones where the admissible load on piles may be the same.

Application of Bayes formula is not new. For years it has been used in different spheres (Morris 1971, Zelner 1980, Hey 1987). In resolving geotechnical problems the approach has been applied in non-Russian practice since 1970s...1980s (Kay 1977, Cividini et al. 1983, Baecher & Rackwitz 1982). Probabilities of hypotheses are approximated by one or the other classical laws of distribution – normal, lognormal, etc. However, the technique allows simpler approach to be applied; in here, one can perform operations distributing the "errors" without compulsory specification of their analytical law (Ryzhkov 1992, 2000). Hence, distributions are considered as discrete, truncated in the form of histograms or tables. If necessary, "leveling" of the histograms apart from approximation by analytical laws may be done (Ryzhkov 2000). This simplifies mathematical calculations, but given the "errors" assessment accuracy over 0.1, calculations are to be done applying software ("BAYES") or tables (e.g. 6.3, 6.4) prepared in advance. It is important to underline that introduction of "errors" as the truncated distributions excludes the situations when the design parameter decreases with the increase of the number of tests. If to apply analytical distributions, the paradoxical situations are possible to occur.

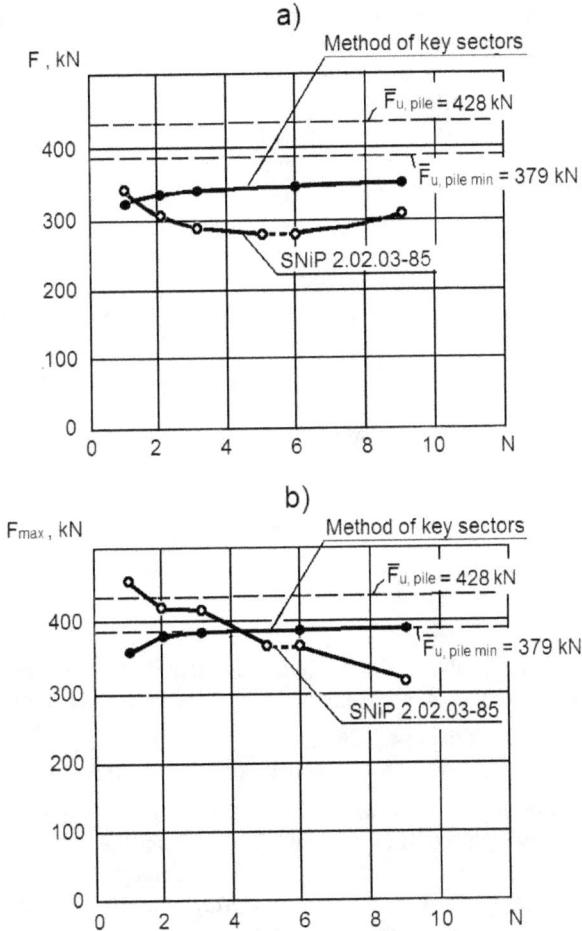

Fig. 6.6 Schematics of dependence of specified load on a pile F upon the number of static tests of piles (key sectors) N which helped the adjustments to be done (neighborhood "L", Ufa, sector150×200 m):

a – mean values of specified admissible load on piles F according to combinations of the tested piles (one, two, three piles, etc., i.e. $N = 1, 2, 3...9$), b – maximum values of pile bearing capacities F after the same variants;

6.3 CPT APPLICATION WITH OTHER INDEPENDENT SOIL INVESTIGATION TESTS

Section 6.2 deals with CPT application with only one "accurate" test with further adjustment of the results obtained. However, there are situations when several approximate methods (types) of tests are used, "accurate" tests necessary for the adjustment to be carried out may be absent. For example, besides CPT one can perform drilling, select monoliths subjected to laboratory tests or carry out dynamic probing tests, rotary shear test of soil, etc. As it was mentioned in section 6.1, incomplete use of the information obtained is typical for these cases; it is due to traditional orientation towards any one method. Attempts have been taken to abandon the traditional approach passing to the unified complex of "unequal measurements" (Rodionov & Ivanov 1967, Rats 1973, Guidance 1977, etc.). Here, the design parameter is understood as weight average parameter of results obtained by various tests, reliability of each method being considered by corresponding "weight coefficients". The coefficients may be obtained by the techniques applied in Bayes analysis of continuous random quantities being normally distributed (Hey 1987, Zelner 1980). A number of solutions to the problem of average assessment a posteriori distribution finding of the required index in unequal measurements are known. The "weight" of each method of measurements in such solutions is usually inversely proportional to dispersion of results of the method. In the simplest case it is as follows:

$$\overline{A} = \frac{\sum \overline{A}_i (n_i / \sigma_i^2)}{\sum (n_i / \sigma_i^2)},$$
(6.6)

where:

$\overline{A} =$ maximum probable mean value of the required parameter A;

$\overline{A}_i =$ mean value of A according to the results of i-method;

$\sigma_i =$ dispersion of data (A_i) obtained by i-method;

$n_i =$ number of A evaluated by i-method (quantity of A_i).

In more careful geotechnical research (Rodionov & Ivanov 1967, Rats 1973, Guidance 1977) "weight coefficients" take into account the errors of different origin differentially, e.g. the ones connected with inaccuracy of the applied empirical dependencies or heterogeneity of soils. Corrections to "systematic deviations" characteristic for specified test methods can be introduced.

Drawbacks of such calculations are due to conditionality of the mathematical models. Therefore, one has to apply the normal law of "errors" distribution (or bring asymmetrical distributions to it artificially), allow the required index to be reduced in obtaining available information. As a rule, reduction of the range of "errors" is not considered in limited conditions (at small sites).

325

Bayes approach ensures wider possibilities, as variables applied are considered as the *discrete* random parameters. There appears the possibility to do operations with any distributions, including the truncated ones (see section 6.2).

In occurrence of several approximate test methods another variant seems to be appropriate (different from that one discussed in 6.2). It is as follows:

- It is the average index that is evaluated resulting from measurements and calculations, but not the correction coefficient to the average index.

- Under "errors" of the required index evaluation we understand the differences between the approximate values and the "accurate" ones, but not their ratios (e.g. instead of the parameters E_{CPT}/E_{st}, $F_{u,CPT}/F_{u,pile}$ (see section 6.2), the parameters $E_{CPT} - E_{st}$, $F_{u,CPT} - F_{u,pile}$) are used.

The same formula is used (6.5), but probable *mean values* of the index itself (e.g. of modulus of deformation $\bar{E}$ or pile bearing capacity $\bar{F}_u$, etc.), but not the probable average "errors" are taken as "hypotheses H_i". The number of hypotheses (the same as in problems discussed in sec. 6.2) is determined by the desired accuracy of the required index (e.g. in accuracy of the modulus of deformation ± 1 MPa, the number of hypotheses will be 10 times lower than in accuracy of $\pm 0,1$ MPa).

"Instant α", in which "a priori" probabilities $p(H_i)$ are recalculated into "a posteriori" $p(H_i/\alpha)$, is considered to be the result of any test carried out at the investigated site and being ignored in preceding calculations. Conditional probabilities $p(\alpha/H_i)$, i.e. the probabilities of "instant α" accomplishment given the validity of hypotheses H_i ($i = 1...n$) will show both the reliability of the specified test method and heterogeneity of soil (changeability of its properties in the site). Consideration of both factors may be realized by transformation of histograms with the methods discussed in section 6.2. The histogram is drawn, i.e. the one reflecting the effect of one of the factors – reliability of the method, then transformation of the histogram into larger dispersion (i.e. its "expansion") is done, and thus the effect of the other factor – heterogeneity of soil is taken into account.

Calculation is done in the following way.

First, distribution of probable approximate values of the required index $A_{approx.}$is drawn on condition that its true value is equal to A_i (to be more precise, it is in the interval of $A_i \pm a/2$, where a – range of values of A). Figure 6.7 illustrates the mode of these distributions drawing.

The whole of the range of A alteration is divided into the zones A_1, A_2, A_3 ...with the displacement amplitude a; distribution of the approximate values of $A_{approx.}$ are evaluated for each zone in the same way. For simplification it is allowable to take the identical distributions in all the hypotheses. It seems possible when minimum values of $A_{approx.}$ do not exceed half of the range of their alteration (otherwise, one has to deal with the negative values of the required index).

Standard deviation $\sigma_{heter.}$, showing the effect of heterogeneity of soils may be evaluated by particular values of the required index in different locations of the site:

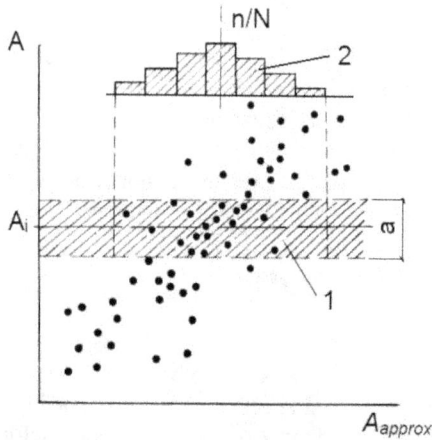

Fig. 6.7 Diagram explaining methods of conditional probability evaluation of approximate values of $A_{approx.}$ in true value of A_i:

1 – zone of considered true values of the required index $A_i \pm a/2$ corresponding to i-hypothesis (H_i), 2 – histogram showing distribution of $A_{approx.}$ in the marked zone of true values $A_i \pm a/2$, i.e. these are the relative frequencies of $A_{approx.}$ given $A = A_i \pm a/2$, approximately corresponding to conditional probabilities $p(a/H_i) = p(A_{approx.,j}/A=A_i)$

$$\sigma_{heter} = \sqrt{\frac{\sum (A_{approx,i} - \overline{A}_{approx})^2}{n-1}}, \tag{6.7}$$

where:

$A_{approx,i}, \overline{A}_{approx}$ = current and mean values of A_{approx};

$n =$ number of locations at the site where $A_{approx,i}$ were evaluated.

It is evident that the approach is approximate, since deviations of pile bearing capacities in the limits of the site show not only heterogeneity of soil but the reliability of pile bearing capacity calculation as well. On the other hand, diagrams of dissipation of the approximate and "accurate" evaluations also reveal both their reliability and heterogeneity of soils (see section 3.2.1). However, the errors caused by these assumptions are taken as a safety factor when choosing the allowable loads on piles.

Standard deviation showing the effects of both factors is evaluated with regard to the properties of dispersions

$$\sigma = \sqrt{\sigma_{rel}^2 + \sigma_{heter}^2} \, , \qquad (6.8)$$

where σ_{rel}^2 & σ_{heter}^2 = dispersions showing reliability of the required index evaluation and heterogeneity of soils correspondingly.

Then, in accordance with the ratio $k_o = \sigma/\sigma_{rel}$ transformation ("expansion") of histograms is carried from σ_{rel} (see fig. 6.6) to σ. As it was mentioned in 6.2, the transformation is in expansion of the histogram intervals in k_o with further regrouping of the "expanded" intervals into the former range (i.e. the number of intervals-"hypotheses" becomes larger). The methods of such transformation are described in (Ryzhkov 2000). These operations are performed by means of software (e.g. SYNT applied in BashNIIstroy). It is evident that transformation of histograms (tables) may be carried out in the traditional way, i.e. by application of the analytical law of distribution and drawing of distribution for new dispersion as well. However, here the procedure of analytical law selection is added and applications of "truncated" distributions are significantly complicated.

In general, calculations prove to be analogous to those described in section 6.2.

The accuracy of calculation when applying SYNT is usually ±1 kN; this presupposes (in application of the diagram of dissipation in fig. 4.10) recording of nearly a thousand of hypotheses. However, in order to characterize the algorithm of this software it seems suitable to consider the principal elements of calculation with the reduced accuracy, e.g. ±25 kN, in which the principal calculations are done "by hand". Below is given an example of such a calculation, i.e. the basic fragments of calculations.

Tables 6.7, 6.8 and 6.9 contain relative frequencies of different "errors" in obtaining pile bearing capacities by the data of three "standard" methods (SNiP 2.02.03-85*):

- Calculations by CPT data $F_{u,CPT}$ (in accordance with the diagram of dissipation, fig. 4.10).
- Calculations by physical-mechanical properties of soils $F_{u,lab}$ (by liquidity index I_L of clays or coarseness and density of sands) (Ryzhkov 1992).
- Dynamic tests of piles $F_{u,dyn}$ (calculations by "sets per blow") (Ryzhkov 1992).

Table 6.7 Relative frequencies n/N ("probabilities") of divergences between particular values of pile bearing capacity by CPT data $F_{u,CPT}$ and its mean "accurate" value $\overline{F}_u$ by fig. 4.10 (in ideally homogeneous soils)

$F_{u,CPT} - \overline{F}_u$ (kN)	-200...-250	-150...-200	-100...-150	-50...-100	0...-50	0...+50	+50...+100	+100...+150	+150...+200
$n/N \approx p(F_{u,CPT} - \overline{F}_u)$	0.06	0.09	0.10	0.12	0.16	0.18	0.13	0.10	0.06

328

Here, under the "errors" we understand the differences between approximate pile bearing capacities $F_{u,CPT}$, $F_{u,lab}$, $F_{u,dyn}$ and their "accurate" values of F_u, i.e. ($F_{u,CPT} - \overline{F}_u$), ($F_{u,lab} - \overline{F}_u$), ($F_{u,dyn} - \overline{F}_u$).

Table 6.8 Relative frequencies n/N ("probabilities") of divergences between particular values of pile bearing capacity by laboratory results of soil analysis $F_{u,lab}$ and its mean "accurate" value $\overline{F}_u$ (in ideally homogeneous soils (Ryzhkov 1992))

$F_{u,lab} - \overline{F}_u$ (kN)	-250...-300	-200...-250	-150...-200	-100...-150	-50...-100	0...-50	0...+50	+50...+100	+100...+150	+150...+200	+200...+250
$n/N \approx p(F_{u,lab} - \overline{F}_u)$	0.05	0.05	0.07	0.09	0.12	0.13	0.15	0.14	0.09	0.06	0.05

Table 6.9 Relative frequencies n/N ("probabilities") of divergences between particular values of pile bearing capacity by dynamic tests of piles $F_{u,dyn}$ and its mean "accurate" value $\overline{F}_u$ (in ideally homogeneous soils (Ryzhkov 1992))

$F_{u,dyn} - \overline{F}_u$, kN	-400...-450	-350...-400	-300...-350	-250...-300	-200...-250	-150...-200	-100...-150	-50...-100	0...-50	0...+50	+50...+100	+100...+150	+150...+200
$n/N \approx p(F_{u,dyn} - \overline{F}_u)$	0.05	0.07	0.07	0.08	0.12	0.13	0.10	0.09	0.07	0.07	0.06	0.05	0.04

Relative frequencies of "errors" given in tables 6.7, 6.8 and 6.9 nearly equal to their conditional probabilities in "true" value of mean pile bearing capacity $\overline{F}_u$. They correspond with $p(\alpha/H_i)$ in the formula (6.5) in ideally homogeneous soil at the site. In real heterogeneity of soils divergences between approximate pile bearing capacities obtained and their *mean* accurate values will be greater than in the tables. For this reason, distribution of "errors" shall be transformed into larger dispersion, i.e. it is necessary to carry out "expansion" of the histogram in σ/σ_{rel} times.

Table 6.10 shows "expanded" distribution of "errors" when the dispersion revealing heterogeneity of soils is $\sigma^2_{heter} = 4900$ kN^2 ($\sigma_{heter} = 70$ kN).

Dispersion of "errors" revealing reliability of calculations (table 6.7) has been $\sigma^2_{rel} = 12374$ kN^2 ($\sigma_{rel} = 111.2$ kN)," total" standard deviation of heterogeneous soil (in $\sigma^2_{heter} = 4900$ kN^2) is as follows (formula 6.7):

$$\sigma = \sqrt{12374 + 4900} = 131.4 \text{ kN.}$$

Correspondingly, $\sigma/\sigma_{rel.} = 131.4/111.2 = 1.18$, i.e. distribution of the table 6.7 "has been expanded" in 1.18 times (by means of SYNT). Values of $\overline{F}_u$ corresponding with the accuracy of the tables or histograms applied are to be considered the hypotheses (table 6.10 presupposes the accuracy to be ±25 kN). The following are considered to be the examples:

hypothesis $H_1 \rightarrow \overline{F}_u = 0...50$ kN ($\overline{F}_u = 25$);

hypothesis $H_2 \rightarrow \overline{F}_u = 50...100$ kN ($\overline{F}_u = 75$);

hypothesis $H_3 \rightarrow \overline{F}_u = 100...150$ kN ($\overline{F}_u = 125$);

hypothesis $H_4 \rightarrow \overline{F}_u = 150...200$ kN ($\overline{F}_u = 175$);

......

hypothesis $H_{20} \rightarrow \overline{F}_u = 950...1000$ kN ($\overline{F}_u = 975$).

Table 6.10 Relative frequencies n/N ("probabilities") of divergences between particular values of pile bearing capacity by CPT data $F_{u,CPT}$ and its mean "accurate" value $\overline{F}_u$ with regard to heterogeneity of soils ($\sigma_{heter} = 70$ kN)

$F_{u,CPT} - \overline{F}_u$, kN	-250...-300	-200...-250	-150...-200	-100...-150	-50...-100	0...-50	0...+50	+50...+100	+100...+150	+150...+200	+200...+250
$n/N \approx p(F_{u,CPT} - \overline{F}_u)$	0.05	0.06	0.07	0.08	0.12	0.15	0.16	0.12	0.08	0.06	0.05

"A priori" probabilities of theses hypotheses may be taken to be equal, i.e. $p(H_1)$ $= p(H_2) = p(H_3) = ... = p(H_{20}) = 0.05$.

Let us assume that the following ultimate pile bearing capacities ($F_{u,CPT}$) in kN: 510; 580; 500; 610; 640; 550; 660; 690 have been obtained by CPT data. When considering each value of $F_{u,CPT}$ successively, one can recalculate the probabilities of hypotheses applying the formula (6.5). According to table 6.10, the first value of pile resistance $F_{u,CPTI} = 510$ kN excludes the validity of hypotheses corresponding to $F_{u,CPT} > 800$ kN and $F_{u,CPT} < 250$ kN, since in here the "errors" $F_{u,CPT} - \overline{F}_u$ become either smaller -300 kN or greater +250 kN, i.e. they appear in the zone of "zero probabilities". Thus, for first five hypotheses ($H_1 ... H_5$) and last four ones ($H_{17}...H_{20}$) "a posteriori" probabilities equal to zero, while hypotheses $H_6...H_{16}$ remain "non-zero".

For hypothesis H_6 (i.e. $\overline{F}_u = 275$ kN) the "error" will be 235 kN, and conditional probability of its occurrence will be 0.03. According to the formula (6.5) a posteriori probability of this hypothesis validity will be

$p(H_6/F_{u,CPT} = 510\text{kN}) = (0.05 \cdot 0.05)/(0.05 \cdot 0.05 + 0.05 \cdot 0.06 + ... 0.05 \cdot 0.16 + ... + 0.05 \cdot 0.05) = (0.05 \cdot 0.05)/[0.05(0.05 + 0.06 + ...0.16 + ... + 0.05)] = 0.05 \cdot 0.05/0.05 \cdot 1 = 0.05$

For hypothesis H_7 a posteriori probability will be

$p(H_7/F_{u,CPT} = 510$ kN$) = 0.05 \cdot 0.06/0.05 \cdot 1 = 0.06$.

It is possible to calculate a posteriori probabilities of the rest hypotheses analogically (table 6.11).

Table 6.11 Probabilities of the analyzed hypotheses after recording of the result $F_{u,CPT} = 510$ kN

Hypotheses (values of $\overline{F}_u$ in kN)	H_6 (275)	H_7 (325)	H_8 (375)	H_9 (425)	H_{10} (475)	H_{11} (525)	H_{12} (575)	H_{13} (625)	H_{14} (675)	H_{15} (725)	H_{16} (775)
Probabilities of hypotheses ("a posteriori")	0.05	0.06	0.08	0.12	0.16	0.15	0.12	0.08	0.07	0.06	0.05

As it could be seen from table 6.11, a posteriori probabilities of hypotheses mirror conditional probabilities of table 6.10, since a priori probabilities of hypotheses are equal [$p(H_i) = 0.05$]. In here, originally admitted range of possible values of $\overline{F}_u$ is not of any importance, since probabilities of the whole of hypotheses beyond the range of "errors" (table 6.10) will be equal to zero.

Analogically, one can take into account the following result: $F_{u,CPT2} = 580$ kN. In here, a posteriori probabilities given in table 6.11 are taken as a priori ones and calculations are repeated for new values of conditional probabilities of $F_{u,CPT}$ obtaining, i.e. for $F_{u,CPT2} = 580$ kN. Recalculation results are presented in table 6.12.

From table 6.12 one can see that uncertainty of evaluated values of $\overline{F}_u$ has decreased; hypotheses H_6 & H_7 are excluded from further analysis as their probabilities are equal to zero.

Table 6.12 Probabilities of the analyzed hypotheses after recording of two results $F_{u,CPT} = 510$ kN and $F_{u,CPT} = 580$ kN

Hypotheses (values of $\overline{F}_u$ in kN)	H_6 (275)	H_7 (325)	H_8 (375)	H_9 (425)	H_{10} (475)	H_{11} (525)	H_{12} (575)	H_{13} (625)	H_{14} (675)	H_{15} (725)	H_{16} (775)
Probabilities of hypotheses ("a posteriori")	0	0	0.05	0.08	0.12	0.20	0.22	0.14	0.10	0.05	0.04

Other results – $F_{u,CPT} = 500, 610, 640, 550, 660, 690$ kN are taken into account analogically. Every time a posteriori probabilities of the previous stage of calculation are taken as a priori ones at the next stage. Table 6.13 gives probabilities of hypotheses followed by recording of eight results mentioned earlier ($F_{u,CPT}$). As one can see from the table, only 5 hypotheses with probabilities unequal to zero have remained after the calculations.

Table 6.13 Probabilities of the analyzed hypotheses after recording of eight results $F_{u,CPT} = 510, 580, 500, 610, 640, 660, 690$ kN

Hypotheses (values of $\overline{F}_u$ in kN)	H_{10} (475)	H_{11} (525)	H_{12} (575)	H_{13} (625)	H_{14} (675)
Probabilities of hypotheses ("a posteriori")	0.10	0.16	0.30	0.24	0.21

Pile resistances evaluated by means of other methods are taken into account in the same way. For example, pile resistances $F_{u,lab}$ evaluated by laboratory treatment of monoliths or resistances $F_{u,dyn}$ obtained after the dynamic tests, etc. The technique of calculation is the same, but instead of table 6.7, it is necessary to use other tables showing the specific features of assessment methods, e.g. for $F_{u,lab}$, $F_{u,dyn}$ – tables 6.8 or 6.9 are appropriate.

In small numbers of $F_{u,lab}$ or $F_{u,dyn}$ the dispersion reflecting heterogeneity of soil may be taken by CPT data. Table 6.14 presents distribution of "errors" in evaluation of $F_{u,lab}$ as the same differences, i.e. $F_{u,CPT} - \overline{F}_u$, but with regard to heterogeneity of soils corresponding to the dispersion $\sigma^2_{heter} = 4900$ kN^2 ($\sigma_{heter} = 70$ kN) after CPT data. Calculations reveal that recording of this heterogeneity corresponds with expansion of the range of "errors" $F_{u,CPT} - \overline{F}_u$ in 1.13 times.

For example, if pile bearing capacity was evaluated by the data of laboratory tests of soils and $F_{u,lab} = 600$ kN, calculations would be done in the following way.

"A priori" probabilities $p(H_i)$ are taken to be equal to "a posteriori" ones in table 6.14, beyond the range of the table $p(H_i) = 0$.

Table 6.14 Relative frequencies n/N ("probabilities") of divergences between particular values of pile bearing capacity by laboratory analyses of soil $F_{u,\,lab}$ and its mean "accurate" value $\overline{F}_u$ with regard to heterogeneity of soils ($\sigma_{heter} = 70$ kN)

$F_{u,lab} - \overline{F}_u$ (kN)	-250...-300	-200...-250	-150...-200	-100...-150	-50...-100	0...-50	0...+50	+50...+100	+100...+150	+150...+200	+200...+250	+250...+300
$n/N \approx p(F_{u,lab} - \overline{F}_u)$	0.06	0.07	0.08	0.09	0.10	0.11	0.12	0.12	0.09	0.07	0.06	0.03

For hypothesis H_{10} the "error" is $F_{u,lab} - \overline{F}_u = 600 - 475 = 125$ kN, in accordance with table 6.14 its probability equals to 0.09. "A posteriori" probability H_{10} is calculated in the same way as in recording of $F_{u,CPT}$ i.e. by the formula (6.5):

$$p(H_{10}/F_{u,lab} = 600 \text{ kN}) = (0.10 \cdot 0.09)/(0.1 \cdot 0.09 + 0.16 \cdot 0.12 + 0.3 \cdot 0.12 + 0.24 \cdot 0.11 + 0,21 \cdot 0.10) = 0.08.$$

Correspondingly

$p(H_{11}/F_{u,lab} = 600$ kN$) = (0.16 \cdot 0.12)/(0.1 \cdot 0.09 + 0.16 \cdot 0.12 + 0.3 \cdot 0.12 + 0.24 \cdot 0.11 + 0,21 \cdot 0.10) = 0.17.$

$p(H_{12}/F_{u,lab} = 600$ kN$) = (0.30 \cdot 0.12)/(0.1 \cdot 0.09 + 0.16 \cdot 0.12 + 0.3 \cdot 0.12 + 0.24 \cdot 0.11 + 0,21 \cdot 0.10) = 0.32.$

etc.

Other results obtained by any methods are recorded in the same way. In the end this results in a tiny number of hypotheses and their probabilities left, thus, allowable load on the pile is chosen (with regard to the confidence coefficient).

The calculations discussed take into consideration both required index reliability evaluation and heterogeneity of soils; therefore, one does not need any additional reliability coefficients to be applied.

The approach presupposes the absence of "accurate" tests on the site to adjust the approximate estimates. If these tests are performed, the efficiency of required index evaluation significantly increases. In particular, there is the possibility of recording of the range of alteration reduction of possible "errors" in limited sites (see sec. 6.2).

In availability of the results of "accurate" tests it seems reasonable to point out two cases:
- Location of "accurate" tests does not coincide with location of approximate evaluations, i.e. "key" sectors are absent.
- Location of "accurate" tests coincides with location of separate approximate tests, i.e. there are "key" sectors (e.g. in fig. 6.2).

In the first case there are not any principal alterations in the algorithm considered. "Accurate" tests are considered as the tests possessing very small "errors" (in relation to the true values of the index in specified locations of the site), i.e. low dispersed σ_{rel}^2. Nevertheless, high accuracy of the tests is related to *the particular values* of the required index, while heterogeneity of soil makes the particular values A_i only the approximate estimates of *mean value (at the site)* of this parameter – $\overline{A}$. In other words, the "errors" of these tests are characterized by dispersions σ_{heter}^2 in general. Therefore, the formula (6.7) is as follows: $\sigma = \sqrt{\sigma_{rel}^2 + \sigma_{heter}^2} \approx \sigma_{heter}$. As the "accurate" tests are carried out in little due to their high prices and complexity, it is impossible to judge about closeness of their results to the average mean required parameter ignoring the analysis the results of a number of approximate tests embracing the whole of the terrain under study. The issue was discussed in section 2.5. It was underlined that a small number of "accurate" tests in high heterogeneity of soil are less informative than a large number of approximate tests. It is important to keep in mind that in designing structures on heterogeneous soils averaging of soil properties is usually done on the local zones (sectors), but not on the whole of the site. Nevertheless, in each zone the local design characteristics of the foundation are established following the same principles given above.

Thus, recording of the results of "accurate" tests in absence of "key" sectors is done in the same way as that of the results of approximate tests, distribution of conditional probabilities. Significant increase of definition is attained due to narrow range of "errors" variations in required index evaluation by the data of "accurate" tests, i.e. the parameters $A_u - \overline{A}_u$ depending upon heterogeneity of soils in general.

In the second case when location of "accurate" tests coincides with location of separate approximate tests (i.e. in occurrence of "key sectors") the efficiency of engineering-geological and geotechnical surveys significantly increases. There appears the possibility of *adjustment* of the approximate results; in here, the range of possible "errors" in specified sites may be reduced, as it was mentioned in section 6.2. The methods of this adjustment are analogous to those ones of reliability coefficient evaluation γ_k discussed in section 6.2. For this BashNIIstroy apply the software program SYNT allowing generalization of the results of different tests, either with the adjustment or without it. Differences occur just only due to the fact that under "errors" we understand not the ratios but differences between the approximate and "accurate" values. Thus, divergence between "accurate" (A) and approximate (A_{approx}) results at "key sectors" is estimated not by the ratio $k=A_{approx}/A$, but the difference $k_d = (A_{approx} - A)$. Correspondingly, possible mean values of these differences at the investigated site are taken as the hypotheses H_i. As in estimation of the reliability coefficient γ_k, the "error" in mean values equals to mean "error"

$$\overline{A}_{approx} - \overline{A} = \overline{(A_{approx} - A)} \ . \tag{6.9}$$

"Instants α" being the base to recalculate the probabilities of hypotheses applying the formula (6.5) are the results (as in evaluation of the reliability coefficient γ_k) obtained at "key sectors", but in other form, i.e. not as the ratio $k' = A_{approx}/A$, but as the difference

$$k'_d = (A_{approx} - A).$$

Probabilities of hypotheses distribution $p(H_i)$ and conditional probabilities of "instants α" $p(\alpha/H_i)$ are taken on the basis of histograms' transformation of corresponding total distribution of "errors". In here the operations of "compression" are used. Distributions given in tables 6.7, 6.8 and 6.9 can be applied as total distributions; for clearness they can be presented as histograms. When evaluating pile bearing capacities the histogram of probabilities of hypotheses may be obtained by compression of the total histogram as in evaluation of the reliability coefficient γ_k i.e. in 1.55 times, and conditional probabilities – in 1.3 times. Calculations are to be done as described in section 6.2.

In carrying out several approximate tests at "key sectors" (by different methods) the adjustment may be done with regard to each method. Corrected distributions obtained are generalized applying the described methods in the same way as without the adjustment.

Figure 6.8 illustrates the results of calculations of the allowable loads on piles for the same site where the discussed experiments were performed (see fig. 6.4 and 6.5).

As in the previous case, a great number of "accurate" tests (11 static tests of piles at the site of 150 m^2) allowed evaluating the required index with high accuracy – average pile bearing capacity $\bar{F}$, which (with regard to its dispersion σ^2/n) was 257±19 kN. Variants of allowable loads were considered; in them the following information was used as the initial data:

- CPT (11 locations) and static tests of piles (one pile, two, three... N piles).
- Complex of definitions including CPT (11 locations), drilling of soil with subsequent laboratory treatment of the selected monoliths (6 bore holes, 5 monoliths from each hole), dynamic tests of piles (11 piles), static tests of piles (one, two, three...N piles).
- The same complex, but selection of loads on piles is done after SNiP 2.02.03-85*, i.e. after static tests of piles as the most accurate method.

As one can see from figure 6.8, design loads allowable on piles differed up to 15...20 % in one and the same initial data, but different ways of generalization. According to calculations based on Bayes formula, allowable loads on piles gently increased with the increase of number of piles tested statically (curves 1 & 2). In application of all four methods of pile resistance estimation (curve 2) these loads proved to be higher than in smaller number of methods applied (curve 1). Reliability and affordability of solutions based on the mode of calculation studied may be considered the principal results. In fact, on no combination of approximate and "accurate" evaluations the allowable load on piles exceeded the actual average pile bearing capacity (257 kN). Neither did it exceed the assumed minimal value of this average (238 kN). At the same time, the traditional method admitted in the Norms SNiP 2.02.03-85* does not ensure this reliability, since it can lead to dangerous errors, if only one static test of piles is carried out. In unsuccessful selection of the site to carry out this test the load on the pile is possible to be higher than average pile resistance (curve 3 in fig. 6.8b). Therefore, the instruction given in p. 5.2 of SNiP 2.02.03-85* concerning the necessity to perform not less than two static tests of piles needs to be understood as the requirement to foundation reliability. In here, in the traditional method (SNiP 2.02.03-85*) one has to apply lower loads on piles than in the mode with application of Bayes formula.

In general, despite the mentioned advantages of the approach, its practical application is complicated by the advanced requirements to quality of measurements and competence of the principal doer. This is conditioned by the following reasons.

First, in parallel applications of test methods different in their accuracy "the spread" of the results obtained may be rather significant, while the effect of each result on selection of the design index shows itself more than in the traditional approach. Here, awareness of information completeness, ability to distinguish correct results from the false ones, especially when the required index is overrated are of greatest importance. It is obvious that the doer must be a highly-skilled specialist who is aware of local conditions and specific features of the performed tests.

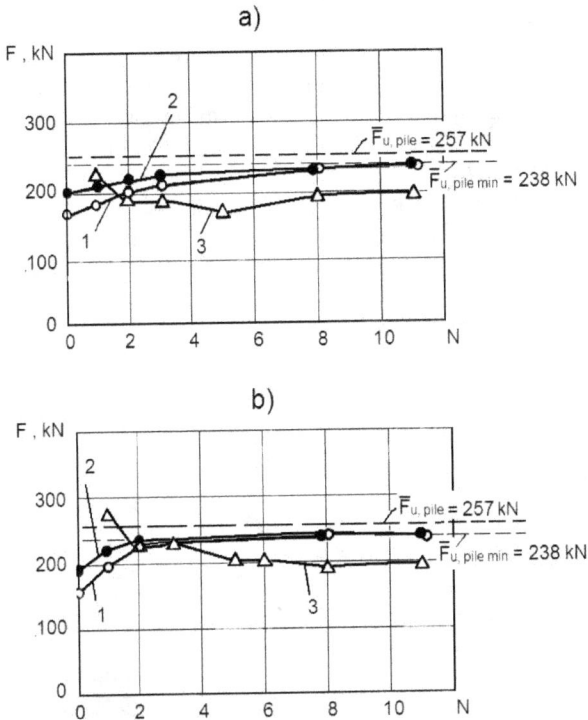

Fig. 6.8 Schematics of design load allowable on piles F in different numbers of static tests of piles (key sectors) N with application of several approximate methods of pile bearing capacities evaluation (the site as well as piles are the same as in fig. 6.4):

a – mean values of design load allowable on piles F by the examined combinations of tested piles (in one tested pile, two, three ...etc., i.e. in $N = 1, 2, 3...11$), b – maximum values of pile resistances F after the same variants. Initial data in calculations: 1 – CPT and static tests of piles; 2 – CPT, laboratory analysis of soil properties, dynamic tests of piles, static tests of piles; 3 – selection of design load allowable on piles in availability of their static tests after SNiP 2.02.03-85* (i.e. after static tests of piles only)

Second, zone sizes where design geotechnical parameters are established (the results of different accuracies are generalized, their adjustment is done, etc.) are determined by either geological factors or mostly engineering ones. These are the total rigidity of the erected structure, local rigidities of its members or ability of the structure to bear one or the other deformations, etc. According to the National Norms of Russia design geotechnical parameters correspond to reduced *mean* values incorporating a share of particular parameters (at local sectors) being lower than the admitted design parameter. These local reductions are compensated by the rigidity of the erected structure, i.e. the possibility of load redistribution onto the adjacent "denser" sectors. Thus, any averaging of test results (with the adjustment or without it) may be acceptable for a small sector of the structure where such redistribution is possible, but it is inacceptable in case the sector is quite large, and hence necessary redistribution should not be anticipated.

It is also important to take into account that the discussed calculations are possible to be done in availability of data characterizing the accuracy of the method of the required index evaluation, i.e. diagrams of dissipation, tables or histograms reflecting the comparative results of approximate evaluations and the "accurate" ones.

CONCLUSION

The analysis of the results obtained by non-Russian and Russian professional engineers during last decades as well as CPT of soils as it is reveal that this method possesses ample opportunities for its vast practical application, but in reality they are not realized on a large scale. One of the reasons is that specialists are not quite aware of the research carried out in Russia and worldwide as well. Moreover, there is a lack of publications on the issue, especially some general ones in Russian language. The authors of this book have tried to fill in the gap. Only the readers can judge whether it has been a success or not.

In general, let us draw the following conclusions:

1. CPT is the method in which inaccuracy of particular estimates of mechanical properties of soil may be compensated by larger number of their evaluations ensuring assessment of the site as a whole and revealing specific features of its structure better than it can be achieved in any other method of soil study. Therefore, increase of efficiency and manufacturability of CPT rigs proves to be one of the most promising trends in CPT development. Intensive exploration of the underground space in cities and megalopolises does require application and improvement of heavy CPT rigs with pushing thrust capacity to be 150...200 kN, capable to carry out CPT in depth of 30...40 m and more. Before, CPT was done up to 15...20 m in depth.

2. During the last two-three decades in non-Russian practice significant increase of nomenclature and the techniques of evaluation of the parameters characterizing the soil under study have been the general tendencies in CPT development. Additional measuring devices are installed in the penetrometer; they allow evaluating the pore pressure, deviation of the penetrometer from the vertical, combining CPT with other tests (pressuremeter, rotary shear, geophysical measurements, etc.), revealing of environmentally unfriendly substances (hydrocarbons, etc.), etc. Independent geotechnical aspect is being formed – "Direct Push Technologies" – DPT. The authors of the book consider this aspect to be the realization of the idea of "CPT equipment application in a new way"; it is of great practical importance, but beyond the problems of CPT applications. That is why the issues are discussed in the book in general.

3. As opposed to non-Russian practice, in Russian one there appear the tendencies to complicate the penetrometers by giving them some new functions. For example, penetrometers equipped with pore pressure sensors are not widely applied in Russian engineering-geological surveys. However, in comparison to the non-Russian approaches greater attention is paid to theoretical issues concerning the connection of CPT results with the properties of soil, application of CPT in specific soils (permafrost, collapsible, etc.), solution of technological problems of pile driving, application of probabilistic models in selection the design geotechnical indexes.

4. CPT proves to be the most efficient in its application with more accurate (and hence, more expensive) methods of soil study, especially in estimation of pile bearing capacity. In here, the "accurate" results shall serve as a means of correction of the approximate data obtained by CPT, i.e. ensure the "anchorage" of these data to the conditions of the specified site. Possible variant of the correction is given in the book; it is based on the existing data application and probabilistic analysis models of the information obtained. In lack of the "accurate" data, correction is necessary to be done proceeding from the least favorable, but possible situations.

5. The results of the done theoretical analysis allow denying the traditional point of view on CPT as a "purely empirical method". Cone resistance may be considered as generalized refection of mechanical processes in soil involving breaking (plastic) and non-breaking (linear) deformations. To a certain extent, the existing empirical dependencies between cone resistance (q_c) and standard properties of soil (φ, c, E) may be considered as particular cases of more generalized dependency showing the process of a rod penetrometer immersion into the elastic-plastic medium. Thus, the variety of the empirical dependencies is easily explained. Thus, different ratios (correlation links) between strengthening and deformation properties of soil in deposits of various genesis explain the variety of simplified dependencies, such as $E \approx f_E(q_c)$, $\varphi \approx f_\varphi(q_c)$, $c \approx f_c(q_c)$. Deviations from Coulomb law occurring in different ways either in clays or sands (in sands flattening of the line "$\tau \sim \sigma$" occurs in σ pressures being many times higher than in clays) explain the differences of the mentioned dependencies in clays and sands.

6. Soil behavior under the cone is of specific character with high speeds of deformations, zones of considerable normal stresses exceeding pressures under shallow foundation footing by one-two orders of magnitude. In here, clayey soil behavior corresponds to the "closed system", i.e. lack of drainage, in high normal stresses clayey soil behaving itself as the ideally coherent thing ($\varphi \approx 0$). There is not any significant consolidation of clays or loams under the cone, i.e. the decrease of the void ratio happens to be insignificant.

7. Theoretical research corroborated by numerous experiments show that there is narrow correlation between cone resistance q_c of clayey soils and their shear stress τ_g in overburden stress. At the same time correlation between q_c and shear parameters φ & c (separately) proves to be weaker, as a rule. Therefore, non-Russian specialists prefer CPT in evaluation of only one parameter of clay strength – "shear strength s_u" ignoring normal pressures. The National Norms of Russia are directed toward each shear parameter (φ & c) to be separately evaluated by CPT data, practice corroborating the approach applicability. This inconsistency is possible to be explained by specific features of Coulomb's model. In different combinations of φ & c corresponding to one and the same value of τ_g, the resistivity of clayey massif (under the foundation footing, in soil creep slope, etc.) almost does not alter. In overstating of φ the value of c is often understated, these factors being

considerably compensated in calculations. For this reason, the parameters φ & c evaluated by CPT data characterize both the investigated soil and its equivalent which will behave in the foundation in the same way as the real one.

8. In spite of significant differences in moduli of deformation E obtained in one and the same soils but by different methods (odometer, stabilometer, stamp, pressuremeter, etc.) correlation between these moduli does exist. That is why, CPT could be "anchored" to any of the moduli of deformation obtaining different empirical formulae. In practice it is necessary to understand clearly what results should be taken as the reference ones and, thus, apply the information obtained.

9. Application of sleeve friction resistance f_s is different in Russia and worldwide. In the National Norms of Russia f_s is considered as the important initial parameter for computation of driven pile side bearing capacity. Non-Russian specialists apply cone resistance q_c here, although they prefer f_s to be applied in identification of soils much wider than Russian specialists. Friction ratio $f_s/q_c \cdot 100\%$ and q_c, are used as the criterion; for this the charts have been developed allowing lithological types of soils to be determined. When using f_s it is necessary to take into account that it is affected by the length of the friction sleeve – the longer the sleeve, the lower f_s is.

10. CPT may be used to study specific soils (permafrost, collapsible, etc.); for this special CPT methods have been developed. Here, "standard" cone penetrometers (without any structural additions) or penetrometers equipped with additional devices (temperature sensors, bulk water transporters, etc.) are used. "Relaxation-creep" CPT may be referred to the simplest special methods of CPT; in here, resistance of soil and other measured parameters are recorded in vanishing velocities of penetrometer movement. Both "standard" and complicated penetrometers are used in these tests, the penetrometers equipped with pore pressure sensors included.

11. Application of CPT in investigation of plastic-frozen soils is one of the most promising trends, since it proves to be rather complicated and laborious when applying the traditional methods then. The research done has revealed that series-produced rigs seem to be applicable for solutions of the emerged problems, and it is desirable to equip the penetrometers with temperature sensors. The empirical formulae for approximate evaluation of strengthening and deformation properties of plastic-frozen soil, pile bearing capacity with regard to their arrangement, etc. have been obtained.

12. Due to the fact that soil behavior around the penetrometer (or driven pile) is practically the same, CPT results happen to be the most appropriate initial data for pile bearing capacity calculation; this ensures the greatest reliability of calculations. Developed in Russia are calculation techniques based on cone resistance application (q_c) and side friction resistance (f_s or Q_s). Transition coefficients "from the penetrometer to the pile" are determined due q_c & f_s (or Q_s) as they are with regard to lithological origin of soil (clayey

soil or sand). For in-situ piles (boring, drilled-in, etc.) in which the similarity is smaller, it is cone resistance q_c that is used. Transition coefficients "from the penetrometer to the pile" are determined due to lithological origin of soil (clay, sand, etc.), q_c and the technology of pile driving as well.

13. CPT may be applied in solution of numerous technological problems of a "zero cycle". This is related to possibility of estimation of pile driving up to the specified depth applying the equipment available, forecasting of pile underdriving in occurrence of coarse boulders, controlling of deep consolidation of soils or liquefaction of sands, etc.

14. The empirical data to estimate mechanical properties of soil by CPT data gained in Russia are not always appropriate in solutions of urgent geotechnical problems. Depth extension of engineering-geological surveys makes it possible to obtain and specify the dependencies relevant to deeply deposited soils (below 20...30 m), non-quaternary deposits included. Mass construction of high and high-rise buildings and complexes in large cities and megalopolises makes the issues concerning CPT results applications for estimation of mechanical properties of soils in high pressures acting on the foundation (above 0.4...0.6 MPa) quite urgent, indeed. The empirical dependencies obtained earlier correspond to the pressures of up to 0.3 MPa.

15. Due to simplicity and quickness of test procedures and CPT data treatment as well as availability of a number of software programs there is the possibility to present the information obtained either as numerous "point" estimations or digital geotechnical 2D or 3D models (sectional views, fields or charts) showing changeability and distribution (in the investigated ground massif) of parameters being necessary in designing – pile resistances, properties of soil, resistances of footings of shallow foundations, etc.

16. CPT is one of the most promising methods of investigation of mechanical properties of space objects, as it gives possibility to carry out measurements "in-situ" and to communicate by means of space communication; this is incomparably easier and cheaper than selecting monoliths and delivering them on the Earth.

REFERENCES

Aas G., Lacasse S., Lunn T., Høeg K. (1986) Use of in situ tests of foundation design on clay. Proc. of the ASCE Specialty Conf. In Situ'86: Use of In Situ Test in Geotechnical Eng, Blackburg, ASCE, pp. 1-30.

Amaryan L.S. (1990) Properties of weak soils and methods of their study. Moscow: Nedra, 220 p.

Andersland O.B., Anderson D.M. (1978) Geotechnical engineering for cold regions. McGraw-Hill Book Company, 551 p.

ASTM Standard. D 3441. Standard Test Method for Mechanical Cone Penetration Tests of Soil.

ASTM Standard. D 5778. Standard Test Method for Electronic Friction Cone and Piezocone Penetration Testing of Soils.

ASTM Standard. D 6067. Standard Practice for Using the Electronic Piezocone Penetrometer Tests for Environmental Site Characterization.

Babkov V.V., Polak A.F. (1965) The issue on pore space geometry. Construction of oil processing and petrochemical plants: BashNIIstroy Trust. – Issue IV. Moscow: Stroyizdat, pp. 483-487.

Baecher G.B., Rackwitz R. (1982) Factor of Safety and Pile Load Test. Int. Journal for Numerical in Geomechanics. Vol. 6, No. 4, pp. 409-424.

Baldi G., Bellotti R., Ghionna V., Jamiolkowski M., Pasqualini E. (1981) Interpretation of CPTs and CPTUs. 2nd part: drained penetration of sand. Proc. of IV Int. Geotech. Seminar, Singapore, pp. 143-156.

Baldi G., Bruzzi D., Superbo S., Battaglio M., Jamiolkowski M. (1988) Seismic cone in Po River sand. Proc. of the Int. Simp. on Penetration Testing, ISOPT-1 Orlando, vol. 1, Rotterdam: Balkema, pp. 643-650.

Been K., Quiñonez A., Sancio R.B. Interpretation of the CPT in engineering practice. Proc. of the 2nd International Symposium on Cone Penetration Testing, Huntington Beach, USA.

Begemann H.K.S. Ph. (1953) Improved method of determining resistance to adhesion by sounding through a loose sleeve placed behind the cone. Proc. of the 3rd Int. Conf. Soil Mech. and Found. Eng. Zürich, vol. 1, pp. 213-217.

Begemann H.K.S. Ph. (1965) The friction jacket cone as an aid in determining the soil profile // Proc. of the 6th Int. Conf. Soil Mech. and Found. Eng. Montreal, vol. 1, pp. 17-20.

Belyaev V.P. (1970) CPT in soil structural properties estimation. TSTISIZ Information bulletin. №4 (21). Moscow: TSTISIZ, pp. 21-28.

Belyaev V.P. (2005) Pile foundations in construction and reconstruction in Samara (1960…2000). Samara: SISU, 141p.

Berezantsev V.G. (1955) Limit equilibrium of consolidated medium under spherical and conical stamps. – News. Academy of Sciences of the USSR, OTN, № 7.

Berezantsev V.G. (1965) Calculation of deep foundation beds. Reports at VI International Congress on Soil Mechanics and Foundation Engineering. – Moscow: Stroyizdat, pp. 119-127.

Berezantsev V.G. (1966) Determination of ultimate pile end bearing capacity in sands by static penetration. Bases, foundations and soil mechanics J. № 4, pp. 1-5.

Berezantsev V.G., Yaroshenko V.A. (1962) Specific features of deforming sandy beds under deep foundations. Bases, foundations and soil mechanics J. №1, pp. 3-7.

Berezina S.L., Enikeyev V.M. (1989) The issue on pile computation by CPT data. Research on advanced pile foundations: NIIpromstroy Trust. Ufa, pp. 93-95.

Bilenko N.Z., Ryzhkov I.B. (1985) Application of multisectional tensopile to study pile behavior in weak clays. Issues on foundation engineering. Soil Mechanics: NIIpromstroy Trust. Ufa, NIIpromstroy, pp. 64-70.

Blouin S., Chamberlain E., Sellmann P. and Garfield D. (1979) Determining subsea permafrost characteristics with a cone penetrometer. Cold Regions Science and Technology. № 1, pp. 3-16.

Bondarick G.K. (1964) Dynamic and static penetration of soils in geotechnical engineering. Moscow, Nedra, 164 p.

Briaud J.L. (1988) Evaluation of cone penetration test methods using 98 pile load tests. Proc. of the Int. Simp. on Penetration Testing, ISOPT-1 Orlando, vol. 2, Rotterdam: Balkema, pp. 687-697.

Building Code MGSN 2.07-01 (2003) Bases, foundations and underground structures. Moscow: GUP "NIATS", 109 p.

Building Code SN-448-72 (1972) Guidance on CPT of soils for construction. – Moscow: Stroyizdat, 33 p.

Building Code SNiP 2.02.01-83* (2002) Foundations of buildings and structures / Gosstroy of Russia – Moscow: GUP TSPP, 48 p.

Building Code SNiP 2.02.02-85* (2004) Foundations of hydraulic structures. Gosstroy of Russia. – Moscow: FGUPP TSPP, 48 p.

Building Code SNiP 2.02.03-85* (2002) Pile foundations. – Moscow, 46 p.

Building Code SNiP 11-02-96 (1997) Investigations for construction. Fundamentals. Minstroy of Russia – Moscow: PNIIS, 44 p.

Building Code SNiP II-15-74 (1975) Foundations of buildings and structures. Gosstroy of the USSR. Moscow: Stroyizdat, 65 p. (functioned until 1983).

Building Code SNiP II-17-77 (1978) Pile foundations. Gosstroy of the USSR. Moscow: Stroyizdat, 48 p. (functioned until 1985).

Building Code SP 11-105-97 (1998) Engineering-geological investigations for construction. Part I. Moscow.

Building Code SP 25.13330.2012 (2012) Soil bases and foundations on permafrost soils. – Moscow, 117 p.

Building Code SP 50-101-2004 (2005) Designing and construction of bases and foundations of buildings and structures. Gosstroy of Russia. – Moscow: FGUP TSPP, 130 p.

Building Code SP 50-102-2003 (2004) Designing and construction of pile foundations. Gosstroy of Russia. – Moscow, 80 p.

Building Code SP 24.13330.2011 (2011) Pile foundations. – Moscow, 85 p.

Building Code SP 47.13330.2012 (2012) Engineering survey for construction. Basic principles. – Moscow, 110 p.

Burns S.E., Mayne P.W. (1998) Penetrometers for Soil Permeability and Chemical Detection. Funding provided by NSF and ARO issued by Georgia Institute of Technology Report No GIT-CEEGEO-98-1, July 1998. Georgia Institute of Technology. 144 p.

Bustamante M., Gianeselli L. (1982) Pile bearing capacity prediction by means of static penetrometer CPT. Proceedings of the 2nd European Symposium on penetration Testing, ESOPT-II, vol. 2, Amsterdam. Rotterdam: Balkema Pub., pp. 493-500.

Campanella R.G. and Howie J.A. (2005) Guidelines for the use, interpretation and application of seismic piezocone test data. The University of British Columbia.

Campanella R.G., Gillespie D., Robertson P.K. (1982) Pore pressure during cone penetration testing. Proc. of the 2nd European Simp. on Penetration Testing, ESOPT-II, Amsterdam, vol. 2, Rotterdam: Balkema Pub., pp. 507-612.

Campanella R.G., Robertson P.K. (1986) Research and development of the UBC cone pressuremeter. Proc. of the 3rd Canadian Conf. on Marine Geotechnical Engineering, St. John's, Newfoundland, vol. 1., pp. 205-214.

Chebotarev G.P. (1968) Soil mechanics, foundations and earth structures. Transl. from English. Moscow: Stroyizdat, 616 p.

Cherkasov I.I., Gromov V.V., Zobachev N.M., Musatov A.A., Mikheev V.V., Petrukhin V.P., Shvarev V.V. (1968) Sampler-penetrometer of automatic lunar station Luna-13. Reports of Academy of Sciences, USSR, volume 179, № 4, pp. 829-831.

Cherkasov I.I., Mikheev V.V., Smorodinov M.I., Shvarev V.V. (1986) 20 years of Soviet research of Lunar soils. Bases, foundations and soil mechanics J. № 6, pp. 17-19.

Cherkasov I.I., Shvarev V.V. (1971) Fundamentals of soil science of the Moon. – Moscow: Nauka, 199 p.

Cherkasov I.I., Shvarev V.V. (1973) Soviet research in mechanics of Lunar soils. Bases, foundations and soil mechanics J. № 4, pp.12-15.

Cividini A., Maier G., Nappi A. (1983) Parameter Estimation of Static Geotechnical Model Using a Bayes Approach. Int. Journal Rock. Mech. and Mining Sci. and Geotech. Abstr. Vol. 20, No. 5, pp. 215-226.

Cosstay J., Sanglerat G. (1981) Soil mechanics. Practical course / Transl. from French into Russian by V.A. Barvashov; edited by B.I. Kulachkin. – Moscow: Stroyizdat, 455 p.

De RuiterJ., Beringen F.L. (1979) Pile foundations for large North Sea structures. Marine Geotechnology, No 3 (3), pp. 267-314.

Douglas B.J., Olsen R.S. (1981) Soil classification using electric con penetrometer. Cone penetration Testing and Experience. Proc. of the ASCE National Convention, St. Louis: ASCE, pp. 209-227.

Dzagov A.M. (2007) On assessment of pile bearing capacity in collapsible soils by CPT data // Bases, foundations and soil mechanics. № 5, pp. 28-30.

ENiR. Collection E2. Excavation works. Issue1 Mechanized and manual excavation works.

Enikeyev V.M. (1980) Research and development of CPT in collapsible soils for pile foundations designing. Thesis...PhD (Engineering). – Dnepropetrovsk, 160p.

Enikeyev V.M., Isaev O.N., Norshayan A.V. (1989) Assessment of possible penetration of piles up to specified depth in soil of Karelia by CPT data. Mechanization and automation of zero cycle works: NIIpromstroy Trust. – Ufa, pp. 60-64.

Enikeyev V.M., Makarov V.N., Ryzhkov I.B., Norshayan A.V., Yurasov N.N. (1986) Research on effect of some structural features of S-832M CPT rig on the results obtained. – Collected papers of NIIpromstroy "Soil Mechanics". – Ufa: NIIpromstroy, pp. 27-36.

Enikeyev V.M., Ryzhkov I.B. (1980) On effect of speed of penetration on resistances of various soils. Collected papers of NIIpromstroy, "Bases and foundations", – Ufa: NIIpromstroy, pp. 71-74.

European prestandard ENV 1997-3 (2000) Eurocode 7: Geotechnical design – Part 3: Design assisted by fieldtesting. BSI.

European standard EN 1997-2 (2007) Eurocode 7: Geotechnical design – Part 2: Ground investigation and testing. BSI.

Ferronsky V.I. (1969) Penetration survey of geological investigations. Moscow: Nedra, 240 p.

Firestein V.D. (1968) Evaluation of pile bearing capacity and selection of optimal pile sizes by CPT results in 3-component scheme of pile behavior. Collection of reports on pile foundations. – Moscow: Stroyizdat, pp. 135-172.

Firestein V.D., Makarov V.N. (1964) S-832 CPT rig for CPT of soils. Equipment for CPT and drilling of soils. – Moscow: TSNTI of Gosstroy, USSR, pp. 1-10.

First panoramas of lunar surface. (1969) Vol. II. After materials of automatic stations "Luna-9" and "Luna-13" – Moscow: Nauka, 70 p.

Fortier, R., Ladanyi, B., and Allard, M. (1993) CPT study of the effect of unfrozen water content on strength of silty permafrost at Kangiqsualujjuaq, Nunavik (Quebec). In Proceedings of the 46th Canadian Geotechnical Conference, Saskatoon, Sask., pp. 307–318.

Gareeva N.B., Gorbatova N. Ya. (1985) On evaluation of design pressure on soil by CPT data. Foundation engineering. Soil mechanics. Collected papers of NIIpromstroy. – Ufa: NIIpromstroy, pp. 111-116.

Gmurman V.E. (2000) Probability theory and mathematical statistics. Moscow: Vysshaya Schola, 479 p.

Goncharov B.V., Ryzhkov I.B., Isaev O.N. (1991) CPT application for pile foundation designing in boulder soils // Pile foundations / Gersevanov VNIIOSP, DalNIIS. – Moscow: Stroyizdat, pp. 4-11.

Guidance on designing and construction of pile foundations in Latvian SSR. (1977) У-1-77, Riga, Gosstroy of Latvian SSR, 115 p.

Heat-and-weight exchange. (1982) Heat engineering experiment: Manual. – Moscow: Energoizdat, 512 p.

Hey J. (1987) Introduction in Bayes statistical resume. Transl. From English. Moscow: Finance and Statistics, 335 p.

Houlsby G.T., Teh C.I. (1988) Analysis of the piezocone in clay. Proc. of the Int. Simp. on Penetration Testing, ISOPT-1 Orlando, vol. 2, Rotterdam: Balkema, pp. 777-783.

Handbook (2004) Cone Penetration Testing (Simplified Description of the Use and Design Methods for CPTs in Ground Engineering). Fugro Engineering Services Limited.

Handbook (2005) Geotechnical & geophysical investigations for offshore and nearshore developments. ISSMGE TC1.

Ignatova O.I. (2009) Deformative properties of Jurassic clays of Moscow. Bases, foundations and soil mechanics J. – Moscow: № 5, pp. 24 - 28.

International Reference Test Procedure for Cone Penetration Test (IRTP): Report of the ISSMFE Technical Committee on Penetration Testing of Soils (1989) – TC 16, with Reference to Test Procedures, ISSMFE, Swedish Geotechnical Institute, Linköping, Information, 7, pp. 6-16.

International Reference Test Procedure (IRTP) for the Cone Penetration Test (CPT) and the Cone Penetration Test with pore pressure (CPTU): Report of the ISSMGE Technical Committee 16 on Ground Property Characterisation from In-situ Testing 1999 (corrected 2001) – TC 16, ISSMGE, 29 p.

Isaev O.N. (1988a) (in co-authorship). Problems of CPT application in permafrost soils. Abstract at All-Union conference "Urgent problems in pile foundation engineering in the USSR". Perm, pp. 92-93.

Isaev O.N. (1988b) (in co-authorship). Assessment of frozen soil durability by CPT data. Minmontazhspetsstroy TSBNTI Express-information "Assembly and special construction works", issue № 7. Moscow, pp. 20-25.

Isaev O.N. (1989a) Principal tasks in CPT development in permafrost soils. Abstract at All-Union Scientific Conference "Problems of engineering-geological investigations in cryolite zone". Magadan, pp.180-182.

Isaev O.N. (1989b) Development of CPT for pile foundation designing in plastic-frozen soils. Thesis...PhD (Engineering). Moscow: MIIT, 148 p.

Isaev O.N., Shvarev V.V., Tikhomirov S.M., Sadovsky A.V., Konstantinov A.V. (1991) Application of CPT for investigations of frozen soils. Bases, foundations and soil mechanics J. № 3, pp. 13-16.

Isaev O.N. (2011a) Identification of the thawed and frozen states of soils by CPT. Proceedings of the IX International symposium on engineering Permafrost. Mirny (Russia), September 3-7, pp. 63-67.

Isaev O.N. (2011b) Methods of frozen soil static sounding. Proceedings of the IX International symposium on engineering Permafrost. Mirny (Russia), September 3-7, pp. 69-73.

Isaev O.N. (2012a) Ground state and determination of the strength and compressibility characteristics of plastic frozen soils with the help of the static probing method. Proceedings of the Tenth International Conference on Permafrost (ICOP), 10, Volume 4, pp. 226-227.

Isaev O.N. (2012b) CPT and TCPT in permafrost: constant rate and relaxation-creep penetrometer test procedures. Proceedings of the 4th International Conference on Site Characterization. ISC-4 – Pernambuco, Brazil.

Isaev O.N. (2014) New in standards for CPT of soils in Russia. Proceedings of the 3rd International Symposium on Cone Penetration Testing. CPT`14, Las Vegas, Nevada, UCA.

Isaev O.N., Ryzhkov I.B. (2010) TCPT in permafrost: penetrometer – soil thermophysical interaction. Proceedings of the Second International Symposium on Cone Penetration Testing. CPT`10, Huntington Beach, UCA.

Isaev O.N., Ryzhkov I.B. (2015) Special probes for Cone Penetration Testing – classification, types and application conditions. Magazine "Metro and tunnels", № 5, pp. 28-30.

Isaev O.N., Shvarev V.V., Konstantinov A.V., Tichomirov C.M., Sadovsky A.V. (1995) Progress of the method of static sounding in the investigation of geotechnical properties of frozen soils. Proc. of the Int. Simp. on Cone Penetration Testing, CPT'95 Linköping, Sweden, October 4-5, , vol. 2, pp. 179-186.

Isaev O.N., Volkov F.E., Minkin M.A. (1987) Pile bearing capacity evaluation in plastic-frozen soils by CPT. Bases, foundations and soil mechanics J. №5, pp. 17-19.

ISO 22476-1 Geotechnical investigation and testing – Field testing - Part 1: Electrical cone and piezocone penetration tests.

ISO 22476-12 Geotechnical investigation and testing – Field testing – Part 12. Mechanical cone penetration test (CPTM).

Jaky J. (1944) The coefficient of earth pressure at rest. Journal of the Society of Hungarian Architects and Engineers, pp. 355-358.

Jamiolkowski M. (1995) Opening Address. Proc. of the Int. Simp. on Cone Penetration Testing, CPT'95 Linköping, Sweden, October 4-5, vol. 3, pp. 7-15.

Jamiolkowski M., Ladd C., Germaine J., Lancelotta R. (1985) New developments in field and laboratory testing of soil. Proc. XI ICSMFE, San Francisco, vol. 1. – pp. 57-154.

Kamp W.G. (1982) The influence of the rate of penetration on the cone resistance in sand. – Proc. ESOPT – II, Amsterdam, 1982, v.2, pp. 627-633.

Kay J.N. (1977) Factor of Safety for Piles in Cohesive Soils. Proc. IX Int. Conf. Soil Mech. and Found. Eng. Vol. 1, pp. 587-592.

Kazantsev V.S., Hamov A.P., Levina R.F., Kolesnik G.S., Ryzhkov I.B. (1974) CPT rigs test results. Designing of foundations for transport buildings and structures on piles and shells in complicated soil conditions: Abstract at scientific seminar in NTO LIZHT. Leningrad, pp. 185-188.

Kolesnik G.S. (1972) Evaluation of pile bearing capacity by CPT results: Thesis...PhD (Engineering). – Ufa, 150 p.

Kolesnik G.S., Ryzhkov I.B. (1977) On immersion of piles up to specified set per blow and specified mark. Collected papers of NIIpromstroy, issue18. Moscow: Stroyizdat, pp.10-18.

Kovalev Yu.I. (1977) Mode of driven pile calculation by mechanical properties of soil and CPT data: Dep. 662. – Moscow: TSNIITEI MPS, 16 p.

Kozlovsky A.D. (1988) Effect of S-832 penetrometer wear on reliability of CPT results. Issues on designing of advanced pile foundations. Collected papers of NIIpromstroy. – Ufa: NIIpromstroy, pp. 23-26.

Krutov V.I., Eiduck R.P. (1971) Determination of relative collapsibility of soil by CPT from pit surface. Bases, foundations and soil mechanics J. – Moscow: №3, pp. 11-12.

Krutov V.I., Kulachkin B.I. (1974) In-situ determination of relative collapsibility of loessial soils by CPT. Bases, foundations and soil mechanics J – Moscow: №3, pp. 29-32.

Kulachkin B.I., Radkevich A.I., Betelev N.P. (2000) Research on soil creep mechanics by means of non-stationary temperature and dynamic (strengthening and hydraulic) fields. Bases, foundations and soil mechanics J. – Moscow: № 1, pp. 16 - 20.

Kulhawy F.H. and Mayne P.H. (1990) Manual on estimating soil properties for foundation design. Electric Power Research Institute. EPRI.

Ladanyi B. (1976) Use of static penetration test in frozen soils. Canad. Geotech. J. № 13 (2), pp. 95-110.

Ladanyi B. (1982) Determination of geotechnical parameters of frozen soils by means of the cone penetration test. Proc. of the Second European Symposium on Penetration Testing. Amsterdam, vol.1, pp. 671-678.

Ladanyi B. (1986) Use of Cone Penetration Test for the Design of Piles in Permafrost. J. of ERT. № 107, pp. 183-187.

Ladd C.C., Foott R., Ishihara K., Schlosser F., Poulos H.G. (1977) Stress-deformation and strength characteristics. State-of-the-art report. Proc. 9th Int. Conf. Soil Mech. and Found. Eng., vol. 2. Tokyo, pp. 421- 494.

Larsson R. (1995) Technical Report: Equipment and testing // Proc. of the Int. Simp. on Cone Penetration Testing, CPT'95 Linköping, Sweden, October 4-5, vol. 3, pp. 83-103.

Lunne T. (2010) The CPT in offshore soil investigations – a historic perspective. 2nd International Symposium on Cone Penetration Testing, Huntington Beach, USA.

Lunne T., Keaveny J. (1995) Technical report of solution practical problems using CPT. Proc. of the Int. Simp. on Cone Penetration Testing, CPT'95 Linköping, Sweden, October 4-5, vol. 3, pp. 119-138.

Lunne T., Lacasse S. Rad N. S. (1989) SPT, CPT pressuremeter testing and recent developments on in situ testing of soils. General report from the 12th Int. Conf.

On Soil Mech. and Found. Eng., Rio de Janeiro, vol. 4, Rotterdam: Balkema Pub., pp. 2339-2403.

Lunne T., Robertson P.K. and Powell J.J.M. (2004) Cone penetration testing in geotechnical practice. London and New York: Spon Press, 312 p.

Lebedev V.I., Ilyichev V.V., Shevtsov K.P., Indukov. A.T. (1988) In-situ methods of engineering-geological investigations. – Moscow: Nedra, 144 p.

Leonychev A.V. (1964) CPT application for evaluation of pile bearing capacity and their length. Applied aggregates and modes of operation. Seminar on exchange of experience in designing and construction of pile foundations (reports). Ufa: BashNIIstroy, pp. 86-91.

Manual on pile foundations designing (1980) N.M. Gersevanov NIIOSP of Gosstroy, USSR. – Moscow: Stroyizdat, 151 p.

Masood T., Mitchell J.K. (1993) Estimation of in situ lateral stresses in soils by cone penetration tests. Journal of Geotechnical Engineering. No. 119(10), ASCE, pp. 1624-1639.

Massarsch K.R. (2014) Cone Penetration Testing – A Historic Perspective. Proc. of the 3rd International Symposium on Cone Penetration Testing, Las Vegas, Nevada, USA, pp. 97-134.

Mayne P. (1992) Tentative method for estimating σ'_{ho} from q_c data in sand. Proceedings of the International Symposium on Calibration Chamber Testing , Potsdam, pp. 249-256.

Mayne P.W. (2014) Interpretation of geotechnical parameters from seismic piezocone tests. Proc. of the 3rd International Symposium on Cone Penetration Testing, Las Vegas, Nevada, USA, pp. 47-73.

Mayne P.W., Rix J.G. (1993) $G_{max} - q_c$ relationships for clays. Geotechnical Testing Journal, ASTM, 16(1), pp. 54-60.

Meigh A.C. (1987) Cone Penetration Testing - Methods and Interpretation", CIRIA, Butterworths.

Meyerhof G.G. (1956) Penetration test and bearing capacity of cohesionless soils. Journal of the Soil Mechanics and Foundations Division, ASCE, No 82 (SMI), pp. 1-19.

Mimura M., Shibata T., Shrivastava A.K., Nobujama M. (1995) Performance of RI cone penetrometers in sand deposits. Proc. of the Int. Simp. on Cone Penetration Testing, CPT'95 Linköping, Sweden, October 4-5, vol. 2, pp. 55-60.

Minkin M.A. (2005) Methodology and methods of engineering-geocryological investigations. Uhta: Institute of Management, Information and Business. 252p.

Mitchell, J.K. and Houston, W.N. (1974) Static cone penetration testing on the moon. Proceedings of the European on Penetration Testing. ESOPT, 2.2, Stockholm: Byggforskningen, pp. 277-284.

Mitchell J.K., Katti R.K. (1981) Soil improvement. State-of-the-Art Report. Proc. X ICSMFE, Stocholm, vol. 4, pp. 567-575.

Morozov A.A., Otrepyev V.P., Motovilov E.A., Favorov A.V., Sheinin V.I. (1999) Development of technique, instruments and software for geotechnical monitoring applying CPT and radioisotopic survey. GEOTECHNIKA-99. Penza, pp. 105-107.

Morris U. (1971) Science on direction. Bayes approach. Transl. From English. Moscow: Mir, 304 p.

Movable laboratory on the Moon "Lunohod-1". (1971) Edited by A.P. Vinogradov – Moscow: Nauka, 79 p.

Möller B., Elmgren K., Hellgren N., Larsson R., Massarsch R., Torstensson B.A., Tremblay M., Viberg L. (1995) National Report for Sweden. Proc. of the Int. Simp. on Cone Penetration Testing, CPT'95 Linköping, Sweden, October 4-5, vol. 1, pp. 221-234.

Norshayan A.V., Ryzhkov I.B. (1998) On optimization of investigations on driven piles as foundations. VI International Conference on Pile Foundation Engineering. Vol. I . M.: RNKMG&F, PGTU, pp. 183-187.

Novozhilov G.F. (1987) Defect-free immersion of piles in thawed and permafrost soils. – Leningrad: Stroyizdat, Lening. Dep. 112 p.

P2-2000 to Building Code of Belarus SNB 5.01-99 (2001) Manual to the Norms of Belarus. Designing of driven and in-situ piles by CPT results. Minsk.

Peuchen J., Heinis F., Graaf H. van de, Staveren M. van. (1995) Cone penetration testing in the Netherlands: State-of-the-art. Proc. of the Int. Simp. on Cone Penetration Testing, CPT'95 Linköping, Sweden, October 4-5, vol. 1, pp. 133-142.

Peuchen J. & Terwindt J. (2014) Introduction to CPT accuracy. Proc. of the 3rd International Symposium on Cone Penetration Testing, Las Vegas, Nevada, USA, pp. 1-45.

Post M.L., Nebbeling H. (1995) Uncertainties in con penetration testing. Proc. of the Int. Simp. on Cone Penetration Testing, CPT'95 Linköping, Sweden, October 4-5, vol. 2, pp. 73-78.

Powell J.J.M. (2010) Session Report 1: CPT Equipment & Procedures. Proc. of the 2nd International Symposium on Cone Penetration Testing, Huntington Beach, USA.

Proposed European Standard of penetration testing (1977) Proc. of the IX Int. Conf. Soil Mech. and Found. Eng. Tokyo, vol. 3, pp. 95-120.

Rad N.S. and Lunne T. (1986) Correlations between piezocone results and laboratory soil properties. Norwegian Geotechnical Institute, Oslo, Report 52155-39.

Ramsey N. (2010) Some issues related to applications of the CPT. Proc. of the 2nd International Symposium on Cone Penetration Testing, Huntington Beach, USA.

Rats M.V. (1973) Structural models in engineering geology. Moscow: Nedra, 216p.

Razorenov V.F. (1980) Penetration tests of soils (theory and practical applications). Moscow: Stroyizdat, 248 p.

Reznikov O.M. (1961) Evaluation of mechanical properties of soil by CPT. Geotechnical issues. Collections of DIIJT. – Issue 5. Dnepropetrovsk, pp. 28-41.

Robertson P.K., Cabal K.L. (2010) Guide to Cone Penetration Testing for Geo-Environmental Engineering. Gregg Drilling & Testing, Inc. USA, 85 p.

Robertson P.K., Cabal K.L. (2010) Guide to Cone Penetration Testing for Geotechnical Engineering. Gregg Drilling & Testing, Inc. USA, 124 p.

Robertson P.K., Campanella R.G., Gillespie D., Greig J. (1986) Use of piezometer cone data. Proc. of the ASCE Specialty Conf. In Situ'86: Use of In Situ Test in Geotechnical Engineering. – Blackburg: ASCE, pp. 1263-1280.

Robertson P.K., Fear C.E. (1995) Application of CPT to evaluate liquefaction potential. Proc. of the Int. Simp. on Cone Penetration Testing, CPT'95 Linköping, Sweden, October 4-5, vol. 3, pp. 57-79.

Robertson P.K., Sasitharan S., Cunning J.C. and Segs D.C. (1995) Shear wave velocity to evaluate flow liquefaction. Journal of Geotechnical Engineering, ASCE, 121 (3), pp. 262-273.

Robertson P.K., Sully J.P., Woeller D.J., Lunn T., Powell J.J.M., Gillespie D.J. (1992) Estimating coefficient of consolidation from piezocone tests. Canadian Geotechnical Journal, 29(4), pp. 551-557.

Robertson P.K., Woeller D.J., Finn W.D.L. (1992) Seismic cone penetration test for evaluating liquefaction potential under cyclic loading. Canadian Geotechnical Journal, 24(4), pp. 686-695.

Rodionov D.A., Ivanov V.A. (1967) Statistical assessments of average contents of block of observations of various presence. Geochemistry, № 1, Moscow, pp. 109-117.

Rodkevich G.S. (1989) Evaluation of soil properties and driven pile bearing capacity by CPT applying penetrometers with friction sleeves: Author's abstract to Thesis...PhD (Engineering). – Perm: PPI, 24 p.

RSN 33-70 (1970) Manual on CPT of soils. Gosstroy RSFSR. – Moscow, 32 p.

Rubinshtein A.Ya., Kulachkin B.I. (1984) Dynamic penetration of soils. – Moscow: Nedra, 92 p.

Ryzhkov I.B. (1970a) Study on soil deformations in cone penetration applying X-ray radiation. Collected papers of NIIpromstroy, issue X. – Moscow: Stroyizdat, pp. 67-69.

Ryzhkov I.B. (1970b) Application of CPT in assessment of soil properties. Thesis...PhD (Engineering). – Odessa: 184 p.

Ryzhkov I.B. (1971) On specific features of interaction between CPT and mechanical properties of soil. Collected papers of NIIpromstroy, issue X, Moscow: Stroyizdat, pp. 69-76.

Ryzhkov I.B. (1973a) Application of elastic-plastic medium model in analysis of CPT procedures. Bases, foundations and soil mechanics J, № 3, pp. 38-40.

Ryzhkov I.B. (1973b) Consideration of random factors in CPT data analysis. Collected papers of NIIpromstroy, issue XI, 1973: Stroyizdat, pp. 51-53.

Ryzhkov I.B. (1988) Correction of approximate pile resistance estimation. Bases, foundations and soil mechanics J, № 2, Moscow: Stroyizdat. – pp. 19-22.

Ryzhkov I.B. (1992) General Methodology and practical applications of CPT of soils for pile foundation designing. Thesis...Doctor of Sciences (Engineering). – Ufa: 552 p.

Ryzhkov I.B. (1995) The approach to application of static CPT together with other methods of soil investigation. Proc. of the Int. Simp. on Cone Penetration Testing, CPT'95 Lincoping, Sweden, October 4-5, vol. 2, pp. 295-300.

Ryzhkov I.B. (2000) A simplified method of operations with the random values in the geotechnical problems. Proc. of III Int. Conf. on advances of computer methods in geotechnical and geoenvironmental engineering. Moscow. A.A.Balkema. Rotterdam. Brookfield, pp. 345-346.

Ryzhkov I.B., Enikeyev V.M. (1979) Selection of pile length with regard to their possible immersion up to specified depth. Organization and technology of building: Abstract Inform. of Minpromstroy, USSR. Series II, issue 6. – Moscow: TSBNTI Minpromstroy, USSR, pp. 18-19.

Ryzhkov I.B., Goncharov B.V. (edited by) (2006). Foundation engineering: BashNIIstroy collected papers. Issue 74. Volume 2. CPT. Technology. Ufa: BashNIIstroy, 162 p.

Ryzhkov I.B., Isaev O.N. (2010) Cone Penetration Testing of soils. Moscow: Publisher ASV, 496 p.

Ryzhkov I.B., Isaev O.N. (2014) CPT sleeve friction resistance – geotechnical practice in Russia. Proceedings of the 3rd International Symposium on Cone Penetration Testing, t.p. № 1-09, Las Vegas, Nevada, UCA. pp. 257 – 262.

Sadovsky A.V., Tikhomirov S.M., Konstantinov A.V., Shvarev V.V., Isaev O.N. (1988) On optimal pile sizes in plastic-frozen soils with regard to their possible driving. Mechanized waste-free technology of prefabricated piles application for pile foundations. Abstract at the second coordination conference. Vladivostok, pp.140-142.

Sandven R. (2010) Influence of test equipment and procedures on obtained accuracy in CPTU. Proc. of the 2nd International Symposium on Cone Penetration Testing, Huntington Beach, USA.

Sanglerat G. (1971) Cone penetration testing of soils. Transl. from French. - Moscow: Stroyizdat, 232 p.

Schaap L.H.J., Zuidberg H.M. (1982) Mechanical and electrical aspects of the electric cone penetrometer tip. Proc. of the 2nd European Simp. on Penetration Testing, ESOPT-II, Amsterdam, vol. 2, Rotterdam: Balkema, pp. 841-851.

Schnaid F. (2010) Session Report 3: CPT Applications. Proc. of the 2nd International Symposium on Cone Penetration Testing, Huntington Beach, USA.

Schneider J.A. (2010) Session Report 2: CPT Interpretation. Proc. of the 2nd International Symposium on Cone Penetration Testing, Huntington Beach, USA.

Shakhirev V.B. (1965) On side pressure of soil on the embedded pile. Construction of oil processing and petrochemical plants: Collected papers of BashNIIstroy. – Issue IV. – Moscow: Stroyizdat, pp. 119-122.

Shibata T., Teparaksa V. (1988) Evaluation of liquefaction potentials of soil using cone penetration tests. Soils and foundations, v. 28, No. 2, pp. 49-60.

Shinn J.D., Bratton W.L. (1995) Innovations with CPT for environmental site characterization. Proc. of the Int. Simp. on Cone Penetration Testing, CPT'95 Linköping, Sweden, October 4-5, vol. 2, pp. 93-98.

Schmertmann J.H. (1978) Guidelines for cone penetration test, performance and design. US Federal Highway Administration, Washington, Report, FHWA-TS-78-209, 145 p.

Schmertmann D. (1967) Cone penetrometers for soil study. Civil Engineering. Journal of ASCE. Transl. from English. № 6. Moscow: Stroyizdat.

Shutenko L.I., Gilman A.D., Lupan Yu.T. (1989) Bases and foundations. Kiev: Вища шк. 328 p.

Shvets V.B., Lushnikov V.V., Shvets N.S. (1981) Determination of structural properties of soils. – Kiev: Будівельник, 104 p.

Solodukhin M.A. (1975) Engineering-geological investigations for industrial and residential construction. Moscow: Nedra, 188 p.

Somerville S.G., Paul M.A. (1986) Glossary on geotechnics. Transl. from English. – L.: Nedra, 240 p.

Standard GOST 5686-94 (1994) Soils. In-situ pile tests. Moscow.

Standard GOST 12248-2010 (2011) Soils. Laboratory methods for determining the strength and strain characteristics. Moscow.

Standard GOST 19912-2012 (2012) Soils. In-situ methods of static and dynamic tests. Moscow.

Standard GOST 20522-96 (1996) Soils. Methods of test results statistic acquisition. Moscow.

Standard GOST 23161-78 (1978) Soils. Laboratory methods of collapsible properties evaluation. Moscow.

Standard GOST 25100-2011 (2013) Soils. Classification. Moscow.

Standard GOST 25260-82 (1982) Rocks. In-situ method of penetration survey. Moscow.

Standard GOST 25358-82 (1982) Soils. In-situ method of temperature evaluation. Moscow.

Standard GOST 27751-88 (1988) Reliability of structures and foundations. Moscow.

Suzuki Y., Taya Y., Tokimatsu K., Kubota Y., Koyamada K. (1995) Field correlation of soil liquefaction based on CPT data. Proc. of the Int. Simp. on Cone Penetration Testing, CPT'95 Linköping, Sweden, October 4-5, vol. 2, pp. 583-588.

Sylvie Buteau, Richard Fortier and Michel Allard. (2005) Rate-controlled cone penetration tests in permafrost. Can. Geotech. J. 42, pp. 184–197.

Tand K.E., Funegard T.G., Warden P.E. (1995) Predicted/measured bearing capacity of shallow footings on sand. Proc. of the Int. Simp. on Cone Penetration Testing, CPT'95 Linköping, Sweden, October 4-5, vol. 2, pp. 589-594.

Tang W.H., Queck S.T. (1986) Statistical Model of Boulder Size and Fraction // Journal of Geotechnical Engineering. No. 112, vol. 1, pp. 79-90.

Temporary manual on pile bearing capacity evaluation by CPT applying S-832 CPT rig (1969) Second edition. Moscow: TSBNTI, Minpromstroy, USSR, 56p.

Terzaghi K. (1933) Structural mechanics of soil based on its physical properties. Transl. from German. Edited by N.M. Gersevanov. – Moscow: Gosstroyizdat, 392 p.

REFERENCES

Terzaghi K., Peck R. (1958) Soil mechanics in engineering practice. Transl. From English by A.V. Sulima-Samuilo; edited by M.N. Goldstein. – Moscow: Gosstroyizdat, 608 p.

Trofimenkov Yu.G. (1995) CPT of soils in construction (foreign experience). Moscow: VNIINTPI, 127 p.

Trofimenkov Yu.G., Mariupolsky L.G. (1975) On determination of soil friction upon pile side surface by CPT. Bases, foundations and soil mechanics J., № 1. – Moscow, pp. 27-28.

Trofimenkov Yu.G., Mariupolsky L.G., Pyarnpuu Z.K. (1977) Evaluation of strengthening properties of clayey soils by CPT data. Bases, foundations and soil mechanics J. № 6, pp. 11-12.

Trofimenkov Yu.G., Minkin M.A., Gvozdick V.I. (1986) Evaluation of pile bearing capacity in permafrost soils by CPT. Bases, foundations and soil mechanics J, № 2, pp. 18-20.

Trofimenkov Yu.G., Vorobkov L.N. (1981) In-situ methods of structural properties of soil study. – Moscow: Stroyizdat, 216 p.

Tsytovich N.A. (1983) Soil mechanics (short course). – Moscow: Vysshaya schola, 288 p.

Tsytovich N.A. (1973) Mechanics of frozen soils. – Moscow: Vysshaya schola, 448 p.

Uhov S.B., Semenov V.V., Znamensky V.V. et al. (2007) Soil mechanics, bases and foundations. Moscow: Vysshaya schola, 566 p.

Van de Graaf H.C., Jekel J.W. (1982) New guidelines for the use of the inclinometer with the cone penetration test. Proc. of the 2nd European Simp. on Penetration Testing, ESOPT-II, Amsterdam, vol. 2, Rotterdam: Balkema, pp. 1027-1034.

Vesic A.S. (1965) Cratering by explosives as an earth pressure problem. Proc. VI Int. Conf. Soil Mech. and Found. Eng., vol. 1, pp. 427-431.

Vesic A.S. (1975) Principles of Pile Foundation Design. Soil Mechanics Series №. 38, Duke University, Durham.

Volkov (1836) (a passageway engineer) Notes on soil investigation done in architecture / Work of lieutenant colonel, – S. Petersburg: Printing House of Passageway and Public Houses Headquarters, 62 p.

Volkov F.E., Isaev O.N. (1983) Possibility of CPT application in plastic-frozen soils by S-832M CPT rig. Pile Foundations: Collected scientific papers. NIIpromstroy, Ufa, pp. 90-94.

Volkov F.E., Isaev O.N. (1985) Application of S-832M CPT rigs for controlling and estimation of temperature changes in permafrost soils. Geocryological forecasting in construction and exploration of terrains: Abstract at All-Union Conference. Vorkuta, pp. 224-226.

Volkov F.E., Rozhnova T.N., Mudarisov M.K. (1985) Increase of bearing capacity of weak water-saturated clayey rocks by alkalization – NIIpromstroy collected papers "Foundation engineering. Soil mechanics", Ufa: NIIpromstroy, pp.116-128.

VSN 29-76 Minpromstroy, USSR (1976) Manual on pile foundation applications with penetration of piles up to specified depth Ufa: NIIpromstroy, 20 p.

Vyalov S.S. (2000) Rheology of frozen soils. Moscow: Stroyizdat, 464 p.

Workbook on controlling of properties of sands and clays in foundations of buildings and structures by means of combined penetrometers. M.: Gersevanov NIIOSP, 1988. – 8 p.

Workbook on speed engineering-geological investigations for designing structures on driven piles for mass construction. (1973) Ufa: NIIpromstroy, 19 p.

Workbook on speed engineering-geological investigations for designing structures of mass construction. (1991) Ufa: NIIpromstroy, 30 p.

Yaglom A.M., Yaglom I.M. (1973) Probability and information, – Moscow: Nauka, 511 p.

Yaroshenko V.A. (1964) Decoding of CPT results of sandy soils. Materials on designing of complex foundations and bases and job practices. Fundamentproject. M.: TSBNTI of Gosmontazhspetsstroy, USSR, pp. 14-25.

Zaharov M.S. (2007) CPT in engineering investigations. S. Petersburg: SP GASU, 71p.

Zelner A. (1980) Bayes methods in econometrics. Transl. from English. – Moscow: Statistics, 438p.

Zhou J., Xie Y., Zou Z.S., Luo V.Y., Tang X.Y. (1982) Prediction of limit load of driven pile by CPT. Proc. of the 2nd European Symposium on Penetration Testing, ESOPT-II, Amsterdam, vol. 2, Rotterdam: Balkema Pub., pp. 957-961.

Ziangirov R.S., Kashirsky V.I. (2006) CPT in engineering-geological investigations. Ehgineering Geology J, Nov. M.: PNIIIS, publishing house "Engineering Geology", pp. 13-20.

THE APPLICABILITY AND USEFULNESS OF IN SITU TESTS

[T. Lunne, P.K. Robertson and J.J.M. Powell (2004)]

Group	Device	Soil type	Profile	u	$*\varphi'$	s_u	I_D	m_v	c_v	k	G_0	σ_h	OCR	$\sigma\text{-}\varepsilon$	Hard rock	Soft rock	Gravel	Sand	Silt	Clay	Peat
		3	4	5	6	7	8	9	10	11	12	13	14	15	16	17	18	19	20	21	22
Penetrometers	Dynamic	C	B	-	C	C	C	-	-	-	C	-	C	-	-	C	B	A	B	B	B
	Mechanical	B	A/B	-	C	C	B	C	-	-	C	C	C	-	-	C	C	A	A	A	A
	Electric (CPT)	B	A	-	C	B	A/B	C	-	-	B	B/C	B	-	-	C	C	A	A	A	A
	Piezocone (CPTU)	A	A	A	B	B	A/B	B	A/B	B	B	B/C	B	C	-	C	C	A	A	A	A
	Seismic (SCPT/SCPTU)	A	A	A	B	A/B	A/B	B	A/B	B	A	B	B	B	-	C	-	A	A	A	A
	Flat Dilatometer (DMT)	B	A	C	B	B	C	B	-	-	B	B	B	C	C	C	-	A	A	A	A
	Standard Penetration Test (SPT)	A	B	-	C	C	B	-	-	-	C	-	C	-	C	C	B	A	A	A	A
	Resistivity Probe	B	B	-	B	C	A	C	-	-	B	C	C	C	-	A	B	B	B	A	B
Pressuremeters	Pre-Bored (PBP)	B	B	-	C	B	C	B	C	-	B	C	C	C	A	A	B	B	B	A	B
	Self-Boring (SBP)	B	B	A[1]	B	B	B	B	A[1]	B	A[2]	A/B	B	A/B[2]	A	B	-	B	B	A	B
	Full Displacement (FDP)	B	B	-	C	B	C	C	C	-	A[2]	C	C	C	-	C	-	B	B	A	A
Others	Vane	B	C	-	-	A	-	-	-	-	-	-	B/C	B	-	-	-	-	-	A	B
	Plate load	C	-	-	C	B	B	B	C	C	A	C	B	B	B	A	B	B	A	A	A
	Screw plate	C	C	-	C	B	B	B	C	C	A	C	B	-	-	-	-	A	A	A	A
	Borehole permeability	C	-	A	-	-	-	B	B	A	-	-	-	-	A	-	A	A	A	A	A
	Hydraulic fracture	-	B	B	-	-	-	C	C	C	B	B	-	-	B	B	-	-	B	A	C
	Crosshole /downhole/ surface seismic	C	C	-	-	-	-	-	-	-	A	-	B	-	A	A	A	A	A	A	A

Applicability: A = high; B = moderate; C - low; - = none.
* = will depend on soil type; [1] = only when pore pressure sensor fitted; [2] = only when displacement sensor fitted.
Soil parameter definitions: $u = in$ $situ$ static pore pressure; $\varphi' =$ effective internal friction angle; $s_u =$ undrained shear strength; $m_v =$ constrained modulus; $c_v =$ coefficient of consolidation; $k =$ coefficient of permeability; $G_0 =$ shear modulus at small strains; $\sigma_h =$ horizontal stress; OCR = overconsolidation ratio; $\sigma\text{-}\varepsilon =$ stress-strain relationship; $I_D =$ density index.

357

ADDITIONAL SENSORS AND DEVICES USED IN PENETRATION TESTING

I. GEOTECHNICAL APPLICATIONS

Sensors (devices)	Measurements	Applications
1	2	3
Accelerometer/ Geophone	Velocities of longitudinal and shear elastic waves	• Modulus of deformation in low deformations of soil • Shear modulus in low deformations of soil
Acoustic Sensor	Acoustic emission	• Type of soil • Compressibility of soil • Structure of soil
Electrical Resistivity Sensor	Current strength in soil between isolated electrodes	• Electrical conductivity of soil • Type of soil • Corrosiveness of soil • Porosity of sands • Determination of groundwater level
Gamma Radiation	Intensity of natural gamma radiation	• Natural radioactivity of soil • Shaliness of dispersed rocks
Gamma- Gamma Radiation	Intensity of secondary gamma-radiation	• Density of soil
Inclinometer	Verticality of the penetrometer	• Prevention of the penetrometer from damage • Correction of depth of penetration
Lateral Stress Sensor	Normal stress on penetrometer side surface	• Estimation of original stress of soil
Neutron- Neutron Radiation	Loss of neutron energy during their dissipation in soil	• Water content in soil
Pore Water Pressure Sensor	Pore water pressure	• Pore water pressure • Type of soil • Coefficient of consolidation • Filtration coefficient, etc.
Pressuremeter Module	Radial deformations	• Modulus of deformation • Shear modulus of soil • Horizontal stresses in soil • Shear strength of soil • Strengthening properties of soil

Vane	Torque	• Shear strength of soil
		• Sensitivity of soil (structural strength)
Vibratory Module	Resistivity of soil to CPT in vibrational pushing of the penetrometer	• Possible liquefaction of sands
Video	Video representation of soil during CPT	• Sizes of soil particles
		• Stratigraphy of soil
Temperature Sensor	Penetrometer temperature in its movement and stopping	• Original temperature of soil
		• Estimation of soil type
		• Determination of soil condition (thawed, frozen)
		• Thermal properties of soil
Temperature Sensor and Heat Element	Penetrometer temperature in its movement and stopping, heating and cooling. Heating and measurement of thawed soil resistance to CPT	• Original temperature of soil
		• Estimation of soil type
		• Determination of soil condition (thawed, frozen)
		• Thermal properties of soil
		• Estimation of mechanical properties of frozen soils in their thawing
Time Domain Reflectometer	Dielectric constant of pulsating electromagnetic wave	• Correlation with water content of soil

359

II. GEOECOLOGICAL APPLICATIONS

Sensors (devices)	Measurements	Applications
1	2	3
Electrical Resistivity Sensor	Currents strength in soil between isolated electrodes	• Penetration of salty water • Acid openings • Mineralization of groundwater
Gamma Radiation	Intensity of natural gamma radiation	• Zones of radioactive pollution
HIM-probe	Dielectric permeability of soil in variable electrical field	• Organic liquid contaminants
Integrated Optoelectronic device	Chemical concentration by wave interference	• Exposure of ammonia • Determination of pH
LIF Sensor	Fluorescence of hydrocarbon contamination	• Contamination of oil products (fuel, gasoline, oil, lubricants) capable to fluoresce
pH Sensor	Hydrogen ions concentration	• Acid openings • Origin of openings
Raman Spectroscopy	Argon ions concentration caused by laser fluorescence	• Contamination by organic liquids • Exposure of chlorinated hydrocarbons
Redox Potential Sensor	Redox potential	• Monitoring during biological renewal of hazardous waste products
Temperature Sensor	Temperature of penetrometer	• Endothermic/exothermal activity due to chemical reactions • Exposure of zones of thermal waters influence • Exposure of zones of groundwater flow violation due to leakage from water pipelines

International Society for Soil Mechanics
and Geotechnical Engineering (ISSMGE)

International Reference Test Procedure for the Cone Penetration Test (CPT) and the Cone Penetration Test with pore pressure (CPTU)

Report of the ISSMGE
Technical Committee 16
on
Ground Property Characterisation from In-situ Testing

1999 (corrected 2001)

Soil Characterisation by In Situ Tests:
International Reference Test Procedure for CPT/CPTU

This report contains the International Reference Test Procedure (IRTP) for the Cone Penetration Test (CPT) and the Cone Penetration Test with pore pressure (CPTU). The report was prepared by a working Group of ISSMGE Technical Committee 16, 'Ground Characterisation from In Situ Testing'. The following persons from the working group compiled the document:

Tom Lunne, Norwegian Geotechnical Institute (NGI), Norway
John Powell, Building Research Establishment (BRE), UK
Joek Peuchen, Fugro, Holland
Rolf Sandven, NTNU, Norway
Martin van Staveren, Delft Geotechnics, Holland
Several other members of the working group contributed by commenting on the document.

ABSTRACT: The Cone Penetration Test (CPT) consists of pushing a cone penetrometer using a series of push rods into the soil at a constant rate of penetration. During penetration, measurements of cone resistance and sleeve friction are recorded. The piezocone penetration test (CPTU) also includes the measurement of pore pressures at or close to the cone. The test results may be used for interpretation of stratification, classification of soil type and evaluation of engineering soil parameters. This report presents the recommended guidelines for test equipment, field procedures and presentation of test results. In addition, recommendations for required accuracy, calibration routines and maintenance procedures are outlined. The recommendations are meant to replace international reference test procedures (IRTP) recommended by the International Society for Soil Mechanics and Foundation Engineering (ISSMFE) in 1989 for the electrical CPT/CPTU. This is not a standard but a set of recommendations for good practice. These are meant to form the basis of future efforts for national/international standardisation. For the mechanical CPT the 1989 version will still be valid.

CONTENTS

1 INTRODUCTION

Two categories of the cone penetration test are considered.
1. The electric cone penetration test (CPT) which includes measurement of cone resistance and sleeve friction.
2. The piezocone test (CPTU) which is a cone penetration test with the additional measurement of pore pressure.

Note: This document may also be used for CPT/CPTU without measurement of sleeve friction.

The CPT is performed with a cylindrical penetrometer with a conical tip, or cone, penetrating into the ground at a constant rate of penetration. During penetration, the forces on the cone and the friction sleeve are measured.

The CPTU is performed as the CPT but with the additional measurement of the pore pressure at one or several locations on the penetrometer surface.

Note: Usually, the measurements are carried out using electronic transfer and data logging, with a measurement frequency that can secure detailed information about the soil conditions.

The results from a cone penetration test can in principle be used to evaluate:
- stratification
- soil type
- soil density and in situ stress conditions
- mechanical soil properties
  - shear strength parameters
  - deformation and consolidation characteristics

Note: The results from cone penetration tests may also be used directly in design e.g. pile foundations, liquefaction potential (e.g. Lunne et al. 1997).

Cone penetration testing with pore pressure measurements (CPTU) gives a more reliable determination of stratification and soil type than standard CPT. In addition, CPTU gives a better basis for interpretation of the results in terms of mechanical soil properties.

The purpose of this reference test procedure is to establish definitions and requirements for equipment and test method, which will lead the users to employ the same procedures on an international basis.

The reference test procedure is to a large extent based on testing procedures and guidelines given by the ISSMFE Technical Committee on Penetration Testing (1989) but is updated to include details on measurements of pore pressure i.e. the CPTU. This is not a standard but a set of recommendations for good practice. These are meant to form the basis of future efforts for national/international standardisation. For the mechanical CPT the 1989 version will still be valid.

Note: It is permitted to deviate from the requirements of this document if it can be demonstrated that the deviation(s) in results are not significantly different compared to results of the tests following the IRTP given herein.

2 DEFINITIONS

2.1 Cone penetration test

The pushing of a cone penetrometer at the end of a series of cylindrical push rods into the ground at a constant rate of penetration.

2.2 Cone penetrometer

The cone penetrometer is the assembly containing the cone, friction sleeve, any other sensors and measuring systems as well as the connection to the push rods. Figure 2.1 shows a section through an example of a cone penetrometer.

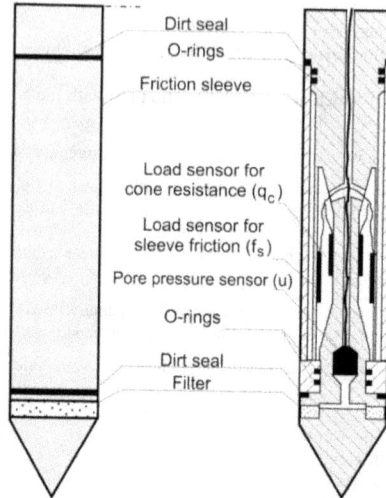

Figure 2.1 Section through an example of a cone penetrometer

The cone penetrometer includes internal load sensors for measurement of force against the cone (cone resistance), side friction against the friction sleeve (sleeve friction) and if applicable pore pressure at one or several locations on the surface of the cone penetrometer. An internal inclinometer is included for measurement of the penetrometer inclination to meet the requirements of the accuracy classes 1, 2 and 3 as given in Table 5.2.

Note: Other sensors can be included in the cone penetrometer.

2.3 Cone

The cone has an apex angle of 60° and forms the bottom part of the cone penetrometer. When pushing the penetrometer into the ground, the cone resistance is transferred through the cone to the load sensor.

Note: In this document it is assumed that the cone is rigid, so that its relative deformation when loaded is very small compared to other parts of the cone penetrometer.

2.4 Friction sleeve

The friction sleeve is the section of the cone penetrometer upon which the sleeve friction is measured.

2.5 Filter element

The filter element is the porous element inserted into the cone penetrometer to allow transmission of the pore pressure to the pore pressure sensor, while maintaining the correct geometry of the cone penetrometer.

2.6 Measuring system

The measuring system includes all sensors and ancillary parts which are used to transfer and/or store the electrical signals which are generated during the cone penetration test. The measuring system normally includes components for measuring force (cone resistance, friction), pressure (pore pressure) and depth.

2.7 Push rods

The push rods are a string of rods for transfer of compressive and tensile forces to the cone penetrometer.

Note: The push rods can also be used for supporting and/or protecting parts of the measuring system. With acoustic transfer of sounding results the rods are also used for transmission of data.

2.8 Thrust machine

The thrust machine is the equipment which pushes the cone penetrometer and rods into the ground along a vertical axis at a constant rate of penetration.

Note: Required reaction for the thrust machine may be supplied by dead weights and/or soil anchors.

2.9 Penetration depth and length

Penetration depth: Depth of the base of the cone, relative to a fixed horizontal plane (Figure 2.2).
Penetration length: Sum of the length of the push rods and the cone penetrometer, reduced by the height of the conical part, relative to a fixed horizontal plane (Figure 2.2).

Note: The fixed horizontal plane usually corresponds with a horizontal plane through the (underwater) ground surface at the location of the test.

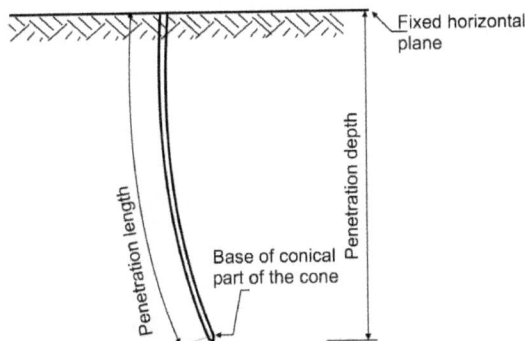

Figure 2.2 Penetration length and penetration depth

2.10 Friction reducer

A friction reducer consists of a local and symmetrical enlargement of the diameter of a push rod to obtain a reduction of the friction along the push rods.

2.11 Cone resistance, q_c

Measured cone resistance, q_c, is found by dividing the measured force on the cone, Q_c, by the cross-sectional area, A_c:

$$q_c = Q_c/A_c$$

2.12 Sleeve friction, f_s

Measured sleeve friction, f_s, is found by dividing the measured force acting on the friction sleeve, F_s, by the area of the sleeve, A_s:

366

$f_s = F_s/A_s$

2.13 Pore pressure, u

The pore pressure, u, is the fluid pressure measured during penetration and dissipation testing. The pore pressure can be measured at several locations as shown in Figure 2.3.

The following notation is used:
u_1: Pore pressure measured on the cone face
u_2: Pore pressure measured at the cylindrical extension of the cone
u_3: Pore pressure measured immediately behind the friction sleeve

Note: The measured pore pressure varies with soil type, in situ pore pressure and filter location on the surface of the cone penetrometer. The pore pressure consists of two components, the original in situ pore pressure and the additional or excess pore pressure caused by the penetration of the cone penetrometer into the ground.

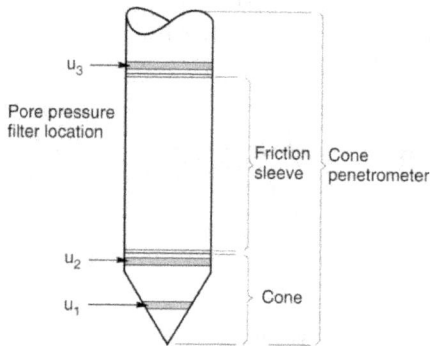

Figure 2.3 Locations of measured pore pressures

2.14 Excess pore pressure, Δu

The excess pore pressure is $\Delta u = u - u_o$, where u_o is the in situ pore pressure existing in the ground at the level of the cone before the penetration starts.

Note: Δu_1, Δu_2 or Δu_3 should be used according to the location at which the pore pressure is measured; see Figure 2.3.

2.15 Net area ratio, a

The ratio of the cross-sectional area of the load cell or shaft of the cone penetrometer above the cone at the location of the gap or groove where pore pressure can act, to the nominal cross-sectional area of the base of the cone.

Note: See section 5.11 and Figure 5.1 for details.

2.16 Corrected cone resistance, q_t

The corrected cone resistance, q_t, is the measured cone resistance, q_c, corrected for pore pressure effects, and is found from:

$$q_t = q_c + (1-a) \cdot u_2$$

Note: Section 5.11 gives more details on this correction.

367

2.17 Friction ratio, R_f

The ratio, expressed as a percentage, of the sleeve friction to the cone resistance both measured at the same depth.

Note: In some cases the inverse of the friction ratio, called the friction index, is used. Whenever possible the corrected cone resistance q_t should be used in calculating R_f.

2.18 Pore pressure ratio, Bq

The pore pressure ratio B_q is defined as:

$$B_q = \Delta u_2/(q_t - \sigma_{vo})$$

where σ_{vo} is the total vertical stress existing in the ground at the level of the cone before the penetration starts.

2.19 Zero reading, reference reading and zero drift

Zero reading: The output of a measuring system when there is zero load on the sensor, i.e. the measured parameter has a value of zero, any auxiliary power supply required to operate the measuring system being switched on.

Reference reading: the reading of a sensor just before the penetrometer is pushed into the soil e.g. in the offshore case the reading taken at the sea bottom - water pressure acting.

Zero drift: Absolute difference of the zero reading or reference reading of a measuring system between the start and completion of the cone penetration test.

2.20 Accuracy, precision and resolution

Accuracy is the closeness of a measurement to the true value of the quantity being measured. It is the accuracy of the measuring system as a whole that is ultimately important not the individual parts.
Precision is the closeness of each set of measurements to each other. It is synonymous with repeatability and can be expressed as a value with say a standard deviation indicating the scatter.

Note: In terms of calibration then if a measurement system shows, for example, a repeatable but non-linear calibration, then the use of a linear approximation for the calibration would immediately result in a loss of accuracy, however the results may still be repeatable and precise. The loss of accuracy would be related to the difference between the true and assumed calibration lines. The use of any incorrect calibration could result in repeatable (precise) results which would have a systematic error and would be inaccurate. Precision or repeatability is not a guarantee of accuracy.

The most desirable situation is to have an instrument that is accurate and precise. This is a prerequisite to obtaining accurate and precise readings in the field where it is then important to record all information such as temperature, wear etc, during the field testing that could influence the accuracy of the final deduced readings.

The resolution of a measuring system is the minimum size of the change in the value of a quantity that it can detect. It will influence the accuracy and precision of a measurement.

2.21 Dissipation test

In a dissipation test the pore pressure change is obtained by recording the values of the pore pressure with time during a pause in pushing and whilst the cone penetrometer is held stationary.

3 METHODOLOGY

The following reference conditions shall be determined:

a) the type of cone penetration test, according to Table 5.1

Note: Filter element location u_1, u_2 or u_3 should be decided upon.

b) the Accuracy Class, according to Table 5.2
c) the required penetration length or penetration depth

Note: The required penetration length or penetration depth depends on the soil conditions, the allowable penetration force, the allowable forces on the push rods and push rod connectors and the application of a friction reducer and/or push rod casing and the measuring range of the cone penetrometer.

d) the elevation of the ground surface or the underwater ground surface at the location of the cone penetration test with reference to a Datum
e) the location of the cone penetration test relative to a fixed location reference point
f) if applicable, the method of back filling of the hole in the soil resulting from the cone penetration test
g) if applicable, the depths and duration of the pore pressure dissipation tests.

Note: The required depth and minimum duration of a dissipation test depends on the soil conditions and the purpose of the measurement. A maximum duration is also a common reference condition for avoiding inappropriately long interruptions.

Note: If the drainage- and/or consolidation characteristics of the soil are to be evaluated, dissipation tests can be carried out at preselected depths in the deposit. In a dissipation test, the pore pressure decay is obtained, by recording the values of pore pressure with time. In fine grained, low permeability soil, the pore pressure record is used to evaluate the coefficient of consolidation, c. In well-draining soils, a dissipation test can additionally be used to evaluate the in situ pore pressure.

The determination of the cone resistance of the soil, the CPT length and, if applicable, the sleeve friction and/or pore pressure of soil and the inclination of the cone penetrometer relative to the vertical axis, shall be according to Section 5, taking into account the Accuracy Class according to Table 5.2, the required depth and the maximum allowable inclination of the cone penetrometer relative to the vertical axis.

The apparatus required to undertake the work shall meet the requirements of Section 4.

4 EQUIPMENT

4.1 *Geometry of the cone penetrometer*

The axis of all parts of the cone penetrometer shall be coincident.

Note: Cone penetrometer design should aim for a high net area ratio and also the end area of the top end of the friction sleeve should preferably be equal or slightly greater than the cross sectional area of the lower end.

4.2 *Cone*

The cone consists of a conical part and a cylindrical extension. The cone shall have a nominal apex angle of $60°$. The cross-sectional area of the cone shall nominally be $1000 \ mm^2$, which corresponds to a diameter of 35.7 mm.

Note: Cones with a diameter between 25 mm ($A_c = 500 \ mm^2$) and 50 mm ($A_c = 2000 \ mm^2$) are permitted for special purposes, without the application of correction factors. The recommended geometry and tolerances should be adjusted proportionately to the diameter.

The diameter of the cylindrical part shall be within the tolerance requirement as shown in Figure 4.1:

$35.3 \ mm \leq d_c \leq 36.0 \ mm$.

The length of the cylindrical extension shall be within the tolerance requirement:

$7.0 \ mm \leq h_e \leq 10.0 \ mm$

The height of the cone shall be within the following tolerance requirement:

$24.0 \ mm \leq h_c \leq 31.2 \ mm$

Note: If a u_2 position filter is included the diameter of the filter element itself may be larger than the steel dimensions given above. See also Sections 4.3 and 4.4.

The face of the cone should be smooth.

Note: The surface roughness, R_a, should typically be less than 5 μm. This is defined as the average deviation between the real surface of the probe and a medium reference plane placed along the surface of the probe. See also note in Section 4.3.

The cone shall not be used if it is asymmetrically worn, even if it otherwise fulfils the tolerance requirements.

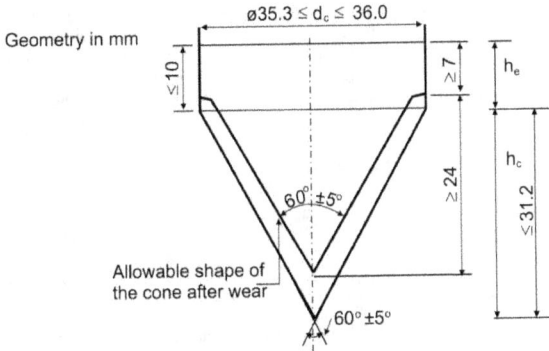

Figure 4.1 Tolerance requirements for use of cone penetrometer

4.3 Friction sleeve

The friction sleeve shall be placed just above the cone. The maximum distance due to gaps and soil seals shall be 5.0 mm.
The nominal surface area shall be 15000 mm². Tolerance requirements are shown in Figure 4.2.

Note: Friction sleeves with an external diameter between 25 mm and 50 mm are permitted for special purposes when used with cones of the corresponding diameter without the application of correction factors. The recommended geometry and tolerances should be adjusted proportionally to the diameter of the base of the cone. The preferred ratio of the length of the friction sleeve and the diameter of the base of the cone is 3.75, but values between 3.5 and 4.0 are permissible.

Note: Conical wear affects the measurement of sleeve friction. It should be taken into account for accuracy of the sleeve friction measurements.

The diameter of the friction sleeve shall be equal to the maximum diameter of the cone, with a tolerance requirement of 0 to +0.35 mm.

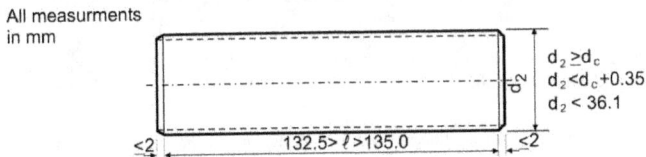

Figure 4.2 Geometry and tolerances of friction sleeve

370

The friction sleeve shall have a surface roughness of 0.4 μm ± 0.25 μm, measured in the longitudinal direction.

Note: The surface roughness refers to average roughness R_a determined by a surface profile comparator according to ISO 8503 (1988) or equivalent. Average roughness is "the arithmetic average of the absolute distances for the actual profile to the centreline" and applies to a specified test length (typically in the range 2.0 mm to 4.0 mm, depending on the applied standard). The intention of the surface roughness requirement is to prevent the use of an "unusually smooth" and "unusually rough" friction sleeve. Steel, including hardened steel, is subject to wear in soil (in particular sands) and the friction sleeve develops its own roughness with use. It is therefore important that the roughness at manufacture approaches the roughness acquired upon use. It is believed that the surface roughness requirement will usually be met in practice for common types of steel used for penetrometer manufacture and for common ground conditions (sand and clay). The effort required for metrological confirmation may thus be limited in practice. The use of the R_a parameter may be reasonable for geotechnical applications, but the use of the parameter R_y is possibly more relevant. The surface roughness R_y is the distance between the highest peak and deepest trough within one cut-off length, taken as the maximum of a series of cut-off lengths within a test length. Further research is necessary to define adequate parameters for the effects of geometry on sleeve friction accuracy.

4.4 *Filter element*

A filter position in or just behind the cylindrical extension of the cone is recommended, but other filter locations can be accepted, see Figure 2.3.

Note: Filter locations in addition to the recommended one can give valuable information about the soil conditions.

Pore pressure u_2:
The filter element shall be placed in or just behind the cylindrical part of the cone. The diameter of the filter shall correspond to the diameter of the cone and the friction sleeve, with a tolerance limit 0.0 to +0.2 mm. The filter can be larger, but never smaller than the diameter of the cone. The filter shall not have a larger diameter than the friction sleeve.

Note: The following relation applies:

$$d_{friction\ sleeve} \geq d_{filter} \geq d_{cone}$$

Note: This filter position also gives more consistent results for classification and interpretation purposes.

Note: For correction of cone resistance for pore pressure effects, the best location of the filter element would be in the groove between the cone and the friction sleeve. A location in the cylindrical part of the cone is recommended for obtaining and maintaining saturation of the pore pressure system.

Pore pressure u_1:
The diameter of the filter shall correspond to the diameter of the cone with a tolerance limit 0.0 -0.2 mm. The shape of the filter should fit to the shape of the cone, i.e. the diameter of the filter shall be equal to but not larger than the diameter of the conical part in the position of the filter.

Note: It is recommended to place the filter element within the middle third of the conical part.

Pore pressure u_3:
The diameter of the filter shall correspond to the diameter of the friction sleeve with a tolerance limit 0.0 - 0.2 mm, i.e. the diameter of the filter can be equal to but not larger than the diameter of the friction sleeve.

Note: It is recommended to place the filter element immediately above the groove between the friction sleeve and the shaft of the cone penetrometer.

The filter shall be saturated at the start of the test.

Note: It is important that the filter remains saturated even when the cone penetrometer is penetrating an upper unsaturated layer.

Note: Porous filters should have a pore size between 2 and 20 μm, corresponding to a permeability between 10^{-4} and 10^{-5} m/sec. Filter materials that get clogged by fine particles should be avoided.

Note: The following types of material have been used with good experience in soft normally consolidated clay: sintered stainless steel or bronze, carborundum, ceramics, porous PVC and HDPE.

The cone penetrometer shall be designed in such a way that it is easy to replace the filter and that the liquid chamber is easy to saturate (see Section 5.3.).

Note: With regard to the choice of saturating liquid, saturation of pore pressure measurement system, and use of slot filters, see Section 5.4.

4.5 Gaps and soil seals

The gap between the different parts of the cone penetrometer shall not exceed 5 mm. The gap shall be protected by a soil seal so that soil particles do not move into it.

Note: The soil seal must be easy to deform relative to the load cell and other elements in the penetrometer, so that no significant forces can be transferred through the gap.

4.6 Push rods

Deviation from a straight line through the ends of 1 m long rod shall be within permissible limits. A check of rod straightness shall follow the criteria given below:

- Each of the 5 lower rods shall have a maximum deviation from the centreline of 1 mm.
- Two connected rods (of the lower 5) shall at a maximum have a deviation of 4 mm.

The other rods shall have a maximum deviation of 2 mm. Two connected rods (of the rest) shall have a maximum deviation of 8 mm.

Note: The above requirements are valid for 1m long rods. If other lengths of rod are used for special purposes then the requirements should be adjusted accordingly.

Note: The straightness of the push rods can be checked by holding the rod vertically and rotating it. If the rod appears to wobble, the straightness is not acceptable.

Note: Friction along the push rods can be reduced by a local increase in the rod diameter (friction reducer). The friction could also be reduced by lubrication of the push rods, for instance by mud injection during the test.

Note: Above the ground level the push rods should be guided by rollers, a casing or similar device to reduce the risk of buckling. The push rods may also be guided by a casing in water or soft strata to avoid buckling.

Note: The push rods should be chosen with respect to the required penetration force and the data signal transmission system chosen.

4.7 Measuring system

The resolution of the measuring system shall be better than one-third of the accuracy applicable to the required accuracy class given in Table 5.2.

Note: An electric cable can be used to transfer signals from the sensors to a recording unit at ground level, or alternatively acoustic transmission through the rods, or electronic transmission to a memory unit in the cone penetrometer.

Sensors for cone resistance and sleeve friction

The load sensor shall be compensated for possible eccentricity of axial forces. The sensor for recording the side friction force shall be constructed so that it measures the friction along the sleeve, and not the earth pressure against it.

Note: Normally strain gauged load cells are used for recording cone resistance and sleeve friction.

Sensor for pore pressure

The sensor shall show insignificant deformation during loading. The sensor communicates with a porous filter on the surface of the cone penetrometer via a liquid chamber.

Note: The pore pressure sensor is normally a pressure transducer of the membrane type.

Note: This system measures the pore pressure in the surrounding soil during penetration.

Sensor for inclination

The inclinometer should have a measuring range of at least 20° relative to the vertical axis.

Measuring system for penetration length.

The measuring system shall include a depth sensor for registration of the penetration length.

Note: If relevant, the measurement system for depth should also include a procedure for correction of measurements if upward movements of the push rods occur relative to the depth sensor, caused by a decrease in force on the push rods.

4.8 Thrust machine

The equipment shall be able to penetrate the cone penetrometer at a standard speed of 20 mm/s ±5 mm/s, and it shall be loaded or anchored such that it limits movements relative to ground level while the penetration occurs.

Note: Hammering or rotation of the penetration rods during measurements shall not be used.

Note: The pushing equipment should give a stroke of at least 1000 mm. Other stroke lengths may be acceptable in special circumstances.

5 PROCEDURES

5.1 Selection of cone penetrometer

Select a cone penetrometer to fulfil the requirements of the penetration test according to Table 5.1.

Table 5.1 Types of cone penetration tests

Type of cone penetration test	Measured parameter
A	Cone resistance
B	Cone resistance and sleeve friction
C	Cone resistance and pore pressure
D	Cone resistance, sleeve friction and pore pressure

Note: Cone penetration tests with measurements of pore pressures at more than 1 location are variants of types C or D.

5.2 Selection of equipment and procedures according to required accuracy class

Equipment and procedures to be used shall be selected according to the required accuracy class given in Table 5.2.

If all possible sources of errors are added, the accuracy of the recordings shall be better than the largest of the values given in Table 5.2.

Note: The errors may include internal friction, errors in the data acquisition, eccentric loading and temperature effects.

Table 5.2 Accuracy classes

Accuracy class	Measured parameter	Allowable minimum accuracy*	Maximum length between measurements
1	Cone resistance Sleeve friction Pore pressure Inclination Penetration depth	50 kPa or 3% 10 kPa or 10% 5 kPa or 2% 2° 0.1 m or 1%	20 mm
2	Cone resistance Sleeve friction Pore pressure Inclination Penetration depth	200 kPa or 3% 25 kPa or 15% 25 kPa or 3% 2° 0.2 m or 2%	20 mm
3	Cone resistance Sleeve friction Pore pressure Inclination Penetration depth	400 kPa or 5% 50 kPa or 15% 50 kPa or 5% 5° 0.2 m or 2%	50 mm
4	Cone resistance Sleeve friction Penetration length	500 kPa or 5% 50 kPa or 20% 0.1 m or 1%	100 mm

*   See definitions in Section 2.20

Note: The allowable minimum accuracy of the measured parameter is the larger value of the two quoted. The relative or % accuracy applies to the measurement rather than the measuring range or capacity.

Note: See Appendix B regarding calculation of penetration depth from penetration length and measured inclination.

Note: Class 1 is meant for situations where the results will be used for precise evaluation of stratification and soil type as well as parameter interpretation in profiles including soft or loose soils. For Classes 3 and 4, the results should only be used for stratification and for parameter evaluation in stiff or dense soils. Class 2 may be considered more appropriate for stiff clays and sands.

Note: At extreme air temperatures, the probe should be stored so that its temperature is in the range 0 - 25°C. During the sounding, zero readings should be carried out with the probe temperature as close as possible to the ground temperature, and all sensors and other electronic components in the data acquisition system should be temperature stabilised.

Note: Current thinking is that for Class 1 testing (see Table 5.2) the temperature sensitivity of the probe transducers should be better than:

2.0 kPa/°C   for cone resistance

0.1 kPa/°C   for sleeve friction

0.05-0.1 kPa/°C   for pore pressure (measuring range 1-2 MPa)

These stability requirements are valid for probes with a load capacity of 5 tonnes. For probes with different capacities, the presented requirements can be changed proportionally with due consideration to the effects on the accuracy of the measured value.

Note: For all classes the temperature sensitivity should be an integral part of the CPT Accuracy Classes given in Table 5.2.

Metrological confirmation applicable to a cone penetration test shall be according to ISO 10012-1; 1992 (E).

5.3 Position and level of thrust machine

Position the thrust machine at a distance of at least 1 m from a previous cone penetration test, or at a distance of at least 20 times the borehole diameter of a previous borehole.

Note: Smaller distances may affect the measurements.

The thrust machine shall push the push rods so that the axis of the pushing force is as close to vertical as possible. The deviation from the vertical axis should be less than 2°. The axis of the penetrometer shall correspond to the loading axis at the start of the penetration.

5.4 Preparation of the cone penetrometer

The actual cross-sectional area of the base of the cone and, if applicable, the actual external cylindrical surface area of the friction sleeve shall be determined and recorded as required to achieve the Accuracy Class of Table 5.2.

For cone penetrometers with measurement of pore pressure the filter element and other parts of the pore pressure system shall be saturated with a liquid before field use.

Note: Usually, de-aired, distilled water is used when testing is carried out in saturated soils. When performing penetration tests in unsaturated soils, dry crust and dilative soils (e.g. dense sands), the filter should be saturated with glycerine or similar, which makes it easier to maintain saturation throughout the test. When deaired water is used, the filters should be boiled for at least 15 minutes. The filter should be cooled in the water, before being stored in a sealed container. A larger volume of de-aired water should also be prepared. This water is necessary when mounting before use. Boiling of filters may not be acceptable for some types of filters (e.g. HDPE). If glycerine (or silicone oil) is used, the dry filters are placed directly in the liquid and treated with vacuum for approximately 24 hours. A larger volume of liquid should be treated similarly and stored in a sealed container. The transducer chamber is usually saturated with the same fluid as used for the filter. This can be done by direct injection of fluid into the chamber, or by treatment of the dismantled probe in a vacuum chamber. The vacuum should be applied until no air bubbles escape from the probe (approx. 15-30 mins). The final mounting of filter and seals should be carried out with the penetrometer submerged in the saturation fluid. After mounting, the fitting of the filter should be checked. The height of the filter should be sufficient so that the filter is not loose, but small enough so that the filter can be rotated by the finger tips. This prevents excessive stresses in the joint around the filter, and also reduces any influence on the measurements. After mounting the filter, it is good practice to cover the filter element with a rubber membrane, which will burst when the penetrometer comes into contact with the soil. Other alternatives are also possible. If clogging is suspected then a new filter should be mounted for each test.

Note: During saturation and mounting of the rubber membrane, the penetrometer will be subjected to small stresses, so that the sensors can show values different from zero.

Note: Slot filter

In this system, the pore pressure is measured by an open system with a 0.3 mm slot immediately behind the conical part (e.g. Larsson, 1995). Hence the porous filter between the soil and the pressure chamber becomes redundant. The slot communicates with the pressure chamber through several channels. The pressure chamber is saturated by de-aired water, antifreeze liquid or other liquid, whereas the channels are saturated with gelatine, silicone grease or similar. Both gelatine and silicone grease are well-suited for field use. When silicone grease is used, this is injected into the channels directly from a tube. This can cause insufficient saturation of the pore pressure system, since air bubbles may be entrapped in the grease. This is avoided by using gelatine, but some more time is needed for preparation using this saturation medium.

The use of a slot filter may reduce the time required for preparation of the probe. In addition, this pore pressure system also maintains its saturation better when passing through unsaturated zones in the soil. The pressure changes in the saturated system are recorded by a pressure sensor, similar to conventional porous filter piezocones. As for other cone penetrometers, the requirements for sufficient saturation are the same, so that adequate pore pressure response is obtained during penetration.

Note: Predrilling

When penetrating coarse materials, predrilling may be used in parts of the profile if the penetration stops in dense, coarse or stone-rich layers. Predrilling may be used in coarse top layers, sometimes in combination with casings to avoid collapse of the borehole. In soft or loose soils, predrilling should be used through the crust down to the groundwater table. The predrilled hole should be filled with water if the pore pressure shall be measured by a water-saturated system. If the ground water table is located at large depths, the pore pressure system should be saturated with glycerine. In some cases, the predrilling can be carried out by ramming a dummy-rod of 45 - 50 mm diameter through the dense layer to provide an opening hole and reduce the penetration resistance.

Note: *Temperature stabilisation*

Before commencing testing, zero readings of all sensors should be taken with the cone penetrometer unloaded and temperature-stabilised ideally at ground temperature.

When the cone penetrometer is lowered into the ground, small temperature gradients will occur if the air temperature is different from the ground temperature. This will influence the sensors, and it is therefore important that the penetrometer is left to come to equilibrium so that the temperature gradients can be reduced to zero before the penetration starts. Usually, the largest gradients will occur after 2 - 3 minutes. The cone penetrometer will usually be completely temperature-stabilised after 10 - 15 minutes.
See Table 5.2 for required accuracies and Appendix A for calibration procedures.

The zero readings of the cone resistance and the penetration length and, if applicable, the sleeve friction and pore pressure and the inclination of the cone penetrometer relative to the vertical axis shall be recorded.

Note: Whenever possible the zero readings should be taken when the cone penetrometer is at or near the temperature of the ground.

Note: The reference readings for underwater cone penetration tests are those applicable immediately above the underwater ground surface.

5.5 *Pushing of the cone penetrometer*

During the penetration test, the probe shall be pushed into the ground at a constant rate of penetration 20 ± 5 mm/s. The rate shall be checked by recording time.

Note: The penetration is regarded as continuous even if the penetration is stopped regularly for a new stroke or mounting of a new push rod. Some thrust machines can carry out true continuous penetration without any stops and this can be an advantage, particularly in layered silt- and clay deposits.

Note: The penetration is regarded as discontinuous if larger stops are introduced, such as dissipation tests (see Section 2.21) or due to unforeseen malfunctions of the equipment.

5.6 *Use of friction reducer*

The use of a friction reducer (see definition Section 2) is permissible. The cone penetrometer and if relevant the push rod shall have the same diameter for at least 400 mm before the introduction of the friction reducer if applicable.

5.7 *Frequency of logging parameters*

The minimum logging frequency of parameters shall be in accordance with Table 5.2. Logging shall include (clock)time for Accuracy Classes 1 & 2 of Table 5.2.

Note: The logging interval for the various measured values can also be chosen from a consideration of the detail required in the profile, i.e. detection of thin layers. Usually the same reading interval is used for registration of cone resistance, sleeve friction and pore pressure.

Note: The average measured value over the 20 mm interval may be used, even if the values are measured more frequently. The maximum logging interval should be according to Table 5.2.

5.8 *Registration of penetration depth*

The level of the cone base shall be determined according to Table 5.2, relative to the ground level or another fixed reference system (not the thrust machine). The resolution of the depth sensor shall be at least 0.01 m.

The penetration length shall also be measured and recorded at least every 5 m for tests according to Accuracy Class 1 of Table 5.2, not using the depth sensor.

The penetration of the cone penetrometer and the push rods shall be terminated when the required penetration length or penetration depth according to Section 3 has been reached, or when the inclination of the cone penetrometer relative to the vertical axis has reached 20°. The penetration length shall be measured and recorded not using the depth sensor.

Note: The measured parameters for a cone penetrometer with a large inclination can deviate from the values that would have been measured if the cone penetrometer was vertical. Appendix B gives guidelines on how to calculate penetration depth from penetration length and inclination measurements.

Note: During the cone penetration test, particulars or deviations from this standard should be recorded, which can affect the results of the measurements and the corresponding penetration length.

5.9 *Dissipation test*

Pore pressure and cone resistance shall be measured with time. It is particularly important to take frequent readings at the beginning of the dissipation test.

Note: The logging frequency should be at least 2 Hz for the initial 1st min of the dissipation test, 1 Hz between 1 min and 10 min, 0.5 Hz between 10 min and 100 min and 0.2 Hz thereafter, as applicable.

Note: The duration of the dissipation test should normally correspond to at least the time needed for 50 % pore pressure dissipation ($t_{50} \rightarrow u_t = u_o + 0.5\Box u_i$), since t_{50} is the time used in most interpretation methods.

Note: The procedure for interruption of penetration should aim for constant cone resistance during the dissipation test. Variation in cone resistance is unavoidable in practice and will depend on factors such as type of equipment and soil conditions.

5.10 *Test completion*

The zero readings of the measured parameters shall be measured and recorded after extraction of the cone penetrometer from the soil and, if necessary after cleaning of the cone penetrometer. The zero drift of the measured parameters shall be within the allowable minimum accuracy according to the required accuracy class of Table 5.2.

The cone penetrometer shall be inspected and any excessive wear or damage noted.

5.11 *Correction of measurements*

Recorded values that are not representative due to penetration interruption shall be corrected for. Correction of measured parameters for zero drift shall be done if appropriate for meeting the requirements of the Accuracy Classes according to Table 5.2.

When the probe is subjected to an all-round water pressure, this will influence the cone resistance and sleeve friction. This is explained by the effect of the water pressure in the grooves between the cone and the friction sleeve, and in the groove above the friction sleeve. This effect shall be accounted for cone penetrometer types C and D of Table 5.1 and where the filter element is at the cylindrical extension of the cone (u_2) by using the following correction formula (e.g. Campanella et al., 1982):

Cone resistance:

$$q_t = q_c + u_2 \cdot (1 - a)$$

where:

q_t = corrected cone resistance
q_c = cone resistance
u_2 = pore water pressure in the cylindrical part of extension of the cone (assumed equal to the pore pressure in the gap between the cone and the sleeve)
a = net area ratio = A_n/A_c (see Figure 5.1)
A_c = projected area of the cone
A_n = area of load cell or shaft

377

Note: It is recommended to only carry out this correction if u_2 is measured. Approximate calculation procedures are available in some soil types for the determination of q_t for filter element positions other than the u_2 location (Lunne et al. 1997).

Note: The net area ratio 'a' varies between 0.3 and 0.9 for commonly used cone penetrometers. The area ratio cannot be determined from geometrical considerations alone, but should be determined by tests in a pressure chamber or similar.

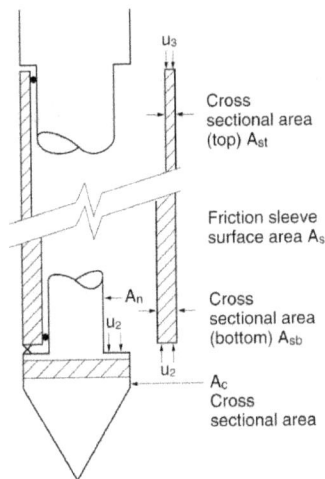

Figure 5.1 Correction of cone resistance and sleeve friction due to the unequal end area effect.

Note: The measured sleeve friction is influenced by a similar effect. However, since it is not usual to measure the pore pressure above the friction sleeve, the uncorrected sleeve friction, f_s, is commonly used. A possible correction method for the recorded sleeve friction is however given below, see Figure 5.1:

Sleeve friction:

$$f_t = f_s - \frac{\left(u_2 . A_{sb} - u_3 . A_{st}\right)}{A_s}$$

where:
f_t = corrected sleeve friction
f_s = sleeve friction
A_s = area of friction sleeve
A_{sb}= cross sectional area of the bottom of the friction sleeve
A_{st}= cross sectional area of the top of the friction sleeve
u_2 = pore pressure measured between the friction sleeve and the cone
u_3 = pore pressure measured above the friction sleeve

This correction should only be carried out if both u_2 and u_3 are measured.

Note: These corrections are most important in fine-grained soils where the excess pore pressure during penetration can be significant. It is recommended to use corrected values of the test results for interpretation and classification purposes.

Correction for inclination, i.e. calculation of penetration depth from penetration length, should also be carried out according to the procedure given in Appendix B to meet the requirements of Accuracy Classes 1, 2 and 3 (see Table 5.2).

Note: Various other corrections may be required to meet the requirements of the Accuracy Classes, e.g. temperature effects, cross sectional area of cone, compression of the push rods, rebound of the thrust machine etc.

6 REPORTING OF TEST RESULTS

6.1 *General reporting and presentation of test results*

The following information shall be reported from a (piezo)cone penetration test (selected information marked with * shall be included on every plot from the test):
- Cone penetrometer type, geometry and dimensions, filter location, net area ratio.

    Note: The actual dimensions of the cone and friction sleeve should be used whenever possible.

- Type of thrust machine used, pushing capacity, associated jacking and anchoring systems
- Use of soil anchors (number and type) if applicable
- Date of test *
- Identification of the test *
- Co-ordinates and altitude of the cone penetration test *
- Reference altitude
- Depth to the groundwater table (if recorded)
- In situ pore pressure measurements (if recorded)
- Depth of predrilling
    - Note: when possible also the type of materials encountered
- If trenching is carried out: trenching depth
    - Note: when possible also type of materials encountered
- Depth of the start of penetration
- Saturation fluid used in pore pressure system (if piezocone)
- Depth and possible causes of any stops in the penetration (e.g. dissipation tests)
- Zero readings of cone resistance and, if applicable sleeve friction and pore pressure before and after the test and zero drift (in engineering units)
- Stop criteria applied, i.e. target depth, maximum penetration force, etc
- Corrections applied during data processing (e.g. zero drifts)
- Reference to this IRTP or other standard
- Test type (Table 5.1) and Accuracy Class (Table 5.2)
- If applicable, the inclination of the cone penetrometer to the vertical axis, for a maximum penetration depth spacing of 1m

Note: In the presentation of test results, the information should be easily accessible, for example in tables or as a standard archive scheme.

Note: In addition to the above it is desirable that the following information is given:
- Manufacturer of cone penetrometer
- Observations done in the test, for example the presence of stones, noise from the pushing rods, incidents, buckled rods, abnormal wear or changes in zero / reference readings
- Identification number of the penetrometer, and measuring ranges of the transducers
- Date of last calibration of sensors

6.2 *Choice of axis scaling*

In the graphical presentation of test results the following axis scaling shall be used when required:

- Penetration depth z: $1 \text{ cm} = 1 \text{ m}$
- Cone resistance q_c, q_t: $1 \text{ cm} = 2 \text{ MPa}$
- Sleeve friction f_s, f_t: $1 \text{ cm} = 0.05 \text{ MPa} = 50 \text{ kPa}$

379

- Pore pressure u: 1 cm = 0.2 MPa = 200 kPa
- Friction ratio R_f: 1 cm = 2 %
- Pore pressure ratio B_q: 1 cm = 0.5

Note: A different scaling may be used in the presentation if the recommended scaling is used in an additional plot. The recommended scaling can for example be used for general presentation, whereas selected parts may be presented for detailed studies, using a different scaling. In clays, and where the test results are to be used for interpretation of soil parameters (Accuracy Classes 1 and 2, see Table 5.2), it is particularly important to use enlarged scaling in the presentation of test results.

The axis scaling for dissipation test results (cone resistance q_c, pore pressure u and time t) shall suit the measured values.

Note: A common presentation format is to use linear scales for q_c and u and a logarithmic scale for t.

6.3 Presentation of test results

The test results shall be presented as continuous profiles of:

- Cone resistance - depth q_c (MPa) - z (m)
- Sleeve friction - depth f_s, (MPa) - z (m)
- Pore pressure - depth u_2 (MPa) - z (m)
- Other pore pressures - depth   u (MPa) - z (m)
(location of pore pressure measurement should be given)

The depth here shall be according to Table 5.2 corrected when necessary for the measured inclination.

Presentation of the results of cone penetration tests according to Accuracy Classes 1 and 2 shall, if required, include tabular data according to Section 6.1. Tabular data per penetration length spacing according to Table 5.2 shall include the time t in s, penetration depth z in 0.01 m, cone resistance q_c in 0.01 MPa and, if applicable, sleeve friction in 1 kPa, pore pressure in 1 kPa, friction ratio R_f in 0.1%, corrected cone resistance q_t in 0.01 MPa, inclination of the cone penetrometer in $^\circ$.

If relevant corrected values of cone resistance (q_t) and sleeve friction (f_t) should be plotted in addition, and should preferably be used in further processing of the data. An exception is made for testing of coarse-grained materials, where the effect of the end area correction is negligible.

Note: In situ pore pressure can be estimated from the location of the groundwater table, or preferably by local pore pressure measurements. It can also be evaluated from the test results by performing dissipation tests in permeable layers. The total overburden stress profile can be determined from density measurements in situ or from undisturbed samples in the laboratory. If adequate information is lacking, an estimate of the density may be obtained by use of a classification chart based on the results from the cone penetration test and local experience.

Note: Further processing of the measured data can be carried out based on the following relationships:

- Excess pore pressure $\Delta u = u - u_o$
- Net cone resistance $q_n = q_t - \sigma_{vo}$
- Friction ratio $R_f = (f_s/q_c) \times 100 \%$
- Pore pressure ratio $B_q = (u_2 - u_o)/(q_t - \sigma_{vo}) = \Delta u_2/q_n$
- Normalised excess pore pressure $U = (u_t - u_o)/(u_i - u_o)$ where u_t is the pore pressure at time t in a dissipation test and u_i is the pore pressure at the start of the dissipation test

Note: In addition the following parameters can be computed for effective stress interpretation:

- Cone resistance number $N_m = q_n/(\sigma_{vo}' + a)$ (a = attraction)

Note: Information of the following parameters is needed in the processing of the test results:

- In situ, initial pore pressure vs depth u_o (MPa) versus z (m)
- Total overburden stress vs depth σ_{vo} (MPa) versus z (m)
- Effective overburden stress vs depth $\sigma_{vo}' = \sigma_{vo} - u_o$

380

Note: These parameters, or additional derived and normalised values, can be used for both identification of strata and classification of soil types, and as basic input values for interpretation of engineering parameters.

7 REFERENCES

Campanella, R.G., Gillespie, D. and Robertson, P.K. 1982 Pore pressure during cone penetration testing. *Proc ESOPT II*, Vol. II: 507-512. Rotterdam: Balkema.

ISO 1988. Preparation of steel substrates before application of paints and related products – Surface roughness characteristics of blast-clean steel substrates. ISO 8503 (1988).

ISO 1992. Quality Assurance Requirements for Measuring Equipment - Part 1: Metrological Confirmation System for Measuring Equipment, ISO 10012-1:1992(E).

ISSMFE Technical Committee on Penetration Testing (1989) *Report on Reference Test Procedures, TC 16*. Swedish Geotechnical Society (SGF), Information No.7.

Larsson, R. 1995. Use of a thin slot as filter in piezocone test. *Proc CPT'95*, 2: 35-40.

Lunne, T., Robertson, P.K. and Powell, J.J.M. 1997. *Cone Penetration Testing in Geotechnical Practice*. London: E & FN Spon, an imprint of Routledge,

APPENDICES:

APPENDIX A- MAINTENANCE, CHECKS AND CALIBRATION

A1   MAINTENANCE AND CHECKS

A1.1   *General*

This Appendix contains informative guidance on maintenance, checks and calibrations. The guidance notes are meant to represent good practice.

A1.2   *Linearity of push rods*

Before the test is carried out, the linearity of the push rods shall be checked. A rough impression of the linearity may be obtained by rolling the rods on a plane surface. If any indications of bending appear, the linearity should be checked according to the procedures outlined in Section 4.6.

A1.3   *Wear of the cone*

The wear of the cone and the friction sleeve shall be checked regularly to ensure that the geometry satisfied the tolerances. A standard geometrical pattern similar to a new or unused probe may be used in this control.

A1.4   *Gaps and seals*

The seals and gaps between the different parts of the probe shall be checked regularly. In particular, the seals should be checked for intruding soil particles and cleaned.

A1.5   *Pore pressure measuring system*

If pore pressure measurements are carried out, the filter should have sufficient permeability for satisfactory response. The filter should be kept saturated between the tests. The pore pressure system should be completely saturated before the penetration starts, and this saturation should be maintained until the cone penetrometer reaches the groundwater surface or saturated soil.

A1.6   *Maintenance procedures*

When maintenance and calibration of the equipment is carried out, the check scheme in Table A1.1 may be used, along with the producer's manual for the particular equipment.

Table A1.1   Control scheme for recommended maintenance routines

Checking Routine	Start of project	Start of test	End of test	Every 3.rd month
Verticality of thrust machine		x		
Penetration rate		x		
Depth sensor				x
Safety functions	x			x
Push rods	x	x		
Wear	x	x	x	
Gaps and seals	x	x	x	
Filter	x	x	x	
Zero drift		x	x	
Calibration	x			x*
Function control	x			x

* and at intervals during long term testing

A2 CALIBRATIONS

A2.1   General procedures

A new cone penetrometer has to be calibrated with respect to:

- the net area ratios, used for correction of measured cone resistance and sleeve friction
- influence of internal friction – restriction to movement of the individual parts.
- possible interference effects (electrical cross talk etc).
- transient temperature effects

The calibrations and checks are specific to each cone penetrometer. They will show variations during a penetrometer's life caused by small changes in the function and geometry of the cone penetrometer. In such cases, a re-calibration of the probe should be carried out. Calibration of the data acquisition system should be carried out regularly, according to the criteria listed below:

- at least every 3 months with the cone penetrometer in continuous use, or after approximately 100 soundings (approximately 3000 m)
- a new calibration should be carried out after soundings under difficult conditions, where the probe has been loaded close to its maximum capacity.

The calibrations should be carried out using the same data acquisition system, including cables, as in the field test, representing a check of possible inherent errors of the system. During the fieldwork, regular function controls of the equipment shall be carried out. These should be carried out at least once per location and/or once per day. Furthermore, a function control and possibly also a re-calibration should be carried out if the operator suspects overloading of the load sensors (loss of calibration).

In general the requirements presented in ISO 10012-1:1992(E) should be followed.

A2.2 Calibration of cone resistance and sleeve friction

The calibration of cone resistance and sleeve friction are performed by incrementally loading and unloading axially the cone and the friction sleeve. When loading the friction sleeve alone, the cone is substituted by a specially adapted calibration unit. This unit is designed so that the axial forces are transferred to the lower end area of the friction sleeve. The calibrations of cone resistance and sleeve friction are carried out separately, but the other sensors are checked to ensure that they are not influenced by the applied load. The calibration is carried out for various measuring ranges, with special emphasis on those ranges relevant for the forthcoming tests. When a new probe is calibrated, the sensors should be subjected to 15-20 repeated loading cycles up to the maximum load, before the actual calibration is carried out. The requirement for separate calibration procedures for cone and friction sleeve is not usually required for subtractive cone penetrometers.

The influence of non axial loading on the cone penetrometer and its effect on the measured parameters should be checked.

A2.3 Calibration of pore pressure and net area ratio

The calibration of the pore pressure measuring system shall be done in a pressure chamber. For pore pressure effects on the cone resistance and sleeve friction, the calibration of the net area ratio a shall be carried out in a specially designed pressure chamber (e.g. Figure A1), constructed so that the lower part of the penetrometer can be mounted in the chamber and be sealed above the friction sleeve. The enclosed part of the probe is then subjected to an incrementally increasing chamber pressure, and cone resistance, sleeve friction and pore pressure are recorded. In this way a calibration curve for the pore pressure transducer is obtained and the net area ratio can be determined from the response curves for cone resistance and sleeve friction. The pressure chamber is also well suited to check the response of the pore pressure sensor to cyclic pressure variations.

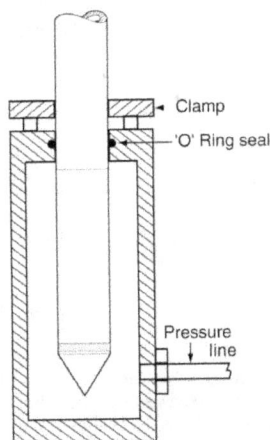

Figure A1 Pressure chamber for determination of the end area ratios a and b (from Lunne et al., 1997)

A2.4 Calibration of temperature effects

The cone penetrometer shall also be calibrated for temperature effects at various temperature levels, for example by lowering the cone penetrometer into water reservoirs at different temperatures. The sensor signals are recorded until the values stabilise. From these results a measure for changes in zero readings per °C is obtained and an impression is gained of the time needed for temperature stabilisation in the field performance. This is important information for a proper preparation of the test equipment before the penetration test starts.

The above applies to ambient temperatures only and not to transient temperatures.

A2.5 *Calibration of depth sensor*

The depth sensor calibration should be calibrated at least every 3rd month or after repair.

APPENDIX B - CALCULATION OF PENETRATION DEPTH

CORRECTION FOR PENETRATION DEPTH DUE TO INCLINATION

The depth of cone penetration tests according to Accuracy Classes 1, 2 and 3 of Table 5.2 can be corrected for inclination by the equation:

$$z = \int_{0}^{1} C_h \cdot dl$$

where:
z   is the penetration depth, in m;
l   is the penetration length, in m;
C_h is a correction factor for the effect of the inclination of the cone penetrometer relative to the vertical axis

Equations for the calculation of the correction factor C_h for the influence of the inclination of the cone penetrometer relative to the vertical axis, on the penetration depth:

a) for an now-directional inclinometer:

$$C_h \quad = \quad \cos\alpha$$

where:
α   is the measured angle between the vertical axis and the axis of the cone penetrometer, in °

b) for a bi-axial inclinometer:

$$C_h \quad = \quad (1 + \tan^2 \alpha + \tan^2 \beta)^{-\frac{1}{2}}$$

where:
α   is the angle between the vertical axis and the axis and the projection of the cone penetrometer on a fixed vertical plane, in°;
β   is the angle between the vertical axis and the axis and the projection of the cone penetrometer on a vertical plane that is perpendicular to the plane of angle α, in°.
Note: It may be necessary to apply additional corrections to the CPT depth.
Note: The determination of the correction factor for the penetration depth should take account of a complex loading sequence. Additional factors include: bending and compression of the push rods and push rods connectors, vertical movements of the ground surface or the underwater ground surface and vertical movements of the depth sensor relative to the ground surface or the underwater ground surface. For some situations, such as penetration interruptions, it is possible to correct for bending and compressing of the push rods and push rod connectors by using a heave compensator.

www.ingramcontent.com/pod-product-compliance
Lightning Source LLC
Chambersburg PA
CBHW061615220326
41598CB00026BA/3770